Proceedings of the 29th International Geological Congress Part C

Also available from VSP

PROCEEDINGS OF THE 29TH INTERNATIONAL
GEOLOGICAL CONGRESS - PART A:
Metamorphic Reaction: Kinetics and Mass Transfer
Edited by T. Nishiyama and G.W. Fisher
Sandstone Petrology in Relation to Tectonics
Edited by F. Kumon and K.M. Yu
Evaporite and Desert Environment
Edited by Y. Watanabe and A. Motamed

PROCEEDINGS OF THE 29TH INTERNATIONAL
GEOLOGICAL CONGRESS - PART B:
Reconstruction of the Paleo-Asian Ocean
Edited by R.G. Coleman
Quaternary Environmental Changes
Edited by E.H. Juvigné

PROCEEDINGS OF THE 29TH INTERNATIONAL
GEOLOGICAL CONGRESS - PART D:
Circum-Pacific Ophiolites
Edited by A. Ishiwatari, J. Malpas and H. Ishizuka

Related titles

Facies Models in Exploration and Development of
Hydrocarbon and Ore Deposits
Edited by A.H. Bouma and R.M. Carter

Regional Metamorphism of Ore Deposits and
Genetic Implications
Edited by P.G. Spry and L.T. Bryndzia

Tectonics of Circum-Pacific Continental Margins
Edited by J. Aubouin and J. Bourgois

PROCEEDINGS OF THE 29TH INTERNATIONAL GEOLOGICAL CONGRESS PART C

Kyoto, Japan, 24 August - 3 September, 1992

Siliceous, Phosphatic and Clauconitic Sediments of the Tertiary and Mesozoic

Editors: A. Iijima, A.M. Abed and R.E. Garrison

Utrecht, The Netherlands, 1994

VSP BV
P.O. Box 346
3700 AH Zeist
The Netherlands

© VSP BV 1994

First published in 1994

ISBN 90-6764-175-8

All rights reserved. No part of this publication may be reproduced, stored in a retrieval system, or transmitted in any form or by any means, electronic, mechanical, photocopying, recording or otherwise, without the prior permission of the copyright owner.

CIP-DATA KONINKLIJKE BIBLIOTHEEK, DEN HAAG

Proceedings

Proceedings of the 29th International Geological Congress.
- Utrecht : VSP
Pt. C / ed.: A. Iijima.
ISBN 90-6764-175-8 bound
NUGI 816
Subject headings: geology.

Printed in The Netherlands by A-D Druk, Zeist.

CONTENTS

Introduction to the symposium on 'Siliceous, phosphatic and
glauconitic sediments of the Tertiary and Mesozoic'
A. Iijima, R.E. Garrison and A.M. Abed   1

Neogene siliceous, phosphatic and glauconitic sediments of the
northwestern Pacific Rim: A comparison with the eastern Pacific Rim
A. Iijima   5

What controls the deposition of bio-siliceous sediments in the Japan Sea?
R. Tada   17

Sedimentary environment of the Onnagawa Sea: Middle Miocene
Japanese Backarc Trough
Y. Watanabe, M. Yamamoto and N. Imai   31

Biomarker geochemistry and paleoceanography of Miocene Onnagawa
diatomaceous sediments, northern Honshu, Japan
M. Yamamoto and Y. Watanabe   53

Occurrence and properties of glauconite in Miocene biosiliceous
sediments of the Noto Peninsula, Hokuriku District, Japan
T. Nishimura   75

Genesis of siliceous deposits in back-arc shelf basins of Kamchatka
A.R. Geptner   89

The origin of chert in the Monterey Formation of California (USA)
R.J. Behl and R.E. Garrison   101

Sedimentation and diagenesis in paleo-upwelling zones of epeiric sea
and basinal settings: A comparison of the Cretaceous Mishash Formation
of Israel and the Miocene Monterey Formation of California
Y. Kolodny and R.E. Garrison   133

Are modern and ancient phosphorites really so different?
*C.R. Glenn, M.A. Arthur, J.M. Resig, W.C. Burnett, W.E. Dean and
R.A. Jahnke*   159

Phosphogenesis and the controls on phosphorus accumulation in
continental margin sediments
G. Filippelli and M. Delaney   189

Shallow marine phosphorite-chert-palygorskite association,
Upper Cretaceous Amman Formation, Jordan
A.M. Abed   205

Lithostratigraphy of Senonian phosphorite deposits in the Palmyridean
region and their general sedimentological and paleogeographic framework
A.Kh. Al Maleh and M. Mouty 225

Sedimentary petrographical, geochemical and sedimentological aspects
of Triassic-Jurassic bedded cherts in Southwest Japan
Y. Kakuwa 233

Introduction to the symposium on "Siliceous, phosphatic and glauconitic sediments of the Tertiary and Mesozoic"

A. Iijima[1], R.E. Garrison[2] and A.M. Abed[3]

[1]*JAPEX Research Center, 1-2-1 Hamada, Mihama-ku, Chiba 261, Japan,* [2]*Earth Sciences Department, University of California, Santa Cruz, California 95064, USA, and* [3]*Department of Geology, University of Jordan, Amman, Jordan*

The twenty-ninth International Geological Congress was held at the International Convention Center in the hilly northern part of Kyoto from August 24 to September 4, 1992. Honorary President of the Congress was the the Crown Prince of Japan whose recent engagement and marriage to Miss Masako Owada, a career diplomat, have been celebrated all over the world. The symposium on "Siliceous, phosphatic and glauconitic sediments of the Tertiary and Mesozoic" was convened by A. Iijima, R.E. Garrison and A.M. Abed as one of the interdisciplinary symposia during the Congress. Originally, twenty two abstracts were submitted for the symposium, but ultimately ten oral and six poster articles were presented at the symposium. This symposium volume contains twelve of these articles as well as two additional papers whose abstracts were submitted for the symposium but whose authors were unable to attend.

Siliceous sedimentary rocks (chert, porcelanite, siliceous shale and etc.) occur in the oldest sedimentary sequences on the Earth. Phanerozoic bedded cherts are found in almost all mountain belts in the world; although they are a rather minor lithological component of "geosynclinal deposits" [1], they have long attracted researchers' attention. Bedded cherts generally outcrop in prominent, large cliffs. They also have a number of very distinctive characteristics such as simple mineralogical and chemical compositions, fine rhythmic bedding, mottled color banding, complex intraformational folding, and close association with submarine mafic volcanics (greenstone, spilite, diabase and schalstein). Before the plate tectonic theory was proposed in the 1960's, the investigation of bedded cherts was mainly related to depositional environment and tectonic framework of the "geosynclinal basins" in the initial stage of the orogenic movements [2]. The bedded cherts were again highlighted in the 1970's by the new plate tectonic theory and the deep-sea cherts recovered from many Deep Sea Drilling Project's (DSDP) cores: the greenstone—radiolarian bedded chert assemblage has been interpreted to represent the uppermost part of the ancient oceanic crust by simply referring it to the modern oceanic basalt—radiolarian ooze association in the Pacific and Indian Oceans.

The symposium reported in this volume was the culmination of a series of activities that focused on siliceous and related lithofacies. The first of these, International Geological Correlation Program's (IGCP) Project No.115, "Siliceous deposits in the Pacific region", was organized in 1976 for comparative investigations of Phanerozoic bedded cherts of the orogenic belts in the Pacific Rim and siliceous oozes and deep-sea cherts of the Pacific Ocean from the viewpoints of biostratigraphy, sedimentology and geochemistry. Project No.115 was succeeded by IGCP Project No.187, "Siliceous deposits of the Pacific and Tethys regions", which terminated in 1987. The research results of these two projects were published in two volumes [3,4]. Among the major ideas emerging from these two projects were the following: (1) Phanerozoic bedded cherts are essentially biogenic sediments; siliceous organisms are radiolarian skeletons and diatom frustules with minor sponge spicules in the Paleozoic—Mesozoic cherts and in the Tertiary, respectively. (2) Radiolarian bedded cherts accumulated in small oceans and continental margins where they were strongly influenced by adjacent land masses; no rhythmic bedding has been found in deep-sea cherts and modern radiolarian oozes. (3) Neogene diatomaceous sediments of continental margin sequences are diagenetically altered to bedded cherts and porcelanites with the same general appearance and composition as the Paleozoic—Mesozoic radiolarian bedded cherts. (4) Rhythmic chert-shale bedding originated from primary sedimentary rhythms, such as periodic paleoceanographic changes related to productivity changes of siliceous organisms, periodic rapid spreading of terrestrial muds into hemipelagic biosediments, or periodic deposition of siliceous turbidites; the cyclicity of each chert-shale unit is estimated to be in the range of 10^3 to 10^4 years. Recently Tada [5] recognized an orbital cycle in bedded siliceous rocks of the Miocene Onnagawa Formation in northern Japan and suggested the possibility of applying it to cyclostratigraphy.

In June of 1989, the US—Japan Joint Seminar on "Neogene siliceous sediments of the Pacific region" was held in California, in order to discuss similarities and dissimilarities of the diatom-rich continental margin sequences in the eastern and northwestern Pacific Rim [6]. Lithologic characteristics of the siliceous facies on the both sides of the Pacific Rim are generally comparable and their geologic age is correlative, although the tectonic situations of the sedimentary basins are quite different, as discussed in this volume. The most remarkable dissimilarity is difference in lithofacies associated with the siliceous facies: phosphatic facies are common but glauconite beds are minor in the eastern Rim, whereas the former is absent but the latter is widespread in the northwestern Rim. These similarities and dissimilarities were likely controlled by world-wide paleoclimatic and paleoceanographic events.

Based on the results of the above noted Joint Seminar, Garrison and Iijima initially made plans to convene a symposium on "Tertiary siliceous, phosphatic and glauconitic sediments" for the twenty-ninth International Geological Congress. Subsequently, Abed proposed to incorporate the Mesozoic phosphatic sediments of the Tethys region which commonly coexist with chert and glauconite. Thus, we finally decided to convene the symposium on "Siliceous, phosphatic and glauconitic sediments of the Tertiary and Mesozoic". This symposium has stressed comparisons and contrasts between different settings of siliceous rocks—phosphorites—glauconitic sediments; e.g.the Neogene of the eastern Pacific Rim versus the Neogene of the western Pacific Rim and the Pacific Neogene versus Cretaceous Middle East systems.

The first part of this volume consists of papers dealing with Tertiary biosiliceous sediments of the Pacific Rim, starting in the northwest. IIJIMA summarizes the sediments of the northwest Pacific Rim which were largely deposited in back-arc basins. He focuses on biosiliceous sediments deposited in the Miocene rifted back-arc trough of northeastern Japan and compares them with correlative deposits in central California, discussing similarities and dissimilarities between the northwestern and eastern Pacific Rim which he interprets as controlled by global paleoclimatic and paleoceanographic events. TADA calculates the mean accumulation rates of biogenic silica and diluting material for each lithostratigraphic unit in Neogene sediments of the Japan Sea basins as well as northeastern Honshu and the northwest Pacific Ocean; he discusses the factors leading to deposition of biosiliceous sediments of the Japan Sea, and maintains that this deposition was essentially controlled by high diatom productivity due to cooling of the world ocean and, locally, by the physiography of the Japan Sea. WATANABE, YAMAMOTO and IMAI present factor analyses of major and minor chemical compositions of biosiliceous sediments of the Middle Miocene Onnagawa Formation deposited in the back-arc trough in northwestern Japan. They quantitatively discuss how the basin configuration acted to accelerate diatom blooms, to prevent coarse-grained terrigenous influx, and to preserve delicate laminations in the sediments. YAMAMOTO and WATANABE through analyses of compounds as well as major and minor elements of Neogene sediments in the Akita oil-producing basin of northeastern Japan, demonstrate high inputs of algal organic matter and low inputs of terrigenous organic matter in biosiliceous sediments of the Onnagawa Formation and discuss the relation of paleoceanographic events to the deposition and early diagenesis of these sediments. Authigenic glauconite facies are commonly associated with Miocene biosiliceous sediments in the northwestern Pacific Rim. NISHIMURA describes the occurrence and properties of such glauconites in Miocene biosiliceous sediments of the Noto Peninsula in central northern Japan, showing that the glauconite beds accumulated on an offshore bank with very slow sedimentation rates. Rhythmically-bedded siliceous rocks of the Tertiary sequence in back-arc shelf basins of the Kamchatka Peninsula are described by GEPTNER who attributes their origin to enrichment of fine-grained terrigenous sediments with fine acid volcanic glass particles; in contrast to prevailing views among other workers, he maintains that the silica cementing the sediments is formed by diagenetic alteration of the glass. BEHL and GARRISON discuss the origin of cherts in the Monterey Formation of central California and maintain that these cherts formed by the diagenetic concentration of silica within very pure biogenic calcareous-diatomaceous pelagic sediments. Combined field, petrographic and isotopic studies shed light on the rate and timing of chertification relative to burial and tectonic deformation.

The second part of this volume is composed of papers dealing with Tertiary and Mesozoic phosphatic rocks and phosphate-bearing sequences, in particular of the eastern Pacific Rim and the Middle East. KOLODNY and GARRISON comparatively summarize the lithologic assemblages and diagenetic histories of the two regions. Though these assemblages, both formed in paleo-upwelling regions, are lithologically similar, their diagenetic histories were sharply different and reflect their divergent tectonic settings including marked differences in geothermal gradients. GLENN et. al. present a similar comparative analysis of Cretaceous phosphorite-bearing successions in Egypt and analogous Cenozoic sequences on the Peru margin. Phosphorites in both regions show clear evidence of winnowing and concentration of phosphate particles, but carbon isotopic data suggests that whereas authigenic carbonate fluorapatite (CFA) in the Egyptian deposits formed within the sediment in the zone of sulfate reduction during shallow burial, much of the Peruvian CFA formed at or near the sediment surface. In order to understand the history of phosphorus accumulation in oceanic sediments and to elucidate the controls on phosphorus accumulation and phosphogenesis, FILIPPELLI and DELANEY examine phosphorus accumulation in a Miocene low-oxygen basin of the Monterey Formation and three phosphorite giant deposits spanning about 250 m.y. They find that phosphorus accumulation and burial rates for phosphorite giants and the modern Peru margin are comparable and conclude that phosphorite giant deposition is not a geochemically-anomalous phenomenon. ABED describes the shallow-marine phosphorite—chert—palygorskite association of the Upper Cretaceous Amman Formation of Jordan where the facies distribution was controlled by basin and swell paleotopography on an epeiric shelf. He maintains that phosphorites were deposited on swells and oyster buildups in a subtidal environment under the influence of upwelling currents, whereas the cherts increase towards the highs; he concludes that the Tethyan phosphorites are much shallower than Miocene phosphorites of the Monterey Formation. AL MALEH and MOUTY describe the lithostratigraphy of Senonian phosphorite deposits in central Syria and show that the uplift and sedimentary basin framework as well as the Senonian transgression onto the Arabian platform controlled phosphorite deposition.

The articles summarized in the paragraphs above serve to emphasize the similarities and differences between the Pacific Neogene successions and the Tethyan Mesozoic sequences of the Middle East. Though both contain similar assemblages of biosiliceous rocks and rocks that are phosphate- and organic-rich, the Middle Eastern successions, which are associated with oyster bioherms and evaporites, are shallow-water, shelf deposits and thus contrast markedly with the mostly deep-water basinal sediments of the Pacific Neogene. Sedimentary reworking of the Cretaceous shelfal successions therefore appears to explain the abundance of large, economically significant phosphorite deposits in the Middle East compared to their scarcity in the Pacific Rim basinal sequences.

Compared to the complex facies associations of the high-fertility successions of the Pacific Rim Neogene and the Middle Eastern Mesozoic, rhythmically bedded radiolarian cherts are much more uniform; their study thus requires more subtle approaches involving detailed geochemical analyses. Mesozoic radiolarian bedded cherts, which are widespread in Japan, were deposited in a marginal sea under strong influences from nearby land areas [3,4]. These bedded cherts are not associated with distinctive phosphatic rocks in contrast to the Miocene bedded cherts of the Monterey Formation. In the last chapter of this volume, KAKUWA presents some aspects of sedimentary petrography and geochemistry of Mesozoic radiolarian bedded cherts in Honshu Island, Japan.

Acknowledgments

We are greatly indebted to Organizing Committee and Scientific Program Committee of the 29th International Geological Congress that supported the symposium by giving funds to support travel by Garrison and Abed. Garrison's travel was also partly supported by a travel grant from the Committee on Research of the University of California, Santa Cruz.

REFERENCES

1. A.B. Ronov, V.E. Khain, A.N. Balukhovsky and K.B. Seslavinsky. Quantitative analysis of Phanerozoic sedimentation, *Sedimentary Geol.* **25**, 311-325 (1980).
2. H.R. Grunau. Radiolarian cherts and associated rocks in space and time, *Eclogae Geol. Helv.* **58**, 157-208 (1965).
3. A. Iijima, J.R. Hein and R. Siever (Eds). *Siliceous deposits in the Pacific region*. Elsevier, Amsterdam (1983).
4. J.R. Hein and J. Obradovic (Eds). *Siliceous deposits of the Pacific and Tethys regions*. Springer-Verlag, New York (1988).
5. R. Tada. Origin of rhythmical bedding in Middle Miocene siliceous rocks of the Onnagawa Formation, northern Japan, *J. Sedim. Petrol.* **61**, 1123-1145 (1991).
6. R.E. Garrison (Ed.). *Japan-US Seminar on Neogene siliceous sediments of the Pacific Region*. University of California, Santa Cruz (1989).

Neogene siliceous, phosphatic and glauconitic sediments of the northwestern Pacific Rim: A comparison with the eastern Pacific Rim

A. IIJIMA
JAPEX Research Center, 1-2-1 Hamada, Mihama-ku, Chiba, 261, Japan

Abstract-- Neogene diatom-rich continental margin sequences around the northwestern Pacific Rim largely developed in back-arc basins, and are characterized by association with authigenic glauconite beds. In contrast, similar sequences around the eastern Pacific Rim formed in pull-apart small basins and fore-arc basins, and are accompanied by distinctive phosphatic rocks. The absence of phosphatic rocks implies that extensive coastal upwelling did not occur on the northwestern Pacific Rim throughout the Neogene. Formation of glauconite beds on banktops, outer shelves and submerged ridges suggests that regional tectonic and topographic conditions balanced with sea level changes. Extensive distribution of Middle Miocene diatomaceous—siliceous facies around the Pacific Rim corresponds roughly to the pattern of modern cold currents, suggesting that modern major circulation patterns of the Pacific Ocean started in the Middle Miocene approximately 14 Ma.

Key words: Neogene, siliceous sediments, phosphorite, glauconite, Pacific Rim, paleoceanographic events.

INTRODUCTION

Neogene marine siliceous rocks are widespread in continental margin sequences of the Pacific Rim. Silica of the siliceous rocks is essentially biogenous. Of siliceous organisms, diatoms are by far most important as a source of the biogenous silica [1-3]. Diatom frustules (opal-A) and their diagenetic alteration products, such as opal-CT, opal-T and microquartz, are the main constituents of the Neogene siliceous rocks [4-6]. Recent developments in the chronostratigraphy using microplankton, particularly diatoms, make possible accurate correlation of the Neogene siliceous facies of the Pacific Rim [7-9].
It is very important to compare the Neogene sequences of the eastern and western Pacific Rim, in order to know how and to what extent major paleoceanographic and paleoclimatic events and regional tectonic and topographic events influenced the Neogene facies. The following characteristic differences in tectonic types of sedimentary basins as well as extents and lithologic associations of the Miocene marine siliceous facies are recognizable between the eastern and western Pacific Rim: (1) The latitudinal extent of the Miocene siliceous facies is much larger in the eastern Rim than in the western Rim where this facies is restricted to the northwestern part. (2) The tectonic types of sedimentary basins are quite distinct on the both sides of the Pacific Rim: pull-apart basins along the San Andreas transform fault and fore-arc basins are the main types in the eastern Rim, whereas back-arc basins are dominant in the western rim. (3) Phosphatic facies are commonly associated with the siliceous facies in the eastern Rim, whereas glauconitic facies are usually associated with it in the northwestern Rim. In this paper, I intend to summarize tectonic types of sedimentary basins with Neogene siliceous facies around the northwestern Pacific Rim, and briefly describe the Miocene Tohoku Trough of Northeast Japan as an example of the back-arc basins. Then, I attempt to compare the Miocene sequences of northeastern Japan with those of central California, and discuss the differences on the both sides of the Pacific Rim.

NEOGENE SEDIMENTARY BASINS AROUND THE NORTHWESTERN PACIFIC RIM

The northwestern Pacific Rim forms the active continental margin in which trenches, island arcs, and back-arc marginal seas have been produced by subduction of the Pacific and Philippine Sea plates under the Asian Continent. Neogene marine siliceous facies were reported from many continental margin sequences of both onshore and offshore sedimentary basins around the northwestern Pacific Rim (Figure 1).

Ten sedimentary basins containing Neogene siliceous facies have been recognized around the northwestern Pacific Rim, and they are classified into three tectonic types. Of these sedimentary basins, seven belong to back-arc basins. Western Kamchatka [10] and Eastern Hokkaido are situated on the back-arc side of the Kurile—Kamchtka arc, and Northern Sakhalin [11] is possibly one of the back-arc basins of the Kurile—Kamchtka arc. The Tohoku Trough of Northeast Japan, the Hokuriku—Saninoki Trough of Southwest Japan, southeastern Korea (Tsushima Basin) and the Sea of Japan are located on the back-arc side of the Honshu arc [12]. Two belong to fore-arc basins. Central Hokkaido is situated on the fore-arc side of the Hidaka orogenic belt running from north to south in central Hokkaido [12]; Southern Sakhalin is regarded as the northern extension of Central Hokkaido. Central Shizuoka is located on the fore-arc side of the Honshu arc [13]. Joban on the outer side of northeastern Honshu Island is an intra-arc sedimentary basin of the Honshu arc [12,14].

In summary, back-arc basins are representative of the sedimentary basins with Neogene siliceous facies around the northwestern Pacific Rim. Below, I intend to exemplify the Miocene Tohoku Trough of Northeast Japan as the typical back-arc basin.

Miocene Tohoku Trough
The Tohoku Trough has been named by Iijima and Tada [12] for a narrow, elongated, deepwater sedimentary basin which existed in Middle to Late Miocene time on northeastern Honshu and southwestern Hokkaido Islands. Neogene trend and events of the Tohoku

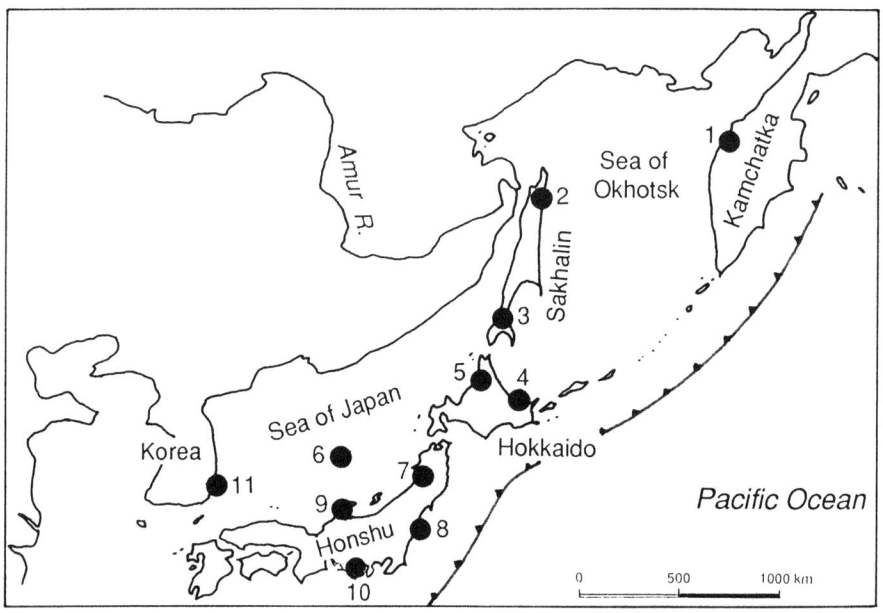

Figure 1. Distribution of Neogene siliceous sediments of the northwestern Pacific Rim. 1 = West Kamchatka, 2 = Northern Sakhalin, 3 = Southern Sakhalin, 4 = Eastern Hokkaido, 5 = Central Hokkaido, 6 = Sea of Japan, 7 = Tohoku Trough, 8 = Joban, 9 = Hokuriku—Saninoki Trough, 10 = Central Shizuoka, 11 = Southeastern Korea (Tsushima Basin).

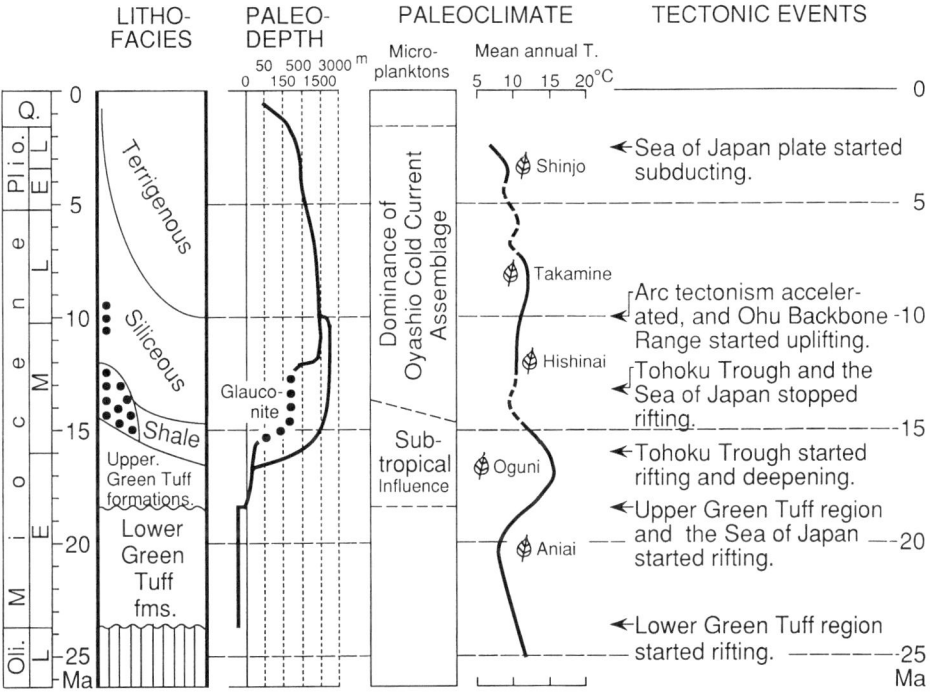

Figure 2. Neogene sequence, paleoclimate and tectonic events of the Tohoku Trough, Northeast Japan.

Trough are shown in Figure 2. Lithofacies are highly diachronous. Estimates of paleo-depth are based on sedimentary textures and structures, benthic foraminiferal assemblages [15], and molluscan assemblages [16]. Paleoclimate is based on microplankton [17] and land floras [18]: the mean annual temperature curve was estimated from the species numbers ratio of leaves with entire margin to leaves with non-entire margin in each flora [18]. Paleoclimates estimated from land plant fossils and microplankton assemblages generally correspond well to one another.

Late Early to earliest Middle Miocene. The margin of the Eastern Asian Continent started rifting, and the Sea of Japan opened not later than about 20 Ma [19]. An extensive volcanic tract, that is called the Upper Green Tuff region, existed on the east of the rapidly rifting Sea of Japan in the late Early to middle Middle Miocene, and was covered by mainly submarine, green-altered, volcanic and volcaniclastic rocks with a total thickness of 1-2 km. The Tohoku Trough was produced by rapid rifting of the Upper Green Tuff region along with two other deep-sea troughs in the time interval of approximately 15-16 Ma. The Tohoku Trough and the Hokuriku—Saninoki Trough extended parallel to the rifting Sea of Japan, while the Fossa Magna Trough bisected these two troughs (Figure 3). This configuration of the three troughs implies an anticlockwise rotation of Northeast Japan and a clockwise rotation of Southwest Japan during the rapid rifting of the Sea of Japan [20], a view supported by paleomagnetic evidence [21,22]. A submerged volcanic ridge (composed of continental and shallow-sea Green Tuff formations), though intercepted by the northern extension of the Fossa Magna Trough, intervened between the Sea of Japan and the Tohoku and Hokuriku—Saninoki Troughs, and was covered with coarse volcanic sediments and shelly limestones with shallow-water large foraminifers and molluscs. In contrast, the Kitakami, Abukuma, Yamizo, Ashio and Kanto Massifs formed small landmasses and surrounding shallow shelves east of the Tohoku Trough, and blocked free circulation of deep waters between the Pacific Ocean and the Tohoku Trough.

Within the Tohoku Trough, intense submarine mafic and silicic volcanism produced volcanic

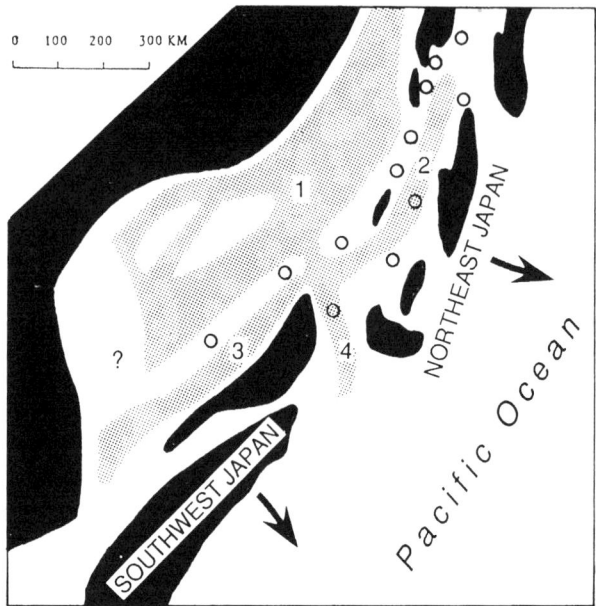

Figure 3. Early Middle Miocene paleogeography and tectonic framework of the Sea of Japan and surrounding areas. Arrows show the clockwise rotation of Southwest Japan and the anticlockwise rotation of Northeast Japan. Gray = rifting deepwater basins (1 = Sea of Japan, 2 = Tohoku Trough, 3 = Hokuriku—Saninoki Trough, 4 = Fossa Magna Trough); black = land; white = shallow sea except the Pacific Ocean; gray = upper to middle bathyal basins; open circle = glauconite bed.

sequences as much as 1500 m thick: the submarine basalts possess chemical compositions of the MORB—back-arc basin basalt type [23,24]. Dark gray shales, over 500 m thick (compacted by a 5800 meters-overburden), and with a total organic carbon content of 0.5—2 percent, accumulated in areas distant from volcanic centers; warm-water planktonic foraminifers and calcareous nannoplankton indicate a subtropical influence.

Early Middle to middle Middle Miocene. The Tohoku Trough continued to rift and subside along with the Sea of Japan in the time interval of 13.2—15 Ma. Extensive glaucogenesis (=glauconite-genesis) is characteristic of this stage. Authigenic glauconite became concentrated on offshore banktops, outer shelves, and upper slopes in the Tohoku Trough as well as on the submerged volcanic ridge between the Sea of Japan and the Tohoku and Hokuriku—Saninoki Troughs, where deposition of terrigenous, volcaniclastic, and even biogenous materials was sparse (Figure 3). Water-depths of the glaucogenesis are estimated to be outer shelfal to upper bathyal [14]. The glaucogenesis resulted in the formation of glauconite beds with a thickness ranging from 0.5 to 10 m. The glauconite beds consist of authigenic glauconite pellets, glauconite aggregates replacing various microfossils and inorganic grains, and clayey—siliceous matrix. Radial cracks of the glauconite pellets caused by dehydration shrinkage suggest that the glaucogenesis occurred in unconsolidated sediments near the water—sediment interface. Such a mode of occurrence is also supported by common presence of reworked and redeposited glauconite grains in the contemporaneous slope and basinal sediments around the sites of glaucogenesis. The glauconite content in the authigenic glauconite beds ranges from 10 to 50 volume percent. This suggests a very slow sedimentation rate: for instance, the sedimentation rate of a 7.3 meters-thick glauconite bed at the base of the Odoshi Formation has been calculated as 2.21 m 10^6 yrs^{-1} [14]. Authigenic glauconite beds of this stage are also widespread in northern Central Hokkaido and Eastern Hokkaido [12].

Within the Tohoku Trough, diatomaceous—siliceous sediments and slightly organic-rich muds were deposited under a transitional subtropical to cold current influence. Frequent occurrences of horizontal laminations in the finer sediments imply that a low-oxygen environment appeared occasionally in the silled basin.

Late Middle to early Late Miocene. The Tohoku Trough and the Sea of Japan are believed to have stopped rifting by about 13.2 Ma [12,19]. Tectonic activity was relatively quiet in the late Middle Miocene, until island-arc tectonism became active at about 10 Ma. The Tohoku Trough as well as the Sea of Japan were deepened by thermal subsidence [25,26]. A restricted sedimentary environment was caused both by blocking of the Tsushima Strait on the southwest of the Sea of Japan, due to regional uplift of Southwest Japan, and by the emergence of the middle part of the Fossa Magna Trough (Figure 4). As a result, the warm Kuroshio Current could not enter the Sea of Japan and the Tohoku Trough. On the contrary, cold currents such as the Oyashio Current flowed into the Tohoku Trough and the Sea of Japan through shallow channels among the basement massifs on the east of the Tohoku Trough. Although free circulation of the Pacific deep waters was obstructed, the Pacific intermediate waters could occasionally invade the Tohoku Trough beyond the upper bathyal swell between central Hokkaido and the Kitakami Massif of northeastern Honshu during the Mid-Miocene high stand.

In the above circumstances, diatomaceous—siliceous facies of the Onnagawa Formation and its equivalent formations were widely deposited under a cold current influence. The diatomaceous—siliceous facies accumulated not only in middle to lower bathyal basins of the Tohoku Trough and the Sea of Japan but also on the submerged volcanic ridge where glaucogenesis had occurred in the previous stage. Frequent occurrences of horizontal laminations as well as a non-calcareous benthic foraminiferal assemblage characterized by *Cyclammina* and *Martinottiella* [15,17] indicate that a restricted and low-oxygen environment prevailed in the deepwater basins. Rhythmical bedding is characteristic in the Miocene siliceous facies: e.g., "dark (clayey and organic)-and-light (siliceous) bands of the Onnagawa Formation, show the orbital forcing cycle and have been interpreted as suggesting that the

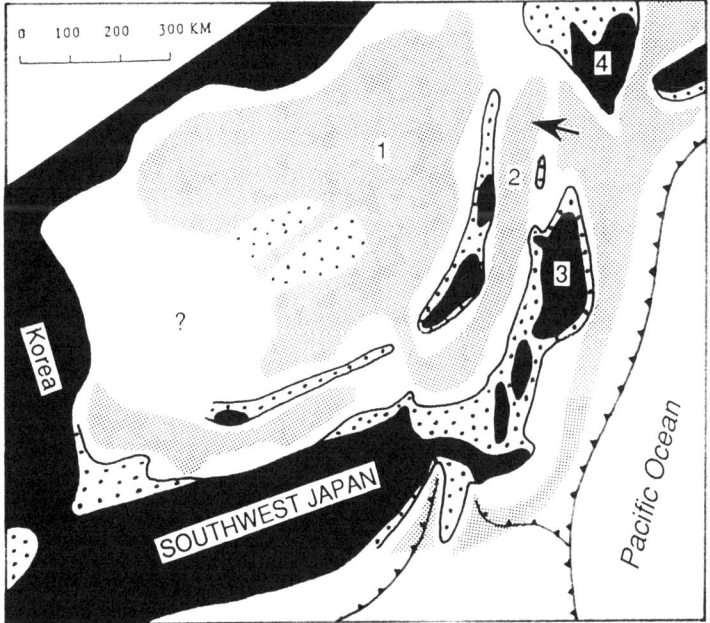

Figure 4. Late Middle Miocene paleogeographical and paleobathymetric map of Japan and the Sea of Japan. Arrow shows the pathway of the Pacific intermediate water to the Tohoku Trough. Gray = middle to lower bathyal basins (1 = Sea of Japan, 2 = Tohoku Trough); white = upper bathyal excepting the Pacific Ocean; dots = shallow sea; black = land (3 = Kitakami Massif, 4 = central Hokkaido).

upper limit of the oxygen minimum zone of the Pacific intermediate water periodically crossed over the swell and topped the Tohoku Trough [27].

Late Miocene. Starting in Late Miocene time, arc tectonism became very active in the Japanese Islands, particularly in Northeast Japan. The eastern part of the Tohoku Trough was converted into the uplift site accompanied by violent, silicic to intermediate, subaerial arc-volcanism, until the Ohu Range emerged embryonically. The emerging Ohu Range resulted in the westward shift of the deepwater basin. Simultaneously the swell that bordered the Tohoku Trough on the west subsided, and the deepwater basin became the eastern part of the Sea of Japan basin. Thus the Tohoku Trough came to an end.
In the late Late Miocene between 5.3 and 8 Ma, a great peninsula extended from Korea to northeastern Honshu and separated the Sea of Japan from the Pacific Ocean. Dark brownish-gray mudrocks and terrigenous and volcanic gravity-flow sediments accumulated in the closed deepwater basin, indicating that the uplift of the Japanese Islands as well as the low stand of this stage accelerated the sedimentation rate of terrigenous material, which surpassed the sedimentation rate of diatoms.

COMPARISON OF MIOCENE SEQUENCES OF CENTRAL CALIFORNIA WITH THOSE OF NORTHEAST JAPAN

Miocene diatom-rich continental margin sequences around the eastern and northwestern Pacific Rim are typically developed and well documented in central California and Northeast Japan, respectively. It is very important to compare the Miocene sequences on the both sides of the Pacific Rim, in order to know how and to what extent major paleoceanographic events and local tectonic and topographic events mutually influenced the distinct facies. The comparison of Miocene sequences of central California [28-30] with those of the Tohoku Trough of northeastern Honshu [12,14,20] and their correlation with major paleoceanographic events are shown in Figure 5, which portrays the evolution of Antarctic ice sheets and eustatic sea level changes, after Tada [31] and Haq *et al.* [32], respectively.

Early Miocene faices. Early Miocene facies of central California and northeastern Honshu are quite different from one another, this being probably derived from differences in regional tectonic situations.
In central California [28-30], terrigenous sediments and hemipelagic calcareous—siliceous facies accumulated diachronously under a subtropical influence in small, pull-apart basins developing along the San Andreas transform fault system. The terrigenous sediments graded upward into the hemipelagic facies as the basin-forming movements progressed with descending age.
In northeastern Honshu [12,14,20], on the other hand, altered volcanic and volcaniclastic rocks of the Green Tuff formations filled the initial stage of back-arc rift basin development, in which the continental and cool temperate environment of the lower Green Tuff (approximately 23—20 Ma) changed into the continental to shallow-sea and subtropical to warm temperate environment of the upper Green Tuff (approximately 18—16 Ma). Otherwise, Early Miocene hemipelagic siliceous facies are locally found in the Setogawa—Mineoka fore-arc basin of central Honshu [13], the Joban intra-arc basin of northeastern Honshu, northern Central Hokkaido and Eastern Hokkaido. Diatomaceous—siliceous rocks of the latter two basins were deposited under a cold current environment related to opening of the Sea of Okhotsk .

Early Middle Miocene facies. The Early Middle Miocene facies of central California and northeastern Honshu are rather distinct from one another, although they were deposited under a transitional subtropical to cold currents influence.
In central California [28-30], prominent phosphogenesis occurred in weak coastal upwelling areas of the pull-apart small basins. Phosphatic and organic-rich shales (as much as 20 percent total organic carbon content) were deposited in silled distal basins, while pelletal-oolitic phosphorites with sparse glauconite grains formed on banktops and outer shelves.
In northeastern Honshu [12,14,20], prominent glaucogenesis occurred widely on banktops, outer shelves and upper slopes in the rifting Tohoku Trough as well as on the submerged

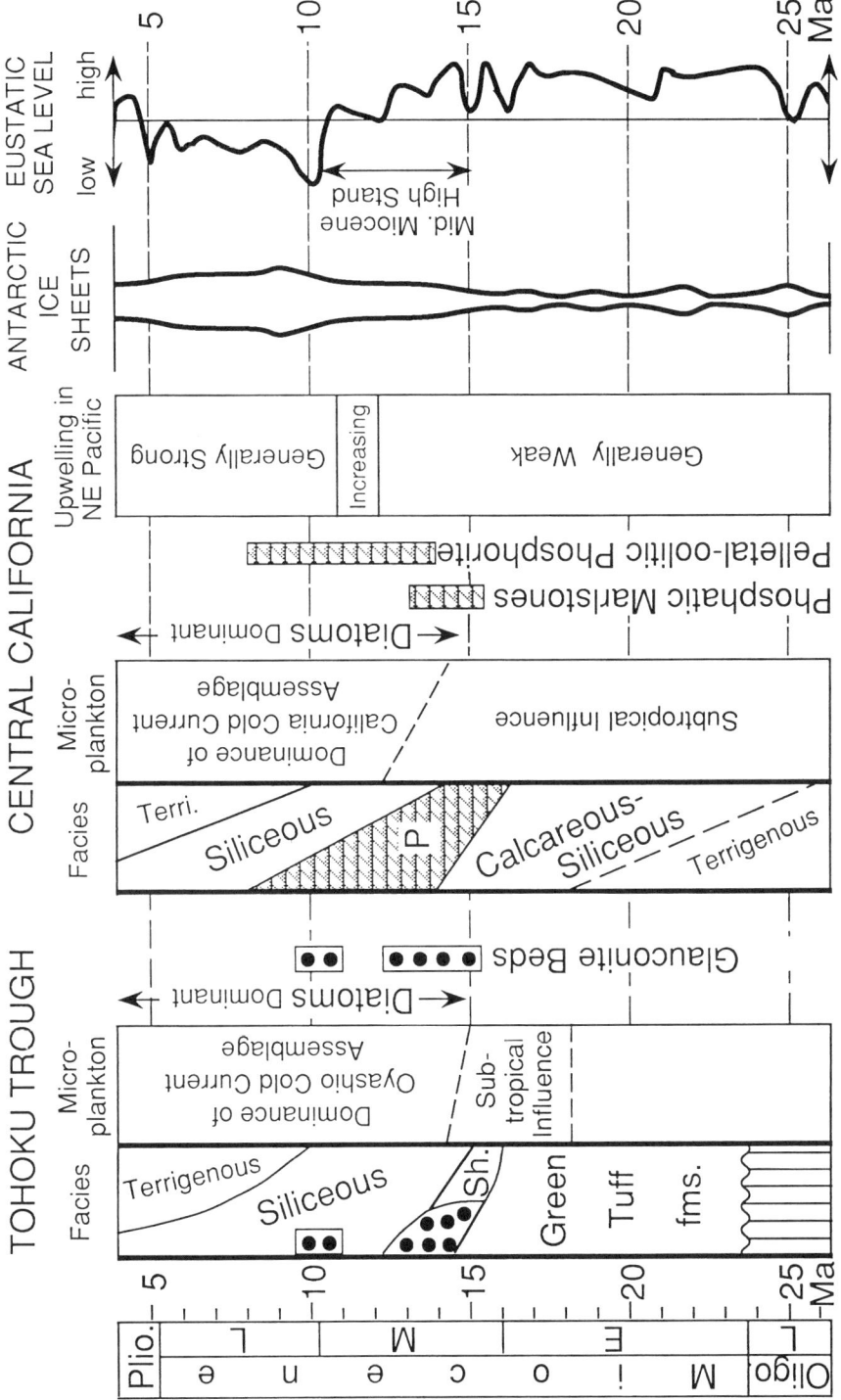

Figure 5. Comparison of Miocene trends in the Tohoku Trough of northern Japan with those of California, and correlation with major paleoceanographic events. Dotted area = authigenic glauconite beds; shaded area = phosphorite beds.

volcanic ridge between the trough and the rifting Sea of Japan. In contrast, slightly organic-rich shales (0.5—2 percent total organic carbon content) accumulated on the lower slope to bottom of the trough associated with extensive submarine volcanism. Within the Tohoku Trough, the slightly organic-rich shales graded into siliceous shales with the increases of cold current influence and water depth.

Late Middle to middle Middle Miocene facies. Late Middle to middle Middle Miocene facies of central California and northern Japan are characterized by unique hemipelagic diatomaceous—siliceous facies, regardless of tectonic types of sedimentary basins. The diatomaceous—siliceous facies accumulated in the small pull-apart basins in central California [6, 28], whereas they were widely deposited not only in the back-arc basins of the Tohoku Trough and Eastern Hokkaido but also in the fore-arc basin of Central Hokkaido in northern Japan [12,14,20]. Also, the Sea of Japan was the site of sedimentation of diatomaceous—siliceous facies [19,26].

Late Miocene facies. Late Miocene diatomaceous—siliceous facies of central California [28,29] and northern Japan [12] were gradually masked by increasing terrigenous clastics. This increasing supply of terrigenous material for sedimentary basins was partly caused by regression of the Late Miocene low stand, but more effectively by increasing tectonic activities in both regions.

PALEOCEANOGRAPHIC ASPECTS OF THE MIOCENE SILICEOUS, PHOSPHATIC AND GLAUCONITIC SEDIMENTS AROUND THE PACIFIC RIM

Marine diatom-rich continental margin sequences of the Neogene are widely distributed around the Pacific Rim [33] (Figure 6): in particular, Late to Middle Miocene siliceous facies are the most widespread. It is worthy of note that the distribution of Miocene siliceous facies of the Pacific Rim is very asymmetric between the east and the west. On the eastern Rim, the siliceous facies are widespread both in the northern and southern hemispheres with the exception of low latitudes areas, whereas on the western Rim, they are recognized only in the northern hemisphere and moreover restricted to the north of about 34.5°N. This distribution pattern corresponds roughly with the present cold currents pattern around the Pacific Rim: i. e., the California and Peru Currents on the eastern Rim and the Oyashio Current on the northwestern Rim [34]. In the Sea of Japan, as already stated, a cold current environment prevailed in the middle Middle to Late Miocene due to the blocking of the Tsushima Straight, though at present the warm Tsushima Current, a branch of the Kuroshio Current, enters the Sea of Japan and flows along the Japanese Islands. Consequently, the distribution pattern of Miocene siliceous facies around the Pacific Rim supports the theory that the recent major circulation of the Pacific Ocean began in the Middle Miocene (approximately 14 Ma) when the Antarctic ice sheets expanded significantly, and the eastern part of the warm Tethyan seaway to. the Pacific Ocean was blocked [35].

There are some common features in the Middle Miocene hemipelagic diatomaceous—siliceous facies around the Pacific Rim. (1) Siliceous rocks accumulated under influences of cold currents. It is noteworthy that the mass accumulation rates of biogenous silica in the siliceous rocks of the Monterey Formation in the Santa Barbara basin, central California and of the Onnagawa Formation in the Tohoku Trough, Northeast Japan are of the same order: i.e., 1.3-6.3 $g \cdot cm^{-2} \cdot 10^3$ yrs^{-1} in the former [36] and 5.7-7.1 $g \cdot cm^{-2} \cdot 10^3$ yrs^{-1} in the latter [37]. However, differences in the sources of the vast amounts of nutrients digested by diatoms are evident: strong coastal upwelling supplied the nutrient-rich Pacific deepwater for the basins of central California [28,29], whereas the fertile Oyashio Current would be the main source of nutrients to the basins of northern Japan. Local contribution of silica by submarine hydrothermal activities are also possible in the Tohoku Trough. (2) Diatomaceous—siliceous facies generally show a deeper environment relative to the preceding early Middle Miocene facies and occasionally onlap against them. These phenomena are explained by the transgression of the world-wide Mid-Miocene high stand, which was superposed upon regional subsidence of the basins, as pointed out by Garrison *et al.* [28]. (3) Rhythmical bedding

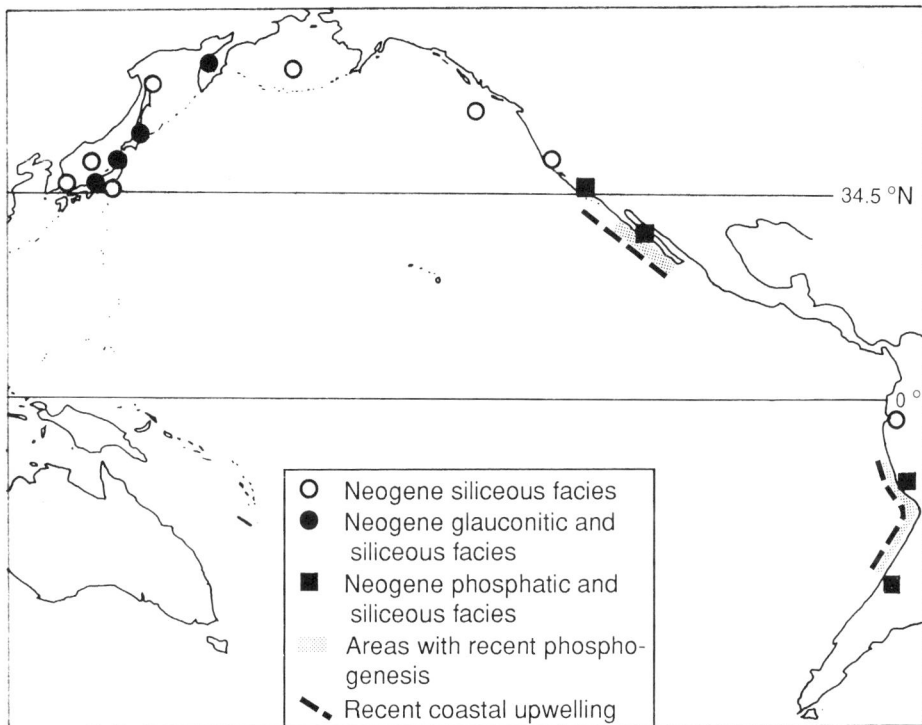

Figure 6. Distribution of biosiliceous, phosphatic and glauconitic facies in Neogene continental margin sequences. Stippled = areas with modern phosphogenesis; broken line = modern coastal upwelling around the Pacific Rim.

characterizes the hemipelagic siliceous sequences in different tectonic types of sedimentary basins. The orbital forcing cycle was reported from the Onnagawa Formation and has been interpreted to be caused by fluctuations of sea level related to short-range growth and retreat cycles of the Antarctic ice sheets in the Middle Miocene [27]. If so, the same cyclic bedding should be sought in other areas.

Lithologic distinctions are recognized in the Middle Miocene hemipelagic siliceous facies between the eastern and northwestern Pacific Rim. Biogenous calcareous (dolomitic) components are much more abundant in central California than in northern Japan. The frequent alternation of calcareous (dolomitic) beds in the calcareous-siliceous member of the Monterey Formation of California is never seen in the siliceous facies of northern Japan. Calcareous tests of foraminifers are frequently dispersed in siliceous rocks of the Monterey Formation [6,38], but they are rarely seen in those of northern Japan. These facts are interpreted to mean that subtropical influences appeared more frequently and strongly in central California than in northern Japan. In contrast, argillaceous components are generally more abundant in siliceous sequences of northern Japan relative to those of central California. Differences in diagenetic changes of the siliceous sediments are inherited from these lithologic distinctions. In central California, early-cemented chert and dolostone beds are common and make fractured reservoirs [6,39,40; Behl and Garrison, this volume]. In northern Japan, on the other hand, late-cemented chert beds are common and porous quartz porcelanites possess reservoir potentiality [5,41].

Quite different lithofacies are associated with the Middle Miocene siliceous facies on the opposite sides of the Pacific Rim, as stated before: i.e., phosphatic facies in the eastern Rim and glauconitic facies in the northwestern Rim (Figure 6). The problem is what controls such difference. Depositional conditions of the glaucogenesis in northern Japan [12,14] and the phosphogenesis in central California [28-30] are shown in Table 1. The phosphorites of

Table 1.
Comparison of depositional conditions of glaucogenesis in northern Japan with those of phosphogenesis in central California

	Glaucogenesis in Northeast Japan	Phosphogenesis in central California
Site of deposition	Offshore Banktop, outer shelf and submerged offshore ridge	P Offshore banktop and outer shelf M Outer shelf, slope, distal basin floor and submerged offshore ridge
Water-depth	Outer shelfal to upper bathyal	P Outer shelfal to upper bathyal M Outer shelfal to bathyal
Upwelling	Insignificant	P Intense coastal upwelling M Weak coastal upwelling
Sedimentation rate	Very slow, occasional reworking	P Very slow, occasional reworking M Slow and little detrital input; rapid deposition of event organic-rich sediments
Bottom water condition	Oxygenated to suboxic	P Oxygenated to suboxic conditions favorable to cyanobacterial coatings M Anoxic to suboxic conditions favorable to microbial mats

P Pelletal-oolitic phosphorite facies, M Phosphatic marlstone facies.

central California are classified into pelletal—oolitic phosphorites and phosphatic marlstones [28,30]. From this comparison chart, we can easily realize that the depositional conditions of glaucogenesis are very similar to those of the pelletal-oolitic phosphorites rather than the phosphatic marlstones, with respect to the site of deposition, water-depth, sedimentation rate, and bottom water environment. In fact, glauconite grain are frequently contained in the pelletal—oolitic phosphorites [28,30]. The formation of authigenic glauconite beds as well as pelletal oolitic phosphorites on banktops, outer shelves and submerged ridges requires a balance of regional tectonic and topographic conditions with sea level changes over a long duration of time. There is, however, a great difference in the condition of water circulation between the phosphogenesis of central California and the glaucogenesis of northern Japan; i.e., regional coastal upwelling existed in the former, whereas it did not occur in the latter. A steady supply of phosphorus dissolved in the Pacific deepwater by significant coastal upwelling is indispensable for the growth of cyanobacterial coatings and microbial mats related to phosphogenesis [28-30]. A recent analog has been found in the Peru margin, where phosphatic sediments and rocks have accumulated since the Miocene [42]. Recent regional coastal upwelling around the Pacific Rim is restricted to the central California—Baja California margin and the Peru—northern Chile margin, in which Miocene phosphorites are distributed [29,35].
Such regional coastal upwelling apparently did not occur around the western Pacific Rim throughout Neogene time, judging by the lack of phosphorites there.

CONCLUSIONS

Neogene diatom-rich continental margin sequences of the northwestern Pacific Rim largely developed in back-arc basins, and are characterized by association with authigenic glauconite beds. In contrast, similar sequences of the eastern Pacific Rim formed in small pull-apart basins and fore-arc basins, and are associated with distinctive phosphatic rocks. The absence of phosphatic rocks implies that extensive coastal upwelling did not occur in the northwestern Pacific Rim throughout the Neogene. Formation of glauconite beds on banktops, outer shelves and submerged ridges suggests that regional tectonic and topographic conditions were balanced with sea level changes. Extensive distribution of Middle Miocene diatomaceous—

siliceous facies in the Pacific Rim corresponds roughly to modern cold currents patterns, thus indicating that modern patterns of major circulation of the Pacific Ocean began in the Middle Miocene at approximately 14 Ma.

Acknowledgments

I am grateful to Prof. R. E. Garrison for invaluable discussion and critical reading of the manuscript. I benefitted from the US—Japan Joint Scientific Seminar on Neogene siliceous sediments of the Pacific region in California, June 11-17, 1989 in improving the basic ideas for this thesis.

REFERENCES

1. M.N. Bramlette. The Monterey formation of California and the origin of its siliceous rocks, *U.S. Geol. Survey Prof. Paper* **212**, 1-57 (1946).
2. J.C. Ingle. Origin of Neogene diatomites around the north Pacific Rim. In: *The Monterey Formation and related siliceous rocks of California*. R.E. Garrison and R.G. Douglas (Eds). pp.159-180. S.E.P.M. Pacific Section, Los Angels (1981).
3. A. Iijima, J.R. Hein and R. Siever (Eds). *Siliceous deposits in the Pacific region*. Elsevier, Amsterdam (1982).
4. K.J. Murata and R.R. Larson. Diagenesis of Miocene siliceous shales, Temblor Range, California, *U.S. Geol. Survey J. Res.* **3**, 553-566 (1975).
5. A. Iijima and R. Tada. Silica diagenesis of Neogene diatomaceous and volcaniclastic sediments in northern Japan, *Sedimentology* **28**, 185-200 (1981).
6. K.A. Pisciotto and R.E. Garrison. Lithofacies and depositional environments of the Monterey Formation, California. In: *The Monterey Formation and related siliceous rocks of California*. R.E. Garrison and R.G. Douglas (Eds). pp.97-122. S.E.P.M. Pacific Section, Los Angels (1981).
7. F. Akiba. Middle Miocene to Quaternary diatom biostratigraphy of Leg 87 in the Nankai Trough and Japan Trench, Deep Sea Drilling Project, and modified Lower Miocene through Quaternary diatom zones for middle-to-high latitudes of the North Pacific — Part 1. Biostratigraphy, *Tech. Center JAPEX Res. Rep.* no. 1, 1-44 (1984).
8. I. Koizumi. Diatom biostratigraphy for late Cenozoic northwest Pacific, *J. Geol. Soc. Japan* **57**, 143-156 (1985).
9. J.A. Barron. Updated diatom biostratigraphy for the Monterey Formation of California. In: *Siliceous microfossil and microplankton of the Monterey Formation and modern analogs*. R.E. Casey and J.A. Barron (Eds). pp.105-119. S.E.P.M. Pacific Section, Los Angels (1986).
10. A.R. Geptner. Genesis of siliceous deposits in back-arc shelf basins of Kamchtka. In: *29th International Geological Congress Abstracts Vol.ume 1*. p.99 (1992).
11. Y. Gladenkov. Paleogeographic and climatic environments in the North Pacific Cenozoic. In: *29th International Geological Congress Abstracts Volume*. 2. p.247 (1992).
12. A. Iijima and R. Tada. Evolution of Tertiary sedimentary basins of Japan in reference to opening of the Japan Sea, *J. Fac. Sci. Univ. Tokyo, Sec. II* **22**, 121-171 (1990).
13. Y. Watanabe and A. Iijima. Evolution of the Tertiary Setogawa—Kobotoke—Mineoka forearc basin in central Japan with emphasis on the Lower Miocene terrigenous turbidite fills, *J. Fac. Sci. Univ. Tokyo, Sec. II* **22**, 53-88 (1989).
14. A. Iijima, R. Tada and Y. Watanabe. Developments of Neogene sedimentary basins in the north-eastern Honshu Arc with emphasis on Miocene siliceous deposits, *J. Fac. Sci. Univ. Tokyo, Sec. II* **21**, 417-446 (1988).
15. S. Hasegawa, K. Akimoto, H. Kitazato and Y. Matoba. Late Cenozoic paleobathymetric indices based on benthic foraminifers in Japan, *Geol. Soc. Japan Memoir* no. 32, 241-253 (1989). *(in Japanese)*.
16. K. Chinzei. Neogene molluscan faunas in the Japanese Islands: An ecologic and zoogeographic synthesis, *The Veliger* **21**, 155-170 (1978).
17. S. Maiya. Neogene events as revealed by changes of foraminiferal assemblages from central and northern Japan. In: *IGCP Project 246—Pacific Neogene events in time and space*. R. Tsuchi, M. Chiji and Y. Takayanagi (Eds). Special Publication, pp.31-48. Osaka Museum of Natural History, Osaka (1988). *(in Japanese)*.
18. T. Tanai. Tertiary vegetational changes in East Asia, *Rep. Mizunami Municipal Fossil Museum* **7**, 117-122 (1990). *(in Japanese)*.
19. R. Tada and A. Iijima. Lithostratigraphy and compositional variation of Neogene hemipelagic sediments in the Japan Sea. In: *Proceedings of the Ocean Drilling Program, Scientific Results Volume.127/128, Part 2*. K. Tamaki, K. Suehiro *et al.* (Eds). pp.1229-1260. College Station, Texas (1992).

20. A. Iijima. Opening of the Sea of Japan and evolution of Neogene sedimentary basins, *J. Japan. Asoc. Petrol. Technol.* **57**, 171-179 (1992). *(in Japanese)*.
21. Y. Otofuji and T. Matsuda. Paleomagnetic evidence for the clockwise rotation of Southwest Japan, *Earth Planet. Sci. Lett.* **70**, 373-382 (1984).
22. T. Tosha and Y. Hamano. Paleomagnetism of Tertiary rocks from the Oga Peninsula and the rotation of Northeast Japan, *Tectonics* **7**, 653-662 (1988).
23. K. Shuto. Miocene abyssal tholeiite-type basalts from Tobishima Island, eastern margin of the Japan Sea, *J. Japan. Petro. Miner. Econ. Geol.* **83**, 252-272 (1988).
24. N. Tsuchiya. Distribution and chemical composition of the Middle Miocene basaltic rocks in Akita—Yamagata Oil Fields of northeastern Japan, *J. Geol. Soc. Japan* **94**, 591-608 (1988).
25. U. Suzuki. Basin subsidence model and structure based on geohistory analysis — The cases of the eastern part of the Sea of Japan and the Beibu Gulf of China, *J. Japan. Asoc. Petrol. Technol.* **56**, 51-63 (1990). *(in Japanese)*.
26. R. Tada and K. Tamaki. Scientific results of ODP Japan Sea Legs, and their implication for stratigraphy, *J. Japan. Asoc. Petrol. Technol.* **57**, 103-111 (1992). *(in Japanese)*.
27. R. Tada. Origin of rhythmical bedding in Middle Miocene siliceous rocks of the Onnagawa Formation, northern Japan, *J. Sedim. Petrol.* **61**, 1123-1145 (1991).
28. R. E. Garrison, M. Kastner and C.E. Reimers. Miocene phosphogenesis in California. In: *Phosphate deposits of the world Volume 3: Neogene to modern phosphorites.* W.C. Burnett and S.R. Riggs (Eds). pp.285-299, Chap. 23. Cambridge University Press, Cambridge (1990).
29. K.B. Föllmi and R.E. Garrison. Phosphatic sediments, ordinary or extraordinary deposits? The example of the Miocene Monterey Formation. In: *Controversies in modern geology, Symposium in honour of K.J. Hsu.* J. Mackenzie, D. Mueller and H. Weissert (Eds). pp.55-84, Chap. 5. Blackwell Scientific Publisher, Oxford (1991).
30. R. E. Garrison, M. Kastner and Y. Kolodny. Phosphorites and phosphatic rocks in the Monterey Formation and related Miocene units, coastal California. In: *Cenozoic basin development of coastal California, Rubey Volume 6.* R.V. Ingersoll and W.G. Ernst (Eds). pp.348-381, Chap. 15. Prentice-Hall, Englewood Cliffs (1987).
31. R. Tada. Change of surface environment through Cenozoic, *J. Geography* **100**, 937-950 (1990). *(in Japanese)*.
32. B.U. Haq, J. Hardenbol and P.R. Vail. The chronology of fluctuating sea level since the Triassic, *Science* **194**, 1121-1132 (1987).
33. R.E. Garrison (Ed.). *Japan-US Seminar on Neogene siliceous sediments of the Pacific Region.* University of California, Santa Cruz (1989).
34. *The live atlas of the world*, Times—Kodansha, Tokyo (1992). *(in Japanese)*.
35. F. Woodruff and S.M. Savin. Miocene deepwater oceanography, *Paleoceanography* **4**, 87-140 (1989).
36. C.M. Isaacs. Compositional variation and sequence in the Monterey Formation, Santa Barbara coastal area, California. In: *Cenozoic marine sedimentation, Pacific margin, USA.* D.K. Larue and R.J. Steel (Eds). pp.117-132. S.E.P.M. Pacific Section, Los Angels (1983)
37. R. Tada, Y. Watanabe and A. Iijima. Accumulation of laminated and bioturbated Neogene siliceous deposits in Ajigasawa and Goshogawara areas, Aomori Prefecture, Northeast Japan, *J. Fac. Sci. Univ. Tokyo, Sec. II* **21**, 139-167 (1986).
38. C.M. Isaacs. Lithostratigraphy of the Monterey Formation, Goleta to Point Conception, Santa Barbara coast, California. In: *Guide to the Monterey Formation in the California coastal area, Ventura to San Luis Obispo.* C.M. Isaacs (Ed.). Pacific Sec., Amer. Assoc. Petrol. Geol. Pub. **52**, 9-24 (1981).
39. C.M. Isaacs. Outline of diagenesis in the Monterey Formation examined laterally along the Santa Barbara coast, California. In: *Guide to the Monterey Formation in the California coastal area, Ventura to San Luis Obispo.* C.M. Isaacs (Ed.). Pacific Sec., Amer. Assoc. Petrol. Geol. Pub. **52**, 25-38 (1981).
40. C.M. Isaacs. Influence of rock composition on kinetics of silica phase changes in the Monterey Formation, Santa Barbara area, California, *Geology* **10**, 304-308 (1982).
41. R. Tada and A. Iijima. Petrology and diagenetic changes of Neogene siliceous rocks in northern Japan, *J. Sedim. Petrol.* **53**, 911-930 (1983).
42. R.E. Garrison and M. Kastner. Phosphatic sediments and rocks recovered from the Peru margin during ODP Leg 112. In: *Proceedings of the Ocean Drilling Program, Scientific Results Volume 112.* E. Suess, R. von Huene *et al.* (Eds). pp.111-134. College Station, Texas (1990).

What controls the deposition of bio-siliceous sediments in the Japan Sea?

R. TADA

Geological Institute, University of Tokyo, 7-3-1 Hongo, Tokyo 113, Japan

Abstract-- It is well known that Neogene fine-grained bio-siliceous sedimentary sequences are widely exposed in northern Japan. Results of recent ODP drilling in the Japan Sea revealed that similar sedimentary sequences are also distributed in the deeper part of the Japan Sea. Two intervals characterized by high contents of biogenic silica are recognizable during 15.5 to 10.5 Ma and 6.5 to 3.5 Ma at the ODP sites in the Japan Sea as well as the onland sites in northern Japan. However, MARs of biogenic silica, ranging from 1.5 to 2.5 $g/cm^2/ky$, for the sediments from the Japan Sea are almost an order of magnitude lower compared to 6 to 20 $g/cm^2/ky$ for the sediments from onland sites in northern Japan. The latter values are comparable to the rate of 15 to 45 $g/cm^2/ky$ observed in modern high productivity areas. The strong positive correlation between MARs of biogenic silica and diluting material found for Neogene bio-siliceous sediments from the Japan Sea, onland sites in northern Japan, and the northwestern Pacific suggests the possibility of enhancement of preservation of biogenic silica at the sediment/water interface with the increase in MAR of diluting material. When taking this effect into account, three intervals, >15.5 Ma, 10.5 to 6.5 Ma, and <3.5 Ma, which are characterized with lower biogenic silica contents, are interpreted as the periods of lower surface biogenic silica productivity within the Japan Sea. The lower surface productivty before 15.5 Ma seems to reflect a global cooling event, whereas the other two low surface productivity intervals seems specific to the Japan Sea and may record semi-isolation of the sea during the global sea level drop along with re-opening of the southern strait which reduced the inflow of cold water from the north.

Key words: Neogene, bio-siliceous sediments, Mass Accumulation Rate, the Japan Sea, paleoceanography

INTRODUCTION

Neogene fine-grained bio-siliceous sediments are widely distributed in the circum-Pacific region including the Japanese Islands and the Japan Sea. It is believed by many that timing of deposition as well as distribution of these bio-siliceous sediments have paleoceanographic implications, especially related to reorganization and intensity changes of global deepwater circulation [1,2]. The bio-siliceous sediments exposed in northern Japan are one of the most intensively studied examples from which paleoceanographic interpretation of regional and global significance could be made [3,4]. However, some of the characteristics of these bio-siliceous sediments could be specific to the Japan Sea because the Japan Sea has been a semi-enclosed marginal sea from the begining of its history, and its paleoceanographic conditions could have been differernt from the open Pacific. If one wishes to extract paleoceanographic information of regional or global significance from the Japan Sea sediments, it is neccesary to identify the local phenomena specific to the Japan Sea.
In this paper, I will discuss timing and extent of deposition of the bio-siliceous sediments in

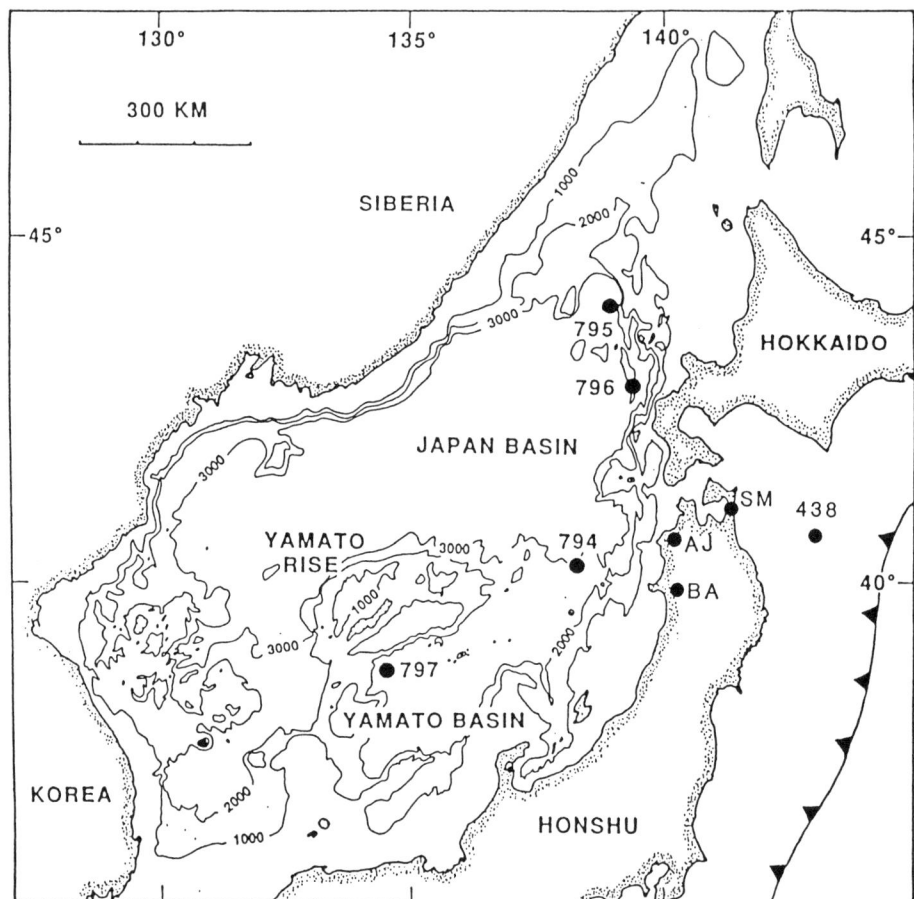

Figure 1. Map of the Japan Sea and its surrounding region showing DSDP, ODP, and onshore sites referred in this paper. AJ=Ajigasawa site, BA=Babame site, SM=Shimokita site.

the Japan Sea based mainly on the results of ODP drilling. Then, I will examine possible controlling factors of their deposition. Finally, I will compare the results with those from the Pacific side of northern Japan so as to differentiate the phenomena specific to the Japan Sea from those of regional or global significance.

RESULTS OF THE JAPAN SEA DRILLING

In 1989, Ocean Drilling Program (ODP) legs 127/128 drilled at six sites within the Japan Sea, of which Sites 794, 795, and 797 are located in the basinal part of the sea (Figure 1). Also shown in the figure is the location of onland studied sites and Deep Sea Drilling Project (DSDP) Site 438, of which results are also discussed here. The acoustic basement, which is composed of basaltic lava and sill complex with ages of 17 to 24 Ma, was penetrated at these three sites [5,6]. Based on the shipboard results of ODP legs as well as those of complementary shorebased studies carried out afterwards, the age of Japan Basin is proved to

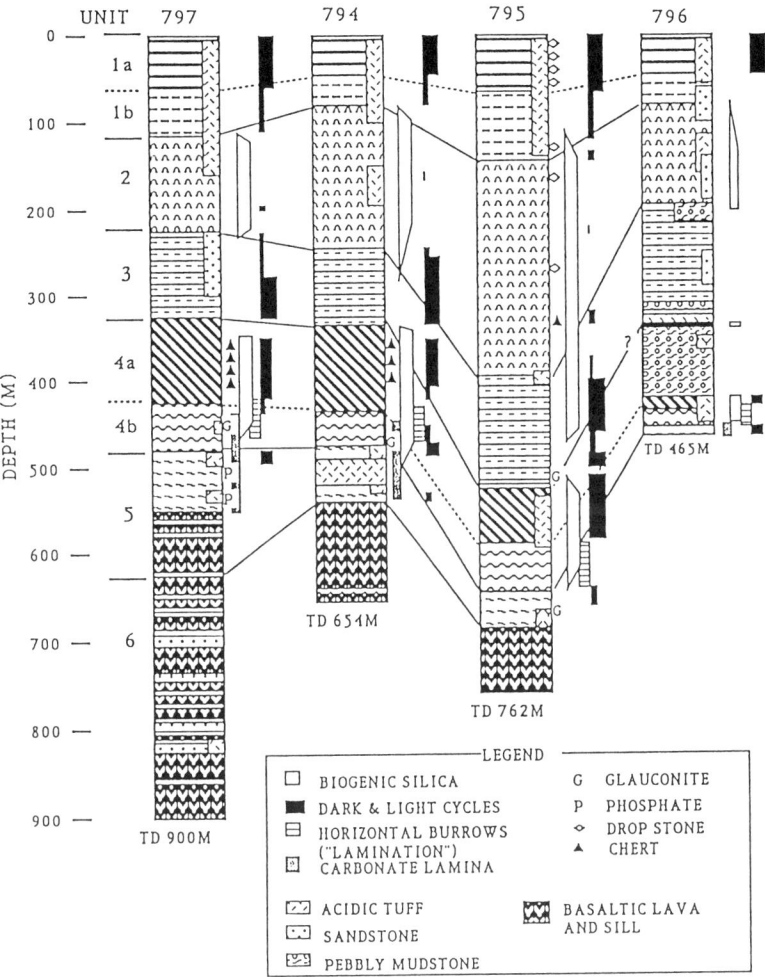

Figure 2. Stratigraphic sections and correlation of the lithological units among sites drilled during Leg 127. Minor lithology is also shown.

be as old as 28 Ma whereas that of Yamato basin is 19 to 20 Ma [7]. Nearly complete hemipelagic sedimentary sequences, which accumulated since the establishment of the Japan Sea as a wide and deep marginal basin, were recovered at Sites 794, 795, and 797. The sequences are characterized by Neogene bio-siliceous sediments which are typical in the circum Pacific rim region.

Lithostratigraphy

Tada and Iijima [8] divided the sedimentary sequence above the basement into five lithological units, which are correlatable basinwide. In addition, sedimentary layers intercalated within the basement are identified as an independent lithological unit at Sites 794 and 797. These lithological units are numbered from 1 to 6 in descending order as is illustrated in Figure 2. The following is a brief summary of the lithology of each units. Complete description of lithology could be found in Tada and Iijima [8].

Unit 6 is composed of decimeter-scale rhythmical alternation of light gray sandstone, reddish brown siltstone, and olive black claystone. Sandstone and siltstone are abundant in sedimentary structures suggestive of rapid sedimentation, whereas claystone is bioturbated suggesting its hemipelagic origin.

Unit 5 is composed of black gray, slightly calcareous claystone with discontinuous dolomitic lamina and micronodules and occasional horizontal burrows. Acidic to intermediate tuff beds are frequently intercalated.

Unit 4 is characterized by highly siliceous rocks and further subdivided into Subunits 4B and 4A. Subunit 4B is composed of gray to dark gray, highly siliceous and occasionally calcareous claystone with parallel lamination. Glauconite pellets are concentrated in the basal part of this subunit. Subunit 4A is characterized by rhythmical alternation of black gray less bioturbated chert and light gray more bioturbated siliceous mudstone. Due to the presence of opaline chert layers, recovery of this interval is poor, and detail observation of the lithology is difficult. However, the results of logging, especially of Formation Microscanner (FMS) images, shows the presence of high contrast rhythmical bedding with the average spacing of 60 cm within this subunit [9]. Based on the limited observation of a small amount of recovered sample, higher resistivity layers are interpreted as chert and lower resistivity layers as siliceous claystone.

Unit 3 is composed of rhythmical alternation of non-bioturbated black gray diatomaceous claystone and bioturbated light gray claystone, and their diagenetic equivalents with occasional intercalations of Mn-carbonate layers in the lower part of the unit [8]. Nature and frequency of rhythmical alternation in this unit are similar to that of Subunit 4A but claystones are less siliceous and the dark and light color contrast is more distinct. At Site 797, thin bioclastic sand layers occur within the basal part of the dark layers in the upper part of this unit.

Unit 2 is composed of gray to dark gray, heavily bioturbated, homogeneous diatom ooze, diatomaceous claystone and their diagenetic equivalents. Dropstones are present in several stratigraphic levels at Site 795. Thin acidic to intermediate ash layers occur in the upper part of the unit.

Unit 1 is composed of rhythmical alternation of dark (olive black) and light (light gray) layers of clay to silty clay which are slightly diatomaceous. Thin ash layers of acidic to intermediate composition are frequent throughout the unit. The unit is further subdivided into Subunits 1B and 1A. In Subunit 1B, light layers are a few meters thick, gray to light gray, heavily bioturbated clay to diatomaceous clay, whereas dark layers are a few decimeters thick, dark brownish gray to black gray, slightly bioturbated silty clay to clay. In Subunit 1A, alternation of dark and light layers is more frequent, more distinct, and dark layers are better laminated. Light layers are a few decimeters to a meter thick, light gray, bioturbated clay to diatomaceous clay, whereas dark layers are a few centimenters to a few decimeters thick, black gray, well laminated silty clay.

In summary, deposition of fine-grained hemipelagic sediments continued from Unit 5 to Unit 1. A switch from bio-calcareous to bio-siliceous accumulation occurred at the begining of Unit 4 deposition. Units 4 and 2 are characterized by bio-siliceous sediments whereas Units 3 and 1 are less siliceous sediments with distinct dark and light layers.

Sediment age and Mass Accumulation Rate (MAR)

Preliminary age of the sediments was determined onboard, and the result was reported in the ODP Initial Reports volume [5]. However, additional stratigraphic and radiometric age data became available afterwards through shorebased investigations. Tada [10] recently reviewed these new results to revise the age model for the sedimentary sequences at Sites 794, 795, and 797. According to his age model, the ages of lithological boundaries at the three sites are approximately synchronous except for that of Units 4/5 at Site 795; namely the ages of Units

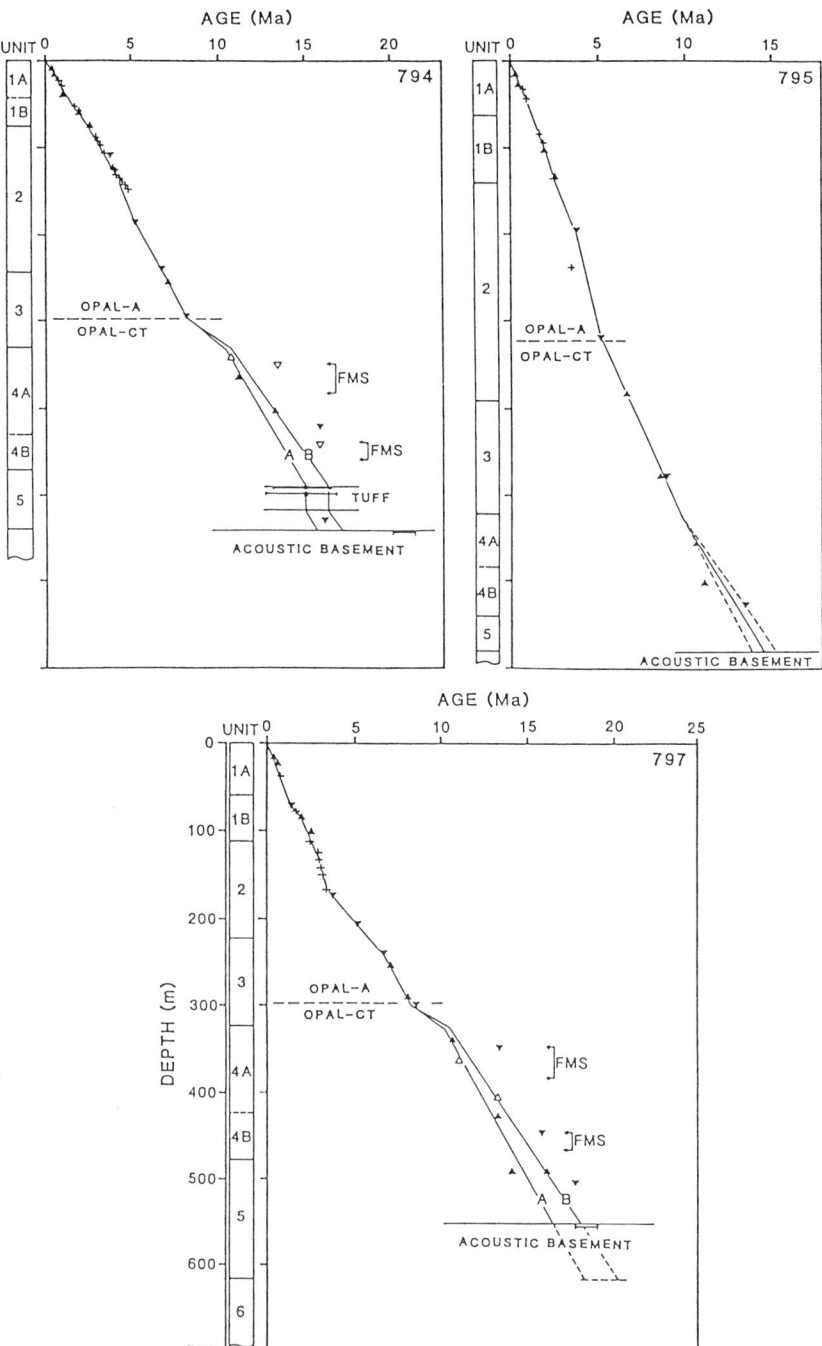

Figure 3. Sedimentation rate diagrams for Sites 794, 795, and 797. Biostratigraphic (normal triangles=last appearance datums, reverse triangles=first appearance datums) and magnetostratigraphic (plus marks) datums, radiometric ages (stars), and FMS correlation intervals are also shown in the figure. Age model B is used in this paper. Further details are described in Tada [10].

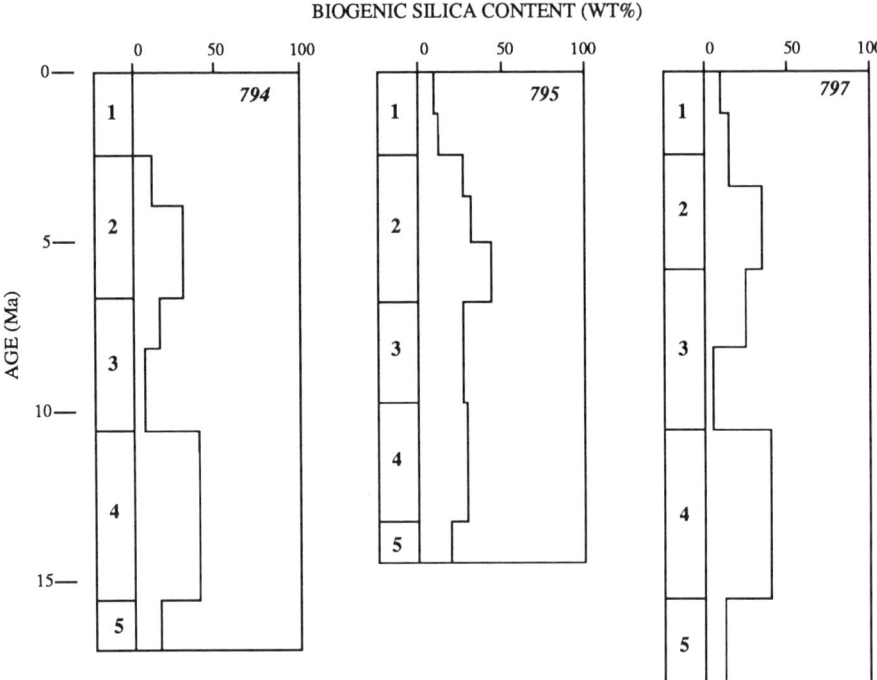

Figure 4. Variation of biogenic silica content with age in hemipelagic sediments at Sites 794, 795, and 797.

1/2, 2/3, 3/4 and 4/5 boundaries are 2.5, 5.9 to 6.8, 9.8 to 10.6, and 13.3 to 15.5 Ma.
Figure 3 shows a sedimentation rate diagram for the three sites after Tada [10]. The linear sedimentation rates at the three sites range from 1.2 to 8.9 cm/ky with generally higher rates at Site 795. Table 1 lists the average linear sedimentation rates after Figure 3, average dry bulk densities based on shipboard measurements [5], and the calculated bulk mass accumulation rate (MAR) for stratigraphically constrained intervals at the three sites. The bulk MAR ranges from 1.2 to 6.3 $g/cm^2/ky$ with generally higher rates at Site 795, especially before 3.7 Ma.

Mineral and chemical compositions
Tada and Iijima [8] analysed chemical and mineral compositions of the hemipelagic sediments recovered from Sites 794, 795, and 797 by X-ray fluorescence analysis [XRF] and X-ray diffraction analysis [XRD]. The result of their mineralogical analysis shows that the hemipelagic sediments are generally composed of detrital minerals such as quartz, feldspar, and clay minerals, biogenic and diagenetic silica minerals such as opal-A, opal-CT, and quartz, and a minor amount of diagenetic pyrite. Carbonate minerals are rare to absent in general. The result of their chemical analysis shows that SiO_2 content of the sediments is variable but inversely correlated with the contents of other major elements suggesting that the sediments are basically mixtures of biogenic (or diagenetic) silica and detrital material with minor contributions of pyrite, organic matter, and calcareous materials.
Based on their chemical and mineral analyses, they estimated the content of biogenic (or diagenetic) silica in the sediments. The average contents of biogenic silica and diluting material for each stratigraphic intervals at Sites 794, 795, and 797 are listed in Table 1. Here, diluting

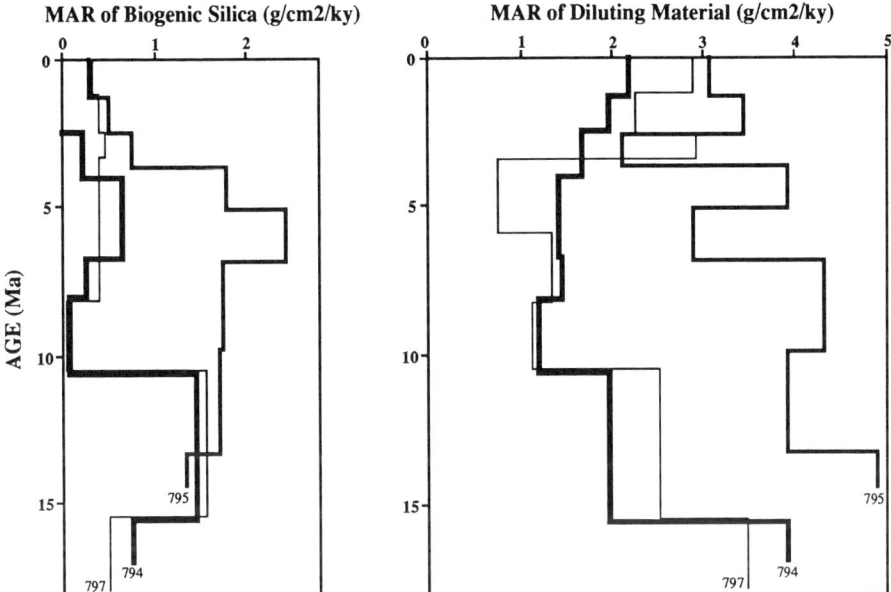

Figure 5. Variation of mass accumulation rates (MARs) of biogenic silica and diluting material with age at Sites 794, 795, and 797.

mateiral simply means the material which dilute biogenic sillica in the sediments, and is calculated as 100 minus biogenic silica content in percent. It basically consists of detrital material with a minor amount of organic matter and pyrite (generally <5 %). Figure 4 shows the vertical profiles of biogenic (and/or diagenetic) silica content at Sites 794, 795, and 797. Biogenic silica content is high between 31 and 46 wt% in Unit 4 and the main part of Unit 2, it is moderate between 6 and 29 wt% in Units 5 and 3, and it is low between 0 and 27 wt% in the upper part of Unit 2 and Unit 1.

MAR of biogenic silica and diluting material
From the bulk MARs, and the contents of biogenic silica and diluting material, it is possible to calculate MARs of biogenic silica and diluting material which are also listed in Table 1. Figure 5 shows variation in MARs of biogenic silica and diluting material with age at Sites 794, 795, and 797. The variation pattern of MAR of biogenic silica is more or less similar between Sites 794 and 797 which are located in the Yamato Basin. At these two sites, MAR of biogenic silica is moderate at 0.6 to 0.8 g/cm^2/ky before 15.5 Ma, highest at 1.5 to 1.6 g/cm^2/ky during 15.5 to 10.5 Ma, lowest at less than 0.1 g/cm^2/ky during 10.5 to 8 Ma, moderate at 0.3 to 0.7 g/cm^2/ky during 8 to 2.5 Ma, and low to moderate at 0 to 0.4 g/cm^2/ky after 2.5 Ma. On the other hand, MAR of biogenic silica at Site 795, which is located in the northernmost part of the Japan Basin, shows quite different patterns. Namely, MAR of biogenic silica is moderate to high between 1.4 and 2.5 g/cm^2/ky from 14.4 to 3.7 Ma with its maximum value between 5.1 and 6.8 Ma, whereas it is moderate to low between 0.3 and 0.8 g/cm^2/ky after 3.7 Ma.

Variation in the MAR of diluting material (dominantly detritus) at the two Yamato Basin sites shows patterns similar to that of MAR of biogenic silica. The pattern is characterized by a distinct maximum at 3.5 to 4.0 g/cm^2/ky before 15.5 Ma, rapidly decreasing to a minimum value of 0.8 to 1.4 g/cm^2/ky during 10.5 to 4.0 or 3.4 Ma, and again increasing to the present

Table 1.
Ages of unit boundaries, linear sedimentation rates, and mass accumulation rates of biogenic silica and diluting material at Sites 794, 795, and 797.

Site	Unit	top depth (mbsf)	bottom depth (mbsf)	interval (m)	top age (Ma)	bottom age (Ma)	duration (m.y.)	LSR (cm/ky)	Av. DBD (g/cm3)	MAR (g/cm2/ky)	Bio.Silica (wt%)	Diluting Material (wt%)	MAR-Silica (g/cm2/ky)	MAR-Diluting Material (g/cm2/ky)
794	1A	0.0	43.8	43.8	0.0	1.3	1.3	3.4	0.65	2.2	0	100	0.0	2.2
	1B	43.8	77.0	33.2	1.3	2.5	1.2	2.8	0.71	2.0	0	100	0.0	2.0
	2	77.0	126.7	49.7	2.5	4.0	1.5	3.4	0.56	1.9	10	90	0.2	1.7
	2	126.7	246.0	119.3	4.0	6.7	2.7	4.4	0.48	2.1	32	68	0.7	1.4
	3 (opal-A)	246.0	298.0	52.0	6.7	8.1	1.4	3.7	0.46	1.7	16	84	0.3	1.4
	3 (opal-CT)	298.0	333.0	35.0	8.1	10.6	2.5	1.4	0.92	1.3	7	93	0.1	1.2
	4	333.0	475.0	142.0	10.6	15.6	5.0	2.8	1.23	3.5	42	58	1.5	2.0
	5	475.0	544.0	41.0	15.6	17.0	1.4	2.9	1.62	4.7	17	83	0.8	4.0
795	1A	0.0	64.6	64.6	0.0	1.3	1.3	5.2	0.65	3.4	8	92	0.3	3.1
	1B	64.6	142.2	77.6	1.3	2.5	1.3	6.2	0.64	4.0	13	87	0.5	3.4
	2 (opal-A)	142.2	200.5	58.3	2.5	3.7	1.2	4.9	0.59	2.9	27	73	0.8	2.1
	2 (opal-A)	200.5	325.2	124.7	3.7	5.1	1.4	8.9	0.64	5.7	31	69	1.8	3.9
	2 (opal-CT)	325.2	395.0	69.8	5.1	6.8	1.7	4.1	1.31	5.4	46	54	2.5	2.9
	3	395.0	525.0	130.0	6.8	9.8	3.0	4.3	1.40	6.1	29	71	1.8	4.3
	4	525.0	645.0	120.0	9.8	13.3	3.5	3.4	1.64	5.6	31	69	1.7	3.9
	5	645.0	684.0	39.0	13.3	14.4	1.1	3.5	1.77	6.3	22	78	1.4	4.9
797	1A	0.0	59.8	59.8	0.0	1.2	1.2	5.0	0.64	3.2	9	91	0.3	2.9
	1B	59.8	112.4	52.6	1.2	2.5	1.3	4.1	0.64	2.7	15	85	0.4	2.3
	2	112.4	167.8	55.4	2.5	3.4	0.9	6.0	0.57	3.4	14	86	0.5	2.9
	2	167.8	224.0	56.2	3.4	5.9	2.5	2.2	0.52	1.2	35	65	0.4	0.8
	3 (opal-A)	224.0	300.0	76.0	5.9	8.2	2.3	3.3	0.54	1.8	23	77	0.4	1.4
	3 (opal-CT)	300.0	327.0	27.0	8.2	10.5	2.3	1.2	1.03	1.2	6	94	0.1	1.1
	4	327.0	480.0	153.0	10.5	15.5	5.0	3.1	1.35	4.1	38	62	1.6	2.6
	5	480.0	554.0	74.0	15.5	18.0	2.5	3.0	1.37	4.1	14	86	0.6	3.5

LSR=linear sedimentation rate, Av. DBD=average dry bulk density, MAR=mass accumulation rate.

value of 2.2 to 2.9 g/cm^2/ky after 4.0 or 3.4 Ma. On the other hand, the MAR of diluting material at Site 795 has been at a higher level and gradually decreased from 4.9 g/cm^2/ky between 13.3 and 14.4 Ma to 3.1 g/cm^2/ky at present with relatively large fluctuations between 6.8 and 2.5 Ma.

DISCUSSION

The biogenic silica content of the sediments from the deeper parts of the Japan Sea shows two maximum peaks of approximately 30 to 45 wt% during Middle Miocene (approximately 15.5 Ma to 10.5 or 9.8 Ma) and late Late Miocene to Early Pliocene (approximately 8.2 or 6.8 Ma to 3.5 or 4 Ma). The minimum between these two peaks is low and distinct at Sites 794 and 797 whereas it is not obvious at Site 795 (Figure 4). The higher contents of biogenic silica could represent higher surface productivity. However, higher contents of biogenic silica in the sediments could be also caused by the decrease in MAR of diluting material. For this reason, the MAR of biogenic silica is generally considered as a better indicator of surface productivity [11,12].

From this view point, the highest MARs of biogenic silica in the Japan Sea sediments, approximately 1.5 g/cm^2/ky during 15.5 to 10.5 Ma at Sites 794 and 797 and approximately 2.5 g/cm^2/ky during 6.8 to 5.1 Ma at Site 795, are still low by an order of magnitude relative to the values of 15 to 45 g/cm^2/ky observed at present high productivity areas such as Antarctica, off Peru, and Gulf of California [13]. Does this indicate that the surface productivity within the Japan Sea never reached levels as high as those observed at the present high productivity areas?

Surface productivity is not the only factor which affects accumulation of biogenic silica. Dissolution of biogenic silica both within the water column and during the early stages of burial (within a few tens of centimenter from the sediment/water interface) could be also important [11,14,15]. Indeed, MAR of biogenic silica for fine-grained siliceous sediments of Middle Miocene to Pliocene age exposed in the Ajigasawa and Babame sites on the eastern margin of the Japan Sea, was 6 to 20 g/cm^2/ky which is 4 to 13 times larger than 1.5 to 1.7 g/cm^2/ky observed at Sites 794, 795, and 797 (Table 2). This difference could be explained in two ways. The first possibility is that higher MAR of diluting material enhanced opal preservation within the sediments through faster burial. The second possibility is that surface silica productivity was higher due to coastal upwelling near the Japanese Islands where higher MAR of detritus (diluting material) would also be expected. The second possibility seems likely especially before Late Miocene when there was a shallow bank between the Japan Sea ODP sites and the onland studied sites in the Japan Sea side of northern Japan [16].

I prefer the first possibility, however, because there tends to be a positive correlation between MARs of biogenic silica and diluting material even for the sediments from adjacent sites which were probably under the similar surface water conditions. One example of this is the Middle Miocene sediments from the Ajigasawa and Babame sites which are only 100 km apart (Table 2). A second example is the Pliocene sediments from Site 794 and the Ajigasawa site which are 200 km apart. The barrier between Site 794 and the Ajigasawa site disappeared, and these two sites became part of a single basin by late Late Miocene [16], thus it is difficult to consieve that the surface productivity differed by as much as factor of 10 between the two sites, although we could not exclude this possibility. On the other hand, the MAR of terrigenous material is more than 10 times larger at Ajigasawa site than Site 794 (Table 2).

The consideration described above leads us to the hypothesis that the increase in MAR of diluting material (dominantly terrigenous material in most cases) may increase MAR of

Table 2.

MARs of biogenic silica and diluting material for Neogene bio-siliceous hemipelagic sediments at DSDP Site 438 and onland sites at Ajigasawa, Babame, and Shimokita.

Site	MAR of Biogenic Silica (g/cm2/ky)				
	(< 3.5 Ma)	(3.5 - 6.5 Ma)	(6.5 - 10.5 Ma)	(10.5 - 15 Ma)	(> 15 Ma)
Ajigasawa		5.7	7.1	5.7 / 0.1	
Babame				20.0	
Shimokita				4.1 / 7.5	2.9
Site 438	1.2	3.0 / 0.8	1.5	0.3	2.6

Site	MAR of Diluting Material (g/cm2/ky)				
	(< 3.5 Ma)	(3.5 - 6.5 Ma)	(6.5 - 10.5 Ma)	(10.5 - 15 Ma)	(> 15 Ma)
Ajigasawa		23.0	8.8	4.3 / 0.2	
Babame				9.8	
Shimokita				4.6 / 8.7	23.5
Site 438	3.4	5.4 / 1.5	4.2	0.5	6.3

biogenic silica though enhancement of preservation of biogenic silica. To test this possibility, MARs of biogenic silica are plotted against MARs of diluting material for fine-grained bio-siliceous sediments from the Japan Sea ODP sites and the onland sites in northern Japan. The result is shown in Figure 6.

As is obvious from the figure, there is a clear positive correlation between the MARs of biognenic silica and diluting material. If we accept the assumption that the surface productivity was more or less constant within the eastern part of the Japan Sea during the late Late Miocene to Early Pliocene (6.5 to 3.5 Ma), then the clear positive correlation shown in the figure supports the hypothesis that higher MAR of diluting material really increases the MAR of biogenic silica through the enhancement of biogenic silica preservation. In addition, the trend for the late Late Miocene to Early Pliocene plots in the figure roughly agrees with the line representing constant biogenic silica content of 30 to 40 wt%. This implies that the dilution effect due to higher MAR of the diluting material is cancelled by the enhancement of biogenic silica preservation due to faster burial, at least within the observed range of MAR.

The Middle Miocene (15.5 to 10.5 Ma) data points also lie on the same trend as the late Late Miocene to Early Pliocene data, suggesting that surface productivity during this period was similar to that of the late Late Miocene to Early Pliocene. The MARs from present high productivity areas [13] are also plotted in the figure. These data tend to have higher MARs of biogenic silica and diluting material, but they are continuous from the trend of the Middle Miocene and the late Late Miocene to Early Pliocene plots from the eastern Japan Sea area.

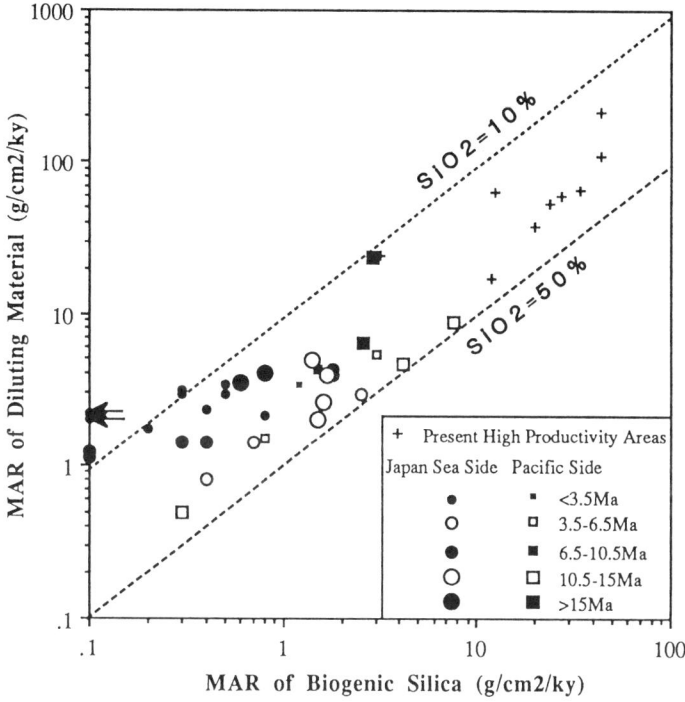

Figure 6. Diagram showing the relationship between MARs of biogenic silica and diluting material in the Neogene hemipelagic bio-siliceous sediments from the Japan Sea, onland northern Japan, and northwestern Pacific Site 438. Data from present high productivity areas [13] are also shown in the figure.

This agreement suggests that surface productivity in the eastern Japan Sea area during these periods could have been comparable to that of present high productivity areas. The higher slope of the trend for present high productivity areas may imply that the dissolution effects of biogenic silica near sediment/water interface becomes insignificant when the MAR of diluting material exceeds 10 g/cm^2/ky. On the other hand, Early Miocene (18 to 15.5 Ma), Late Miocene (10.5 to 6.5 Ma), and Late Pliocene to Pleistocene (3.5 to 0 Ma) data from the eastern Japan Sea area tend to lie to the left of the Middle Miocene and late Late Miocene to Early Pliocene trend ("high productivity" trend), suggesting that the surface productivity during these periods was lower in the eastern Japan Sea area.

Are these lower surface productivity periods during the Early Miocene, Late Miocene, and Late Pliocene to Pleistocene local phenomena specific to the Japan Sea or regional phenomena common in the northwest Pacific? In order to answer this question, we will compare the MARs data from the Japan Sea area with those from the Pacific side of the northern Japan. Figure 6 shows the data from ODP Site 438 and the onland Shimokita site which are located on the Pacific side of northern Japan; and these are compared with the data from the Japan Sea area. As is obvious from the figure, nearly the data from the Pacific side since Middle Miocene (after 15.5 Ma) all plot on the "high productivity" trend, and no deviation from the "high productivity" trend is observed for samples from the Late Miocene and Late Pliocene. This observation suggests that surface productivity in the northwestern Pacific region has been continuously high since Middle Miocene, and the probable drop of the surface productivity within eastern Japan Sea area during Late Miocene and Late Pliocene to Pleistocene could have been the local phenomena specific within the Japan Sea.

Reconstruction of paleoceanographic conditions within the Japan Sea since Early Miocene was recently made by Tada [11] based on the sedimentary records from ODP sites. Based on his reconstruction, the switch from bio-calcareous to bio-siliceous deposition in the Japan Sea area occurred between 15.5 and 14 Ma, which coincides with the global cooling due to build up of the stable Antarctic Ice Sheet [17]. Global cooling caused the southward shift of subpolar front and allowed inflow of the cold current to the Japan Sea. After 14 Ma, deposition of bio-siliceous sediments in the Japan Sea area seems to have been dominated over bio-calcareous deposition until the present.

Tada [10] further argued, based on paleobathymetric reconstruction of the Japan Sea area by Iijima and Tada [16], that the Japan Sea became gradually isolated especially since the Late Miocene probably due to the tectonic uplift in northern Japan caused by east-west compression which started approximately at 7 Ma [18]. Superimposed on this tectonic effect on the shoaling of the sill depths were global eustatic sea level changes. According to Haq et al. [19], a sea level drop of large magnitude occurred at 10.5 Ma and relatively low sea level stands lasted till approximately 6 Ma. Global sea level is considered to have been relatively high until 2.5 Ma when another sea level drop seems to have occurred corresponding to the build up of continental ice sheet in the northern hemisphere [19]. After 2.5 Ma, global sea level oscillated with high frequency and amplitude, but on average the sea level gradually lowered until present.

The sediments deposited within the deeper part of the Japan Sea during 10.5 to 6.5 Ma and 2.5 to 0 Ma are characterized by the presence of dark and light color cycles suggesting periodically restricted circulation within the basin [10]. It is likely that the shallower sill depths and narrower strait widths during these intervals reduced inflow of nutrient-rich Pacific deep water into the sea, thus reduced the surface productivity. Slightly preceding the sea level drop at 2.5 Ma, the Tsushima Strait reopened and the warm current started intruding into the Japan Sea from the south at 3.5 Ma [20]. This event probably weakened the inflow of cold current from the north and helped reduce the surface diatom productivity within the sea.

CONCLUSIONS

1. Significant production and subsequent accumulation of biogenic silica in the Japan Sea area started between 15.5 and 14 Ma. This event coincides with the initiation in upwelling of the nutrient-rich deepwater in the north Pacific as a result of reorganization of the global deepwater circulation as is suggested by Woodruff and Savin [2].
2. Biogenic silica productivity within the Japa Sea was low during 10.5 to 6.5 Ma and 3.5 to 0 Ma when the sill depths of the sea shoaled and inflow of nutrient-rich Pacific deepwater was reduced. Re-opening of the Tsushima Strait and subsequent inflow of the warm current could also have affected surface biogenic silica productivity through reducing inflow of the cold water into the sea. Shoaling of the sill depths was probably related to eustatic sea level drop during these intervals which was superimposed on the gradual tectonic uplift of northern Japan since the Late Miocene.
3. Biogenic silica productivity on the Pacific side of Japan has been high since ca. 15.5 Ma. It did not decrease during 10.5 to 6.5 Ma and after 3.5 Ma. This observation implies that probable decrease in the surface silica productivity recorded in the Japan Sea sediments during these intervals was a local phenomena specific to the Japan Sea.
4. The accumulation rate of biogenic silica, which is often used as a proxy of the surface

biogenic silica productivity, could be affected not only by surface productivity but by MAR of diluting material. Positive relationship between MARs of biogenic silica and diluting material strongly argues for possible enhancement of MAR of biogenic silica by increasing MAR of diluting material through enhancing preservation of biogenic silica at the sediment/water interface. The possibility of this hypothesis should be pursued in the future before using MAR of biogenic silica as a proxy of the surface biogenic silica productivity. Combined research of sediment trap experiments and piston core studies of the surface sediments from various environments with different surface productivities and MAR of diluting material should be the most effective approach to this problem.

Acknowledgments

I express my sincere thanks to Professors A. Iijima and R. E. Garrison for their useful discussions and reading of the manuscript. This study is financially supported by the Grant-in-Aid for Scientific Research of the Ministry of Education, Science and Culture Nos. 03640638 and 04212101.

REFERENCES

1. G. Keller and J.A. Barron. Paleoceanographic implications of Miocene deep-sea hiatuses, *Geol. Soc. Am. Bull.*. **9 4**, 590-613 (1983).
2. F. Woodruff and S.M. Savin. Miocene deepwater oceanography, *Paleoceanography* **4**, 87-140 (1989).
3. R. Tada. Origin of rhythmical bedding in Middle Miocene siliceous rocks of the Onnagawa Formation, northern Japan, *J. Sedim. Petrol.*. **6 1**, 1123-1145 (1991).
4. I. Koizumi. Successional changes of middle Miocene diatom assemblages in the northwestern Pacific, *Palaeogeogr. Palaeoclimatol. Palaeoecol.*. **7 7**, 181-193 (1990).
5. K. Tamaki, K.A. Pisciotto, J. Allan, et al. *Proceedings of the Ocean Drilling Program, Initial Reports vol. 127*. College Station, Texas (1990).
6. I. Kaneoka, Y. Takigami, N. Takaoka, S. Yamashita, and K. Tamaki. ^{40}Ar-^{39}Ar analysis of volcanic rocks recovered from the Japan Sea floor: constraints on the age of formation of the Japan Sea. In: *Proceedings of the Ocean Drilling Program, Scientific Results.* K. Tamaki, K. Suyehiro et al. (Eds). vol. 127/128, part 2, pp. 819-836. College Station, Texas (1992).
7. K. Tamaki, K. Suyehiro, J. Allan, J.C. Ingle, Jr. and K.A. Pisciotto. Tectonic synthesis and implications of Japan Sea ODP drilling. In: *Proceedings of the Ocean Drilling Program, Scientific Results.* K. Tamaki, K. Suyehiro et al. (Eds). vol. 127/128, part 2, pp. 1333-1348. College Station, Texas (1992).
8. R. Tada and A. Iijima. Lithostratigraphy and compositional variation of Neogene hemipelagic sediments in the Japan Sea. In: *Proceedings of the Ocean Drilling Program, Scientific Results.* K. Tamaki, K. Suyehiro et al. (Eds). vol. 127/128, part 2, pp. 1229-1258. College Station, Texas (1992).
9. J.A. Meredith and R. Tada. Evidence for late Miocene cyclicity and broad-scale uniformity of sedimentation in the Yamato Basin, Sea of Japan, from formation microscanner data. In: *Proceedings of the Ocean Drilling Program, Scientific Results.* K. Tamaki, K. Suyehiro et al. (Eds). vol. 127/128, part 2, pp. 1037-1046. College Station, Texas (1992).
10. R. Tada. Paleoceanographic evolution of the Japan Sea, *Palaeogeogr. Palaeoclimatol. Palaeoecol.* (in press).
11. W.S. Broecker and T.-H. Peng. *Traces in the Sea*. Eldigio Press, New York (1982).
12. A.P. Lisitzin. The silica cycle during the last ice age, *Palaeogeogr. Palaeoclimatol. Palaeoecol.*. **5 0**, 241-270 (1985).
13. P.A. Ledford-Hoffman, D.J. DeMaster and C.A. Nittrouer. Biogenic-silica accumulation in the Ross Sea and the importance of Antarctic continental-shelf deposits in the marine silica budget, *Geochim. Cosmochim. Acta* **5 0**, 2099-2110 (1986).
14. L.H. Burkle, A.Sturz and G. Emanuele. Dissolution and preservation of diatoms in the Sea of Japan and the effect on sediment thanatocoenosis. In: *Proceedings of the Ocean Drilling Program, Scientific Results.* K.A. Pisciotto, J.C. Ingle, Jr., M.T. von Breymann, J. Barron et al. (Eds). vol. 127/128, part 1, pp. 309-

316. College Station, Texas (1992).
15. L.D. White and J.A. Alexandrovitch. Pliocene and Pleistocene abundance and preservation of siliceous microfossil assemblages from Site 794, 795, and 797: implications for circulation and productivity in the Japan Sea. In: *Proceedings of the Ocean Drilling Program, Scientific Results*. K.A. Pisciotto, J.C. Ingle, Jr., M.T. von Breymann, J. Barron et al. (Eds). vol. 127/128, part 1, pp. 341-357. College Station, Texas (1992).
16. A. Iijima and R. Tada. Evolution of Tertiary sedimentary basins of Japan in reference to opening of the Japan Sea, *J. Fac. Sci. Univ. Tokyo, Sec II* **22**, 121-171 (1991).
17. K.G. Miller, J.D. Wright and R.G. Fairbanks. Unlocking the ice house: Oligocene-Miocene oxygen isotopes, eustacy, and margin erosion. *J. Geophys. Res.* **96**, 6829-6848 (1991).
18. L. Jolivet and K. Tamaki. Neogene kinematics in the Japan Sea region and volcanic activity of the Northeast Japan Arc. In: *Proceedings of the Ocean Drilling Program, Scientific Results*. K. Tamaki, K. Suyehiro et al. (Eds). vol. 127/128, part 2, pp. 1311-1348. College Station, Texas (1992).
19. B.U. Haq, J. Hardenbol and P.R. Vail. The chronology of fluctuating sea level since the Triassic. *Scinece* **235**, 1156-1167 (1987).
20. I. Koizumi. Diatom biostratigraphy of the Japan Sea: Leg 127. In: *Proceedings of the Ocean Drilling Program, Scientific Results*. K.A. Pisciotto, J.C. Ingle, Jr., M.T. von Breymann, J. Barron et al. (Eds). vol. 127/128, part 1, pp. 249-289. College Station, Texas (1992).

Sedimentary Environment of the Onnagawa Sea: Middle Miocene Japanese Backarc Trough

Y. WATANABE[1], M. YAMAMOTO[2] and N. IMAI[3]
[1,2,3]*Geological Survey of Japan, 1-1-3 Higashi, Tsukuba, Ibaraki 305, Japan*

Abstract: The middle Miocene organic-rich diatomaceous argillite facies is prominent in the backarc trough of northeastern Japan, and serves as a source rock for several commercial oil and gas fields such as Akita and Niigata. This facies was deposited in an N-S trending, narrow, marginal trough, here called the Onnagawa Sea, where its configuration was important in accelarating diatom productivity, in preventing coarse-grained terrigenous influx, and in preserving delicate laminations in the sediments.
Geochemical characteristics indicate that these deposits were spatially controlled by the influxes of siliceous biogenous, terrigenous, carbonate, organic (hydrocarbon-related), and phosphatic components and by bottom water redox conditions. These geochemical factors helped to display the depositional setting of the Onnagawa Sea where the typical laminated diatomaceous argillites with high organic carbon content are concentrated along the western side of the axis of the basin. The accumulation of siliceous biogenic and organic materials were high, and very fine-grained terrigenous fraction was very low under anoxic to euxinic consition which were favorable for developments of organic-rich laminated diatomaceous deposits.

Keywords: diatomaceous deposits, middle Miocene, sedimentary environment, factor analysis

INTRODUCTION

Miocene laminated siliceous sediments are well developed in the northwestern side of Japanese Islands, which represent a counterpart of the Monterey Formation in the western coast of California [1]. They are typically called the Onnagawa Formation in Akita [2] and its equivalents are observed onshore from the northwestern part of Hokkaido [3] to Noto Peninsula [4] on the eastern margin of the Japan Sea. The basin of the Onnagawa Formation contains several oil and gas fields particularly in the Akita and Niigata districts where the source of the hydrocarbons has been attributed mainly to this horizon [5].
The origin and mechanism of sedimentation of the Onnagawa Formation have attracted the attention of many sedimentologists and petroleum geologists because evidence of sea level fluctuations and bottom water conditions could be detected in these sediments [6, 7], which in turn explain the genesis of organic-rich source rocks favorable for oil genesis. Detailed stratigraphic analyses of the sedimentological and geochemical properties led to the formulation of several models to explain environmental changes of the Onnagawa Sea. Tada [6] presented a dilution model where siliceous biogenic productivity and detritus supply are both cyclic, and bottom-water redox also fluctuates in phase with the detritus dilution. He sujested that the detritus supply and bottom-water redox fluctuations were caused by global sea-level oscillations and that productivity was changed more frequently by upwelling intensity in the basin. Fukusawa [7] proposed a seasonal productivity fluctuation model based mainly on analyses of facies and diatom assemblages in the Tempoku Basin, northern Hokkaido. He

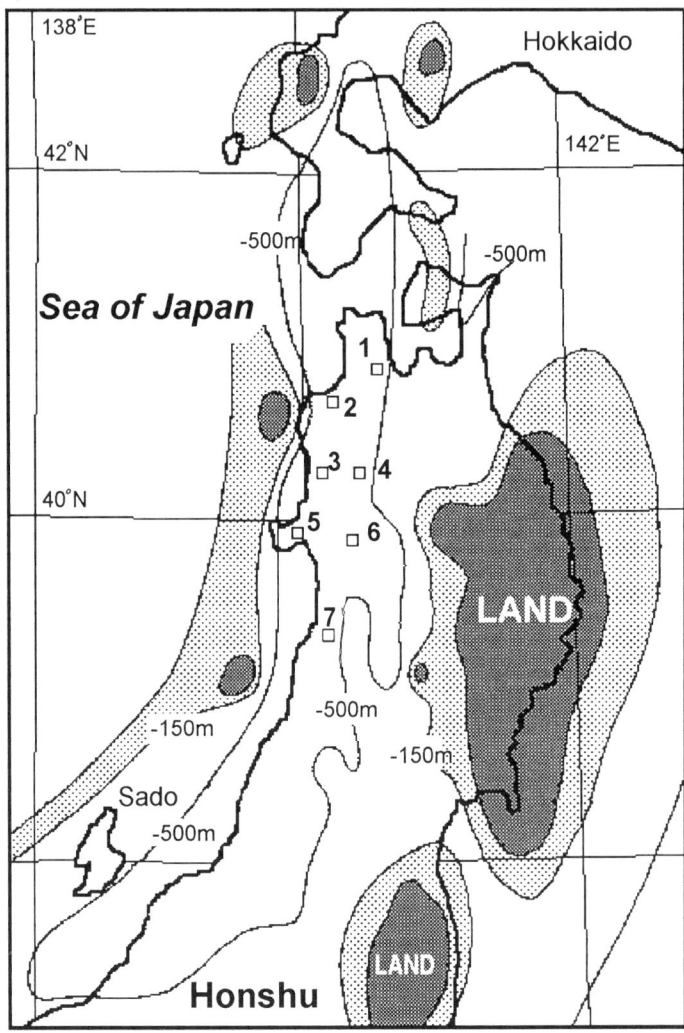

Figure 1. Paleogeographic map of Northern Honshu during middle to late Miocene time (after Iijima et al. [8]). The Onnagawa laminated diatomaceous sediments were deposited in the central part of the N-S trending basin at an approximate batymetrical depth of more than 500 m. Both sides of the basin were fringed by shallow (lightly dotted) and land (heavily dotted) areas. The seven areas of this study are: 1: Goshogawara, 2: Ajigasawa, 3: Noshiro, 4: Takanosu, 5: Oga, 6: Gojonome, and 7: Yashima areas.

explained the fluctuations in the local upwelling system were controlled by the wind system. It should be also emphasized, however, that the basin configuration sometimes plays an important role in the spatial distribution of laminated biosiliceous sediments [8, 9]. Among the several paleogeographic reconstructions of the Onnagawa Basin for the Miocene period, Iijima et al. [8, 10] presented detailed basin outlines and suggested that submarine topography controlled the influx of detritus as well as the locations of upper batyal sills which determined bottom-water oxygenation. Hence, the origin and mechanism of these sediments were controlled by both the time-sequential fluctuation of environments and the paleobathymetrical settings.

For the understanding of the sedimentary environment of the homogenous Miocene Onnagawa siliceous deposits, measurement of the variance of the geochemical characteristics is one of the most useful techniques. The purpose of this study is to geochemically deduce the factors controlling the sedimentation of the typical diatomaceous Onnagawa Formation. More detailed studies with integrated minor and trace element investigations will be presented separately.

GEOLOGICAL BACKGROUND AND SAMPLES

The middle Miocene Onnagawa Basin comprises one of the subsided rifts that followed the rapid opening stage of the Japan Basin during approximately 22 to 15 Ma [11]. These highly siliceous, laminated sediments are observed in a narrow NNE-SSW trending basin, which lay immediately west of the main backbone range from northern Honshu to central Hokkaido during middle to late Miocene time [8]. This basin contained middle bathyal deeps in its center, and extended at least from southern Hokkaido to Niigata (Sado) (Fig. 1). The eastern slopes of the deeps were wide upper bathyal to shallow shelfal seas bounded by large islands (Kitakami and Abukuma massifs), while a line of shallow banks with small islands probably separated the Onnagawa basin from the main part of the Japan and Yamato Basins to the west. The onset of the Onnagawa siliceous sedimentation was rather gradual at most localities, and locally began with deposition of a basal glauconitic facies [8, 10, 12]. Nevertheless, homogeneous siliceous sediments were dominant from 13 Ma until 10 to 8 Ma [13].

The samples for this study were taken from five areas in the Akita Prefecture. Additional data were obtained from Aomori Prefecture, from the Ajigasawa and Goshogawara areas [14, 15 and R. Tada; person. comm.]. The five sample sites with two reference areas are well scattered in the northern part of the Onnagawa Sea (Fig. 1). According to the paleogeographic map of Iijima *et al.* [8], the Ajigasawa, Noshiro and Oga areas represent the western deep part of the basin; the Goshogawara, Takanosu and Gojonome areas are generally located in the central portion of the basin; and the Yashima area represents deposition on the eastern shallow slope of a glauconite-topped bank.

Although biostratigraphic datum could not be obtained from all the areas due to diagenesis, lithostratigraphic correlation gives a good definition of the bottom of the Formation. More problematic is the top of the formation. The lithology generally changes gradually from pure siliceous sediments to argillaceous siliceous mudrocks, but this change was not always synchronous among the localities [8]. As shown in the Yashima area, the lithology changes to more argillaceous from approximately 10 Ma although the top of the formation is estimated as 8.4 Ma [16], whereas the top of the diatomaceous horizon in the Taiheizan area were dated as 10.7 to 9.7 Ma [17]. These ages are rather older than the type locality in the Oga Peninsula, where the top is dated as 5.8 Ma [17]. All sections have not been fully dated, however, and we only included the samples in the main part of the Onnagawa Formation, defined as lying above the locally developed basal glauconite bed and below the first argillaceous horizon, within the age range of 13 to 10 Ma. The main part of the Odoji Formation in the Ajigasawa area and the Umanokamiyama Formation in the Goshogawara area are correlated with a time interval of approximately 12 to 10 Ma [14, 18, 19]. The Onnagawa Formation in the Takanosu area includes diatom biozones of *Denticulopsis praedimorpha* and *Coscinodiscus yabei* zones in its upper part which represent 12.8 to 9.0 Ma in age [20]. Although there is insufficient biostratigraphic data from the Noshiro and Gojonome areas, lithostratgraphic correlations suggest the same duration of the deposition as in with the Takanosu area, i.e. approximately 13 to 10 Ma [13]. In the Oga area, the formation is dated as 12.9 to 5.8 Ma based on diatoms [17, 21]. In the Yashima area, the dates of the unit are approximately as 13 to 8 Ma [16] where the samples were taken below the horizon of ca.10.2 Ma. Thus, 58 rock samples with additional 14 reference data analyzed in this study record the main part of the Onnagawa Formation deposited from approximately 12 to 10 Ma.

ANALYTICAL METHODS

In our study, chemical compositions of the rock samples from the Noshiro, Oga, Takanosu, Gojonome, and Yashima areas were determined by various methods, including X-ray fluorescence analysis for major elements (Si, Ti, Al, Fe, Mn, Mg, Ca, Na, K and P) [22], ICP-OES (inductively coupled plasma optical emission spectrometry) for selected major and minor elements (Ti, Al, Fe, Mn, Mg, Ca, P, Cu, V, and Zn) [23], ICP-MS (inductively coupled plasma mass spectrometry) for selected minor elements (Ba, Co, Cr, Cu, Mn, Ni, P, Sr, V, Zn and Zr) [24] and atomic absorption analysis for Na and K [25]. Total organic and carbonate carbon of some representative samples were determined by CHN analyzer [26]. Petrographic observations and determinations of lamina preservation were done according to Tada *et al.* [14]. Samples with higher contents of carbonate and tuffaceous material were excluded by analyzing only those samples with the carbonate carbon contents less than 0.5 wt % and by omitting the highly tuffaceous samples based on petrographic observation. Other detailed procedures are described in Yamamoto and Watanabe in this volume.

RESULTS AND DISCUSSION

Principal constituents of the siliceous sediment

The chemical composition of the samples are listed in Table 1. As has been discussed by the many authors [6, 14-16], laminated siliceous sediments of the Onnagawa Formation are principally admixtures of various components including siliceous and calcareous biogenic, detrital, organic, phosphatic, diagenetic, etc. The main composition, however, is determined by the terrigenous detrital component and the diluting siliceous biogenic one. This is apparent from the high negative correlation between SiO_2 and Al_2O_3 (r = -0.78) as shown in Fig. 2, where the former represents the siliceous biogenic and the latter the detrital fractions. If we can assume the siliceous biogenic fraction contains no aluminum, most, if not all, of the Al_2O_3 amount can be attributed to the detrital fraction. The average value for the detrital fraction in the Neogene siliceous rocks of the northern Honshu was formerly reported as 16 % for Al_2O_3 [14]. Assuming this value is valid for the samples in this study, the content of the detrital fraction in samples is calculated based on the above Al_2O_3 value, as bulk Al_2O_3 divided by 16 %. Fig. 3 shows the relationship between SiO_2 and the detrital fraction (hereafter called DET; wt %) where the intersection of the regression line is 61.4 % SiO_2 at the 100 % DET value. This value represents the hypothetical SiO_2 value of the terrigenous detrital component. Compared with the previous works by Tada *et al.* [14] and Tada and Watanabe [15], who reported a value of 67.5%, this SiO_2 value is slightly lower.

The siliceous biogenic component, represented as the biogenic silica content (hereinafter called BIO; wt%) is then calculated based on the hypothetical SiO_2 content of the detrital component, as the residue of SiO_2 subtracted by the detrital SiO_2 (61.4 % of DET).

Table 1.
Chemical composition of the Onnagawa siliceous sediments. Data from the Ajigasawa and Goshogawara areas (*) are referred from Watanabe and Tada [15], those from the Noshiro, Takanosu and Yashima areas (**) are refereed from Yamamoto and Watanabe [26]

Sample #	SiO_2	TiO_2	Al_2O_3	Fe_2O_3	MnO_2	MgO	CaO	Na_2O	K_2O	P_2O_5	Cu	V	Zn
Ajigasawa area*													
G-10	80.15	0.30	6.60	3.79	0.03	1.15	0.29	0.46	1.35	0.02	112	27	225
G-9L	89.01	0.15	3.06	1.64	0.00	0.36	0.04	1.13	0.84	0.10	85	12	49
G-0BL	83.82	0.26	5.48	3.07	0.02	0.83	0.06	0.50	1.14	0.00	83	18	167
G-0BD	81.24	0.32	7.09	3.61	0.03	1.06	0.10	0.50	1.37	0.15	116	35	261
K-29A	87.69	0.15	3.68	2.33	0.03	0.79	0.63	0.53	0.73	0.00	115	28	220
K-26	89.02	0.08	1.45	2.19	0.01	0.30	0.29	0.43	0.38	0.01	61	0	157
K-25B	88.10	0.13	2.63	2.75	0.03	0.44	0.39	0.48	0.59	0.01	88	11	215
K-20A	77.29	0.30	6.86	3.79	0.04	1.52	0.96	0.61	1.35	0.31	111	52	233
K-19B	82.27	0.21	4.72	2.54	0.03	1.01	0.69	0.52	0.99	0.00	113	31	243
K-19A	74.07	0.32	8.27	4.05	0.05	1.65	1.17	0.80	1.55	0.00	156	65	283
Goshogawara area*													
M-25B	81.50	0.20	4.43	3.07	0.04	0.73	0.23	0.23	0.92	0.00	78	45	130
M-24	85.30	0.19	4.62	2.34	0.02	1.02	0.15	0.48	0.94	0.00	49	35	110
M-23A	82.19	0.25	6.55	2.81	0.02	2.05	0.64	0.71	0.95	0.00	67	49	209
M-22	82.72	0.31	7.36	3.14	0.03	1.99	0.34	0.45	1.49	0.00	181	45	277
Noshiro area**													
88NSR2	90.80	0.13	4.31	0.80	0.01	0.64	0.26	0.28	0.77	0.20	58	87	58
88NSR4	90.78	0.17	4.50	1.53	0.02	0.58	0.11	0.26	0.77	0.07	50	51	150
88NSR14	85.18	0.31	7.11	1.76	0.02	0.72	0.05	0.34	2.76	0.05	135	137	435
88NSR15	81.49	0.18	4.70	1.94	0.07	2.80	4.93	0.43	1.00	0.51	47	157	93
88NSR16	91.05	0.08	1.90	0.79	0.04	1.53	2.79	0.25	0.33	0.03	38	114	59
Takanosu area**													
88NSR32	89.55	0.18	5.02	1.50	0.02	0.81	0.58	0.38	0.91	0.08	43	62	217
88NSR33	88.67	0.21	4.94	1.98	0.03	0.99	0.23	0.33	1.17	0.07	51	72	252
88NSR35	94.77	0.09	2.14	0.76	0.01	0.22	0.22	0.13	0.33	0.02	25	32	82
88NSR36	90.53	0.17	4.13	1.36	0.01	0.50	0.20	0.34	0.53	0.06	49	70	144
88NSR37	92.67	0.12	3.10	1.13	0.01	0.47	0.44	0.33	0.63	0.06	33	33	80
88NSR38	81.21	0.35	8.06	3.26	0.02	1.67	1.49	1.00	1.60	0.12	57	111	125
Oga area													
SZ7F	97.35	0.01	0.46	0.28	0.00	0.04	0.00	0.03	0.08	0.02	0	1	15
SZ21	98.00	0.05	1.34	0.54	0.00	0.08	0.01	0.29	0.20	0.06	5	23	6
SZ20C	78.84	0.30	6.95	1.99	0.06	1.93	1.54	0.33	1.49	0.05	21	45	104
SZ20B	81.73	0.23	5.21	3.86	0.03	0.63	0.06	0.26	1.15	0.06	61	53	150
SZ20A	78.10	0.21	5.37	1.87	0.03	1.70	2.12	0.57	0.79	0.03	31	43	77
SZ20	78.28	0.34	8.48	3.08	0.03	1.39	0.21	0.34	1.61	0.02	54	65	237
SZ16	76.59	0.30	7.35	2.49	0.02	1.09	0.02	0.40	1.61	0.02	67	66	249
SZ15	82.48	0.29	6.62	1.23	0.01	0.92	0.42	0.34	1.39	0.04	51	44	61
SZ12B	97.64	0.02	0.76	2.00	0.00	0.06	0.02	0.39	0.13	0.02	0	3	13
SZ12A	79.18	0.29	7.39	3.29	0.02	1.02	0.06	0.42	1.45	0.02	59	52	243
SZ11	81.32	0.27	7.19	2.13	0.02	0.82	0.23	0.58	1.29	0.02	48	50	244
SZ10A	71.95	0.43	11.30	3.96	0.03	1.33	0.46	1.64	2.20	0.02	39	77	198
SZ10C	88.40	0.13	3.69	1.18	0.01	0.37	0.08	0.27	0.61	0.63	30	45	56
SZ8	75.13	0.32	9.67	2.92	0.03	1.58	0.45	0.63	1.38	0.02	40	62	208
SZ7C	89.92	0.14	4.08	1.28	0.02	0.55	0.24	0.31	0.71	0.02	28	41	162
SZ5D	83.40	0.19	5.23	1.71	0.03	0.85	0.59	0.44	0.95	0.05	48	47	139
SZ5C	97.81	0.03	0.69	0.26	0.00	0.05	0.02	0.08	0.10	0.02	1	9	21
SZ7B	77.64	0.26	6.67	3.00	0.02	1.06	0.79	0.15	1.83	0.25	60	53	273

Table 1.
Chemical composition of the Onnagawa siliceous sediments (continued).

Sample #	SiO$_2$	TiO$_2$	Al$_2$O$_3$	Fe$_2$O$_3$	MnO$_2$	MgO	CaO	Na$_2$O	K$_2$O	P$_2$O$_5$	Cu	V	Zn
Oga area													
SZ5B	84.50	0.21	5.58	1.86	0.02	1.01	0.27	0.59	0.88	0.02	23	45	94
SZ5A	98.79	0.01	0.32	0.10	0.00	0.05	0.00	0.15	0.08	0.01	0	1	7
SZ3	79.52	0.03	0.82	0.40	0.00	0.08	0.01	0.47	0.19	0.01	9	8	6
SZ2B	83.26	0.21	4.88	1.80	0.09	0.58	0.09	0.34	1.17	0.13	60	55	22
SZ2A	79.58	0.28	4.65	2.56	0.01	0.96	0.02	0.13	1.48	0.02	54	53	93
SZ1	71.47	0.29	7.43	2.68	0.02	2.20	2.27	0.70	1.01	0.03	27	54	89
SZ9	82.54	0.20	5.04	2.11	0.04	0.74	0.41	0.21	1.11	0.04	57	53	153
Gojonome area													
BA-2	83.81	0.19	4.14	1.83	0.00	0.71	0.19	1.01	0.86	0.09	48	50	243
BA-3-1	80.10	0.27	5.52	2.59	0.04	1.08	0.25	0.50	1.11	0.09	30	45	56
BA-4	74.23	0.27	5.36	2.48	0.06	2.02	1.74	0.71	1.11	0.11	40	62	208
BA-5	83.95	0.21	4.73	2.08	0.00	1.03	0.24	0.43	0.87	0.07	51	44	61
BA-6	85.58	0.13	2.74	1.41	0.00	0.94	0.87	0.53	0.68	0.07	39	77	198
BA-8-2	80.96	0.21	5.21	2.17	0.03	1.12	0.30	0.63	0.99	0.07	31	43	77
BA-9-1	86.81	0.16	3.63	1.74	0.00	0.70	0.09	0.42	0.75	0.05	59	52	244
BA-9-2	86.87	0.12	2.82	1.36	0.00	0.54	0.07	0.38	0.57	0.08	67	55	193
BA-10	83.50	0.22	5.20	2.14	0.00	1.05	0.09	0.49	1.00	0.05	60	53	272
BA-11	81.97	0.25	5.54	2.45	0.05	1.08	0.26	0.50	1.05	0.10	67	66	250
BA-12	87.32	0.14	2.85	1.37	0.00	0.50	0.08	0.43	0.69	0.08	0	1	7
BA-13	89.91	0.13	2.29	1.34	0.00	0.44	0.04	0.36	0.58	0.08	9	8	6
BA-15	83.02	0.20	3.61	1.88	0.02	0.97	0.73	0.57	0.90	0.08	51	63	112
BA-16	80.55	0.26	5.75	2.47	0.01	1.40	0.56	0.72	0.75	0.07	27	54	89
BA-18	87.90	0.15	2.22	1.48	0.00	0.39	0.03	1.16	0.67	0.07	0	1	15
*Yashima area***													
927-2	83.52	0.21	5.62	2.05	0.03	0.53	0.64	0.59	0.96	0.09	29	36	79
1003-2	79.92	0.30	8.59	2.58	0.03	0.80	0.81	0.80	1.37	0.09	36	49	125
927-1	88.16	0.26	6.01	1.75	0.02	0.54	0.53	0.52	1.33	0.07	46	62	154
923-4	83.40	0.22	6.14	1.54	0.02	0.60	1.34	0.62	1.31	0.56	40	70	184
921-3	87.39	0.17	4.23	1.24	0.01	0.42	0.38	0.42	0.76	0.07	37	40	122
921-1	86.27	0.35	7.70	3.17	0.02	1.37	0.89	0.67	1.32	0.09	41	56	104
808-1	91.23	0.36	7.91	3.07	0.02	1.36	0.69	0.88	1.50	0.09	44	72	131

Hypothetical end member composition of the detrital component

According to the value of DET, the individual contribution of each element in the hypothetical detrital composition can be calculated as follows. Provided that the concentration of each element in a sample can be attributed to the sum of the detritus and additions from other components, scatter on the element vs. DET plot should be deviated upward from a linear line of the true detrital regression. Although there may be downward deviations due to depletion by post-depositional diagenesis, the detrital regression line can be defined at the bottom of the scatter for most of the elements and should intersect the origin of both axes.

There are three element groups in the element vs. DET plot. The first one includes TiO$_2$ (Fig. 4), K$_2$O and Zr which are tightly concentrated along a straight line with correlation coefficients greater than 0.8. This element group is interpreted as being mostly concentrated

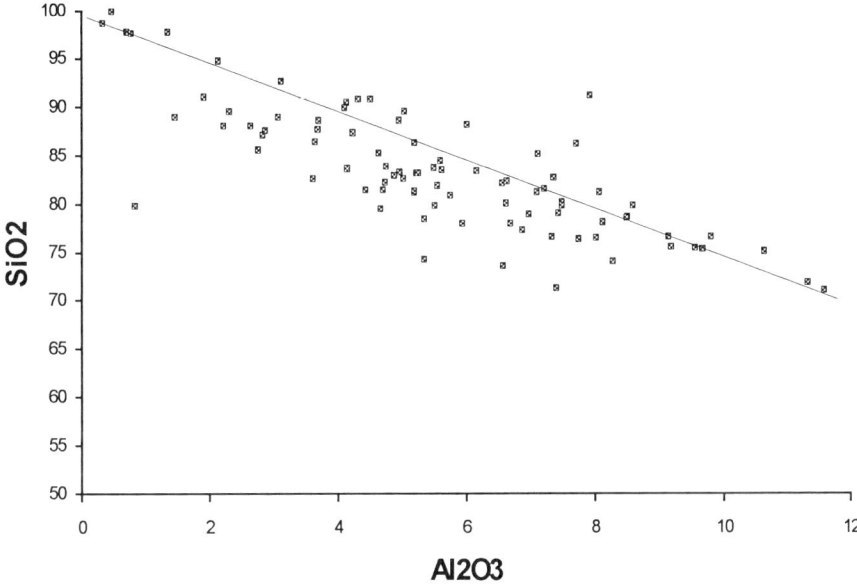

Figure 2. Relationship between Al2O3 and SiO2 in the Onnagawa siliceous sediments.

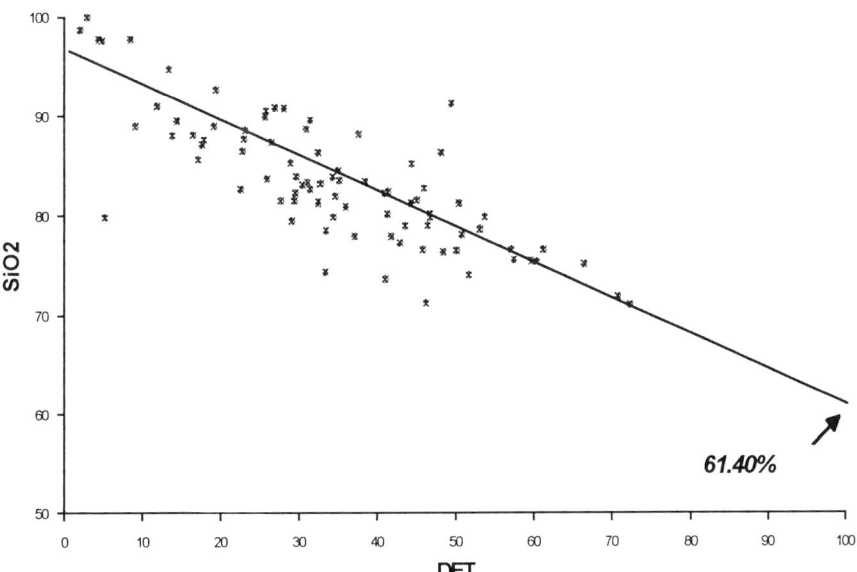

Figure 3. Relationship between SiO2 and DET value calculated from the average Al2O3 content in detrital fraction.

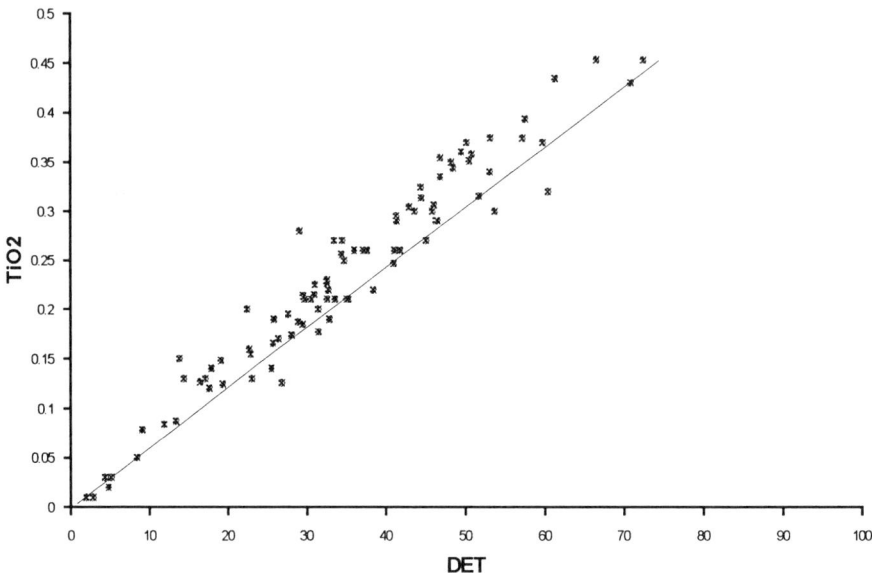

Figure 4. Relationship between TiO2 and DET value, where the hypothetical TiO2 content in detrital fraction is shown by a regression line.

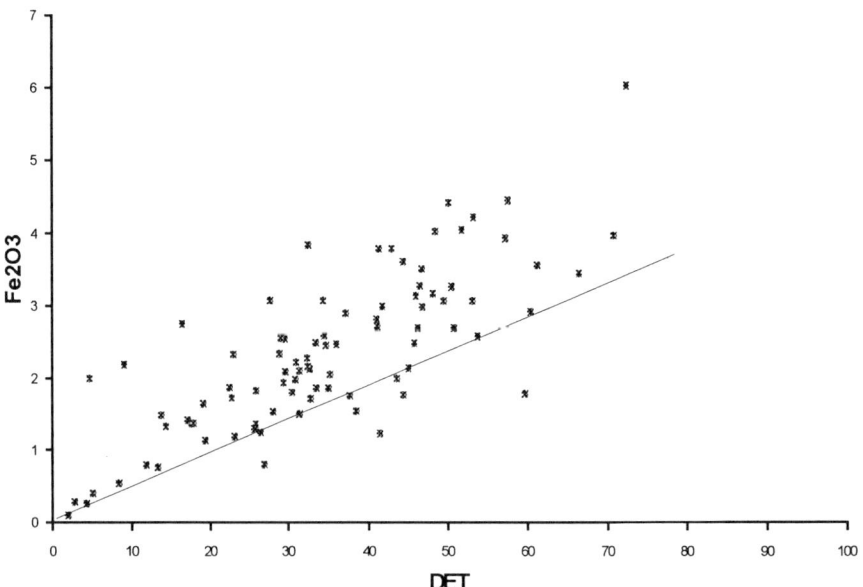

Figure 5. Relationship between Fe2O3 and DET value. Plots are rather scattered indicating a polygenetic nature of iron not only from detrital fraction.

in the detritus components and was least affected by later diagenetic remobilization.

The second group consists of Fe2O3 (Fig. 5), MgO, Na2O, Co, Cu, Ni, Sr, V and Zn. For this group, a schematic regression line can be drawn neglecting some exceptionally low values, although the regression lines for detritus are rather problematic due to a wider scatter of the plots. These elements are polygenic and originally were derived from several components other than the detritus. Mg, Fe and Sr are thought to be contributed from carbonate minerals although significantly calcareous samples were excluded from this study. Iron and other minor elements such as Cu, Ni, Co, V and Zn are generally incorporated into the iron sulfide phase. Cu, Ni and V are sometimes incorporated into organic compounds such as metal porphyrins, and Fe, Co, Cu, Ni and Zn are known to become concentrated in some hydrothermal deposits.

The last group comprises other elements (CaO (Fig. 6), MnO2, P2O5, TOC, S and Ba) which scattered widely in the diagram and have no apparent correlation with DET. They generally have very low values even at the higher DET values. This suggests rather strong depletion due to post-depositional diagenesis because these elements must also occur in detritus fraction in variable amount. These elements are known to be genetically mobile during diagenesis, e.g. Ca occurs in diagenetic carbonate (dolomite), P in phosphate, and Ba in barite. Mn is remobilized in response to changing redox conditions, and organic materials are decomposed by microbial activities, which in turn fixes sulfur in the sediments.

Total organic carbon (TOC) in marine sediments is controlled by the processes of accumulation and preservation. If the organic matter is exclusively of detrital origin, increased detrital fluxes lead to a higher TOC flux in the sediment; increased preservation of organic matter due to rapid burial; and also increased preservation due to higher amounts of clay minerals which prevent organic material from decomposing due to microbial activity [27]. In contrast, if the organic materials accumulate independently of the detrital component, a higher amount of detritus serves only to dilute the bulk amount of TOC in the

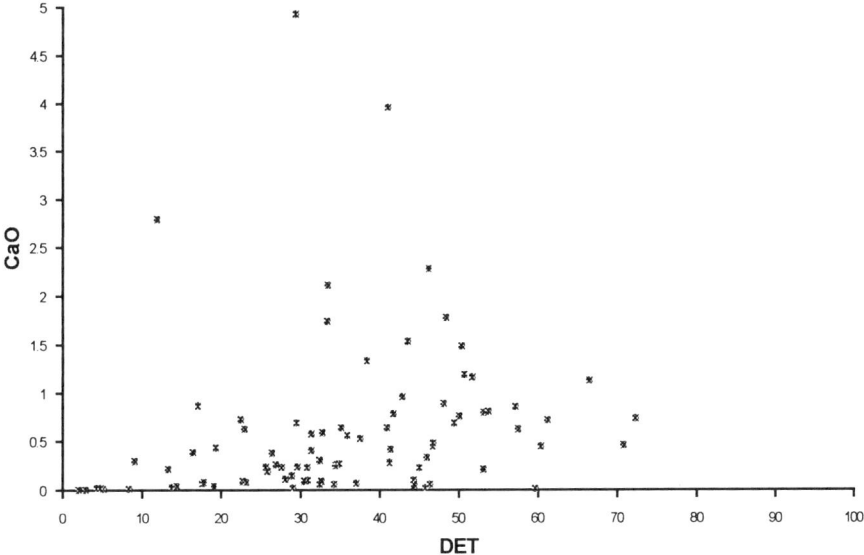

Figure 6: Relationship between CaO and DET value. Plots are very scattered indicating both the polygenetic origin of calcium and post depositional mobilization.

sediments. As shown in Fig. 7, TOC is negatively correlated with DET although the coefficient is very low (r = -0.42), while the biogenic silica (BIO) is positively correlated with TOC. This suggests that the detrital component serves principally to dilute the organic material in the sediment, and that at least some portion of the organic material was incorporated in the biogenic silica fraction.

Consequently we exclude the elements of the third group in our calculation of the hypothetical detrital composition. The hypothetical compositions of the detrital component in the Miocene siliceous sediments are listed in Table 2. This composition seems to represent a mixture of andesitic to dacitic volcanoclastics, argillaceous sedimentary rocks and clay minerals which were also detected under microscopic observation of silt constituents and by X-ray diffraction analysis. Geochemically, the rather high K_2O/Na_2O and FeO/MgO ratios compared with the average composition of modern back-arc mud [28] also suggest the incorporation of dacitic volcanics with higher FeO/MgO ratios and K-rich clay minerals such as illite-smectite mixed layer clays. This lithologic combination is quite reasonable because andesitic to rhyolitic submarine volcanism occurred sparsely in the eastern part of the basin and the main Kitakami landmass, consisting of an older metamorphic and granitic complex, formed the eastern border of the basin [10].

Factors controlling sedimentation of the siliceous sediment

Because the detrital and biogenic components are the by far the largest two constituents of the rocks, statistical analyses are necessary to interpret several minor components. We conducted two fold Q-Mode factor analyses on the elements measured for all samples shown in Table 1. Initially, the bulk chemical composition was used to distinguish all controlling factors, although it was apparent dilution of the biogenic component by detritus was the most important Factor.

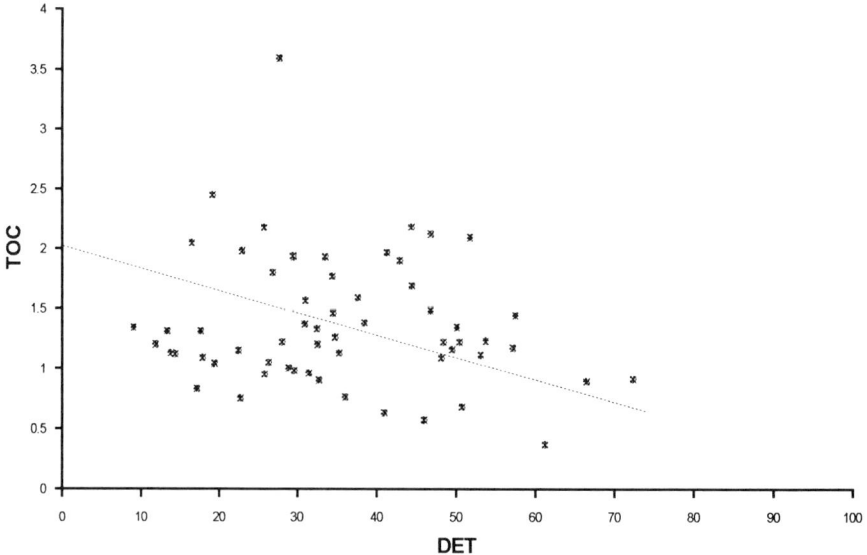

Figure 7. Scatter plot of total organic carbon vs. DET value. A weak negative correlation is observable although the coefficient is low (r = -0.42).

To elucidate the other components (or factors) more clearly by excluding the dilution effect by the detritus, we performed a second factor analysis based on the same data set after all were normalized by the amount of detritus (DET). This analysis focuses on the behavior of excess amounts of components above the detritus contribution, but it may contain some analytical deviation due to the error in the estimate of the DET value. These two factor analyses, however, gave consistent results, as shown below.

The procedures were taken with Varimax rotation for 100 iteration, and the largest of several factors, explaining more than 75 to 80 % of the total sample variance, are extracted for later interpretation. The scores of the extracted factors for each sample were then used for correlation analyses with the value of external variables such as DET, BIO, TOC, S, carbonate carbon and other minor elements.

Factor 1, derived from the first factor analysis and based on the bulk composition data, is 95 %-confidently loaded with silicate-lithogenous major elements such as Al_2O_3, TiO_2, Na_2O, K_2O, and Fe_2O_3, and the scores for each sample positively correlate with Zr, Sr, Ba and DET values (Table 3). This factor apparently represents the terrigenous detrital fraction. It is obvious from the negative relation between SiO_2 and BIO that the detrital fraction is diluting the siliceous biogenic fraction within the limit of the bulk compositional values.

Factor 2 is rich in Fe_2O_3, Cu, Zn and positively correlates with Ni, S and BIO. This combination indicates iron sulfide with subordinate metal elements associated with organic matter. It is known that pyrite contains small amounts of metals such as Cu, Ni, Mn, Mo and Zn [29], particularly in sediments of high organic carbon content such as black shales and oil shales [30, 31]. It is not clear whether the correlation between iron and TOC indicates the coexistence of iron sulfides. Also unknown at this stage of the analyses is the origin of the coexisting metals. As discussed later, the origin of the iron in the sulfides is supposed to be a reactive iron in the detritus. Therefore, the iron loaded in the sulfides component independent of the detritus may reflect post-depositional remobilization of iron, and accompanied metals in this component.

Factor 3 is dominated by carbonate components (CaO and carbonate carbon), where MnO_2, MgO, V and Ba are incorporated. This is concordant with the fact that the carbonate concretions in these horizons are sometimes dolomitic [32].

Factor 4 has high loadings for MnO_2 and small contributions of Fe_2O_3, MgO, Co and Ni. Manganese is often fixed into the sediment primarily as hydroxides unless the bottom water anoxia decomposes the hydroxides and discharges the Mn into the water [33]. Hence the high MnO_2 content is interpreted as the result of bottom water oxygenation and slow sedimentation rate [34]. Co and Ni are also the main hydrogeneous elements which are precipitated from the sea water in oxygenated deep sea pelagic environments with slow sedimentation rate [35].

However this interpretation is only valid assuming no anomalous addition of these elements from other sources such as metallic-rich hydrothermal waters. Chemical and mineralogical suites of submarine hydrothermal deposits also have the same combination of these elements in primary ferromanganese hydroxyoxides along with trace metals such as Ni, Cu and Zn [36].

Table.2
Hypothetical chemical composition of detrital fraction in the Onnagawa siliceous sediments.

SiO_2	TiO_2	Al_2O_3	Fe_2O_3	MgO	Na_2O	K_2O	(%)
61.4	0.59	16	4.6	1.35	1.2	2.16	
Ba	Co	Cu	Ni	Sr	V	Zn	Zr	(ppm)
620	14	58	62	74	80	116	137	

Table. 3
Varimax-rotated factor matrix of the bulk chemical composition of the Onnagawa siliceous deposits. Also shown are the correlation coefficients between each factor score and external variables shown with asterisk (*). Those factor loadings and coefficients with more than 95% significance are shown in bold face. BIO: biogenic silica, DET: detrital fraction, TOC: total organic carbon, CarbC: carbonate carbon content. The "without detritus" value with ** represents % of the variance explained by each factor without the detritus factor (Factor 1), for comparison with Table 4.

	Factor 1	Factor 2	Factor 3	Factor 4	Factor 5	Factor 6
SiO_2	-0.797	-0.127	-0.318	-0.268	-0.024	0.035
TiO_2	**0.867**	0.278	0.124	0.248	0.081	-0.031
Al_2O_3	**0.936**	0.091	0.145	0.100	0.175	-0.019
Fe_2O_3	**0.616**	**0.439**	0.149	**0.458**	-0.224	-0.110
MnO_2	0.201	0.041	0.199	**0.872**	0.109	0.029
MgO	0.281	0.290	**0.746**	**0.384**	0.123	-0.012
CaO	0.181	-0.207	**0.896**	0.044	0.181	0.133
Na_2O	**0.782**	-0.278	0.154	-0.331	-0.024	-0.034
K_2O	**0.770**	0.308	-0.012	0.267	**0.346**	0.004
P_2O_5	-0.075	0.027	0.096	0.012	0.113	**0.976**
Cu	0.148	**0.893**	0.009	0.029	-0.055	0.110
V	0.215	0.138	**0.386**	0.092	**0.799**	0.209
Zn	0.109	**0.782**	-0.051	0.106	**0.436**	-0.110
BIO*	**-0.927**	-0.100	**-0.218**	-0.175	-0.135	0.051
DET*	**0.936**	0.080	0.147	0.108	0.186	-0.046
S*	**0.515**	**0.580**	0.177	0.073	-0.194	0.034
TOC*	-0.154	**0.248**	-0.013	0.177	-0.072	0.180
CarbC*	-0.162	-0.145	**0.377**	0.124	**0.364**	**0.450**
Ba*	**0.399**	0.108	**0.262**	**-0.315**	0.213	0.066
Sr*	**0.654**	-0.032	**0.317**	**-0.266**	0.108	-0.045
Zr*	**0.924**	-0.080	0.221	-0.032	**0.433**	-0.015
Ni*	0.045	**0.759**	0.004	**0.267**	0.005	-0.046
Co*	0.076	0.115	-0.165	**0.685**	0.252	0.031
% variance	43.5	14.0	12.9	8.0	5.6	4.3
without detritus**		24.8	22.8	14.2	9.9	7.6
Interpretation	detritus	metal sulfides	carbonate	hydrothermal	detrital var.	fluorapatite

Northern Honshu was the site of rhyolitic to andesitic submarine volcanic activities along the eastern margin of the basin [8], which suggests some contribution from submarine hydrothermal activities. The scores for this factor are only positive and are high in the Ajigasawa and Goshogawaia areas. Several hydrothermal manganese deposits are distributed around the Ajigasawa area in the lower part of the Onnagawa Formation [37], and oxygen isotope data of some carbonate nodules from both areas were reported to indicate higher temperatures, suggesting hydrothermal precipitation [32]. Also reported from the Ohdate area was submarine hydrothermal activities and their deposits during the Onnagawa time [40]. Therefore this factor is explained as hydrothermal precipitation.

Factor 5 has the heaviest loading of V with subordinate K2O and slight contributions of Zr and carbonate carbon. Presumably this reflects variability in the detrital composition such as variation in feldspar minerals, because K is a major component in alkali feldspars and K2O, Zr, and V are positively loaded, representing lithogenous elements. Positive correlation with DET possibly indicates higher contributions of K-feldspar and heavy minerals, which in turn

Table 4.
Varimax-rotated factor matrix for the biggest five factors based on the DET-normalized data set and the correlation coefficients between each factor score and external variables listed with asterisk (*). Those loadings and coefficients with more than 95% significance are emphasized. BIO: biogenic silica, DET: detrital fraction, TOC: total organic carbon, and CarbC: carbonate carbon content.

	Factor 1	Factor 2	Factor 3	Factor 4	Factor 5
SiO_2	-0.137	-0.080	**0.592**	-0.144	**0.350**
TiO_2	0.123	0.115	0.026	**0.857**	-0.201
Fe_2O_3	0.000	**0.314**	**0.774**	0.038	-0.204
MnO_2	**0.521**	0.128	-0.071	0.063	-0.232
MgO	**0.920**	0.073	0.000	0.152	-0.027
CaO	**0.953**	-0.019	0.013	-0.070	0.038
Na_2O	-0.055	-0.123	**0.882**	0.109	0.137
K_2O	-0.077	0.221	0.090	**0.832**	0.197
P_2O_5	0.161	-0.028	0.034	-0.005	**0.877**
Cu	0.123	**0.866**	0.022	0.211	-0.068
V	**0.788**	0.156	-0.157	-0.027	**0.317**
Zn	0.043	**0.912**	0.052	0.079	0.030
BIO*	0.053	**0.260**	**0.439**	0.156	**0.424**
DET*	-0.111	**-0.269**	**-0.456**	**-0.235**	**-0.385**
S*	-0.117	**0.663**	0.262	**0.358**	-0.249
TOC*	0.007	**0.268**	-0.009	0.100	-0.013
CarbC*	**0.729**	-0.066	0.046	0.271	**0.358**
Ba*	0.165	0.140	-0.202	**-0.283**	-0.142
Sr*	0.124	-0.079	-0.162	**-0.328**	**-0.273**
Zr*	0.115	**-0.321**	**-0.490**	-0.165	**-0.332**
Ni*	0.121	**0.547**	**-0.302**	**0.301**	**-0.302**
Co*	-0.143	0.016	-0.177	0.039	-0.005
% variance	25.5	18.6	15.1	9.3	8.6
Interpretation	carbonate	metal sulfides	biogenic silica	detritus	fluorapatite

represents higher detrital flux.
Factor 6 is primarily phosphorus and carbonate carbon with weak contributions from CaO and V, which reflects phosphatic minerals, probably carbonate fluorapatite.
The second factor analysis yielded five factors derived from the elemental values normalized by DET, which explain 77.1 % of the total variance (Table 4).
Factor 1 is attributed to the dolomitic carbonate component (CaO, MgO, V and carbonate carbon) with some contribution from MnO_2.
Factor 2 represents metal sulfide based on the element combination of Fe_2O_3, Cu, Zn, Ni and sulfur. TOC and BIO have slightly positive correlation.
Factor 3 represents the association of Na_2O, Fe_2O_3, SiO_2, and BIO (biogenic silica) which is interpreted as recording siliceous biogenic accumulation. The sulfur content is weakly correlated with this factor, while TOC shows no obvious correlation. The average chemical composition of Diatomeae is reported as rich in Na (0.43%), Fe (0.41%), P (0.28%), Ba (0.25%), and Ca, Zn, Mg, etc. [38]. This factor was not independent in the first analysis based on the chemistry, because it is primarily dependent on dilution by the largest detrital factor.
Factor 4 has high loadings for TiO_2 and K_2O and correlates with S and Ni. This factor cannot be interpreted unambiguously, but Ti and K are lithogenic elements well regressed with DET values, therefore slight variance of these elements in the detrital components might be

emphasized by this factor.

Factor 5 represents phosphatic minerals incorporated with the P_2O_5, SiO_2, V, BIO and carbonate carbon combination.

These two fold factor analyses produced very consistent results. Several dominant factors controlling the chemical composition of the samples are derived from both analyses, and for the metal sulfide, dolomitic carbonate, phosphatic factors and compositional deviation in detrital materials, the amounts of variance explained by these factors are nearly the same in both of these analyses (Tables 3 and 4). The contribution from the metal sulfide factor is slightly reduced when normalized by DET (22.8 to 18.6 %). This is partly because some part of sulfur is explained by biogenic silica factors which was derived only in the DET normalized analysis. An exception is the hydrothermal factor only obtained from the bulk chemistry data. Variances of the detrital composition from the hypothetical one are represented as higher K_2O and Zr in the Noshiro area by the first factor analysis while TiO_2 and K_2O are higher in the Gojonome areas based on the second factor analysis These factor score distributions indicate both the stratigraphic and geographical variance of the detritus composition, which may urge us to revise our hypothetical average composition of the detritus.

Sedimentary environments of the siliceous sediment

To study the geographical variance of the geochemical characteristics, we plotted the factor scores of the major factors calculated in the previous two fold factor analyses according to the locality of the samples. The mean value and standard deviation for the factors are listed in Tables 5 and 6, which are used to draw contour maps for each factor in Figs. 8 to 10. The latter figures show the control points of the seven localities where the mean values of the

Table 5.
Average (bold face) and standard deviation of factor scores for each area derived from the factor analysis based on the bulk chemical composition.

	Factor 1 detritus		Factor 2 sulfides		Factor 3 carbonate		Factor 4 hydrothermal		Factor 5 detritus var.		Factor 6 fluorapatite	
Ajigasawa	**0.46**	0.19	**0.77**	0.22	**-0.01**	0.10	**-0.08**	0.21	**-0.88**	0.10	**0.04**	0.24
Goshogawara	**0.32**	0.27	**0.47**	0.46	**0.37**	0.29	**0.66**	0.38	**-0.59**	0.09	**-0.47**	0.13
Noshiro	**-0.95**	0.38	**0.09**	0.57	**0.97**	1.18	**-0.12**	0.36	**2.06**	0.72	**0.67**	0.73
Takanosu	**-0.65**	0.32	**-0.11**	0.16	**-0.15**	0.24	**-0.30**	0.13	**0.89**	0.28	**-0.17**	0.11
Oga	**0.01**	0.24	**-0.50**	0.16	**-0.13**	0.22	**0.15**	0.22	**0.14**	0.13	**-0.12**	0.19
Gojonome	**-0.33**	0.10	**-0.12**	0.19	**-0.01**	0.15	**-0.33**	0.25	**-0.02**	0.22	**-0.07**	0.06
Yashima	**0.24**	0.17	**-0.37**	0.11	**-0.33**	0.13	**-0.04**	0.13	**0.41**	0.20	**0.68**	0.57

Table 6.
Average (bold faces) and standard deviation of factor scores for each area derived from the factor analysis based on the chemical composition normalized by DET.

	Factor 1 carbonate		Factor 2 sulfides		Factor 3 biogenic silica		Factor 4 detritus var.		Factor 5 fluorapatite	
Ajigasawa	**-0.23**	0.05	**0.51**	0.30	**0.15**	0.16	**0.11**	0.13	**-0.48**	0.14
Goshogawara	**0.09**	0.09	**-0.02**	0.30	**-0.07**	0.08	**-0.10**	0.16	**-0.86**	0.10
Noshiro	**2.02**	1.56	**0.36**	0.33	**-0.53**	0.31	**0.05**	0.62	**1.14**	0.49
Takanosu	**0.01**	0.09	**0.32**	0.20	**-0.10**	0.09	**-0.34**	0.28	**0.20**	0.06
Oga	**-0.23**	0.11	**-0.39**	0.10	**0.04**	0.27	**-0.35**	0.21	**0.12**	0.23
Gojonome	**0.17**	0.15	**0.00**	0.33	**0.28**	0.19	**0.77**	0.31	**0.23**	0.16
Yashima	**-0.23**	0.08	**-0.39**	0.13	**-0.40**	0.08	**-0.15**	0.19	**0.30**	0.43

respecting factor scores were calculated. Contours are drawn with each standard deviation value taken into consideration to make smooth lines. Also drawn is the limit of shallow sea area based on the paleogeographic reconstruction by Iijima *et al.* [8].

The detrital factor is notably lower (negative) in the Noshiro and Takanosu areas and rapidly increases northward (Fig. 8A). Although the Kitakami landmass is interpreted to have been far to the east or southeast of the basin, detrital influx is apparent in the northern part of the basin. This supports the idea of an uplifted high between the Shimokita and Goshogawara areas which might have provided surrounding areas with detritus. However, a small landmass is interpreted to have existed just to the west of the Goshogawara area (Fig. 1), and it may have been another source of the detritus.

The siliceous biogenic factor is generally low (negative) on both sides of the basin (Fig. 10C). Positive scores occur in most samples from the Noshiro and Gojonome areas with rather lower detrital factor scores. This seems to be indicative of the deposition of pure siliceous sediments.

Diagenetic carbonates are characteristically high at Noshiro, and they develop along the N-S extension of the basin (Figs. 8C and 10A). Matsumoto and Matsuda [32] concluded that the early carbonate concretions in the Ajigasawa and Goshogawara areas were precipitated at rather shallow depths, less than 100 meters. These dolomitic carbonates of early diagenetic origin are often closely associated with anoxic bacterial processes. The metal sulfides are also high along the western side of the basin from Ajigasawa to Noshiro [Figs. 8B and 10B].

High phosphatic factors generally occur in high biogenic silica component and oxygen-deficient areas (Fig. 9C), but they also extend farther to the southeast to Yashima. Yamamoto and Watanabe [26] showed that the existence of locally high productivity and oxygenated bottom-water conditions in the Yashima area. This is different from the general occurrence of diagenetic carbonate fluorapatite associated with pyrite under dysaerobic/anoxic conditions [39].

The oxygenation level in the bottom water is interpreted to have been generally anaerobic to dysaerobic in the main part of the Onnagawa Formation, where minute laminations are preserved in the rocks [6, 14, 41]. Although an extensive study by Tada [6] in the Gojonome area showed that the most laminated horizon of the main part of the formation suffered frequent deviations between aerobic and dysaerobic conditions within the order of few meters in thickness, he also concluded that the bottom water became gradually oxygenated following deposition of the main part of the Onnagawa Formation leading up to deposition of the overlying Funakawa Formation during approximately the time interval of 9 to 6 Ma.

As stated before, the detrital fraction primarily dilutes organic materials in the sediments and the sulfide factor is positively correlated with biogenic silica and TOC. If we can assume that all the sulfur in the sediments is present as iron sulfides, the amount of sulfur is a function of the availability of sulfate, organic carbon and reactive iron [42]. In the case of the marine Onnagawa sediments, sufficient sulfate was available from sea water. The amount of reactive iron was most likely largely derived from the detritus fraction because neither hydrothermal nor hydrogenous iron source were significant. Provided that the sedimentation rate is sufficiently low, the amount of pyrite formed depends largely on the reactive iron from the detrital fraction and to a lesser extent on the amount of locally accumulated organic matter under euxinic conditions, while the reactivity of organic matter in the sediments is the primary controlling factor on the sulfate reduction and pyrite formation in oxygenated environments [43]. In the latter case, the degree of pyritization (ratio of the total iron to pyrite iron) should increase with the flux of reactive organic carbon.

The Fe_2O_3 to sulfur ratio is primarily proportional to the degree of pyritization if all the sulfur exists in pyrite (Fig. 11). Although not completely clear, the samples with lower BIO value

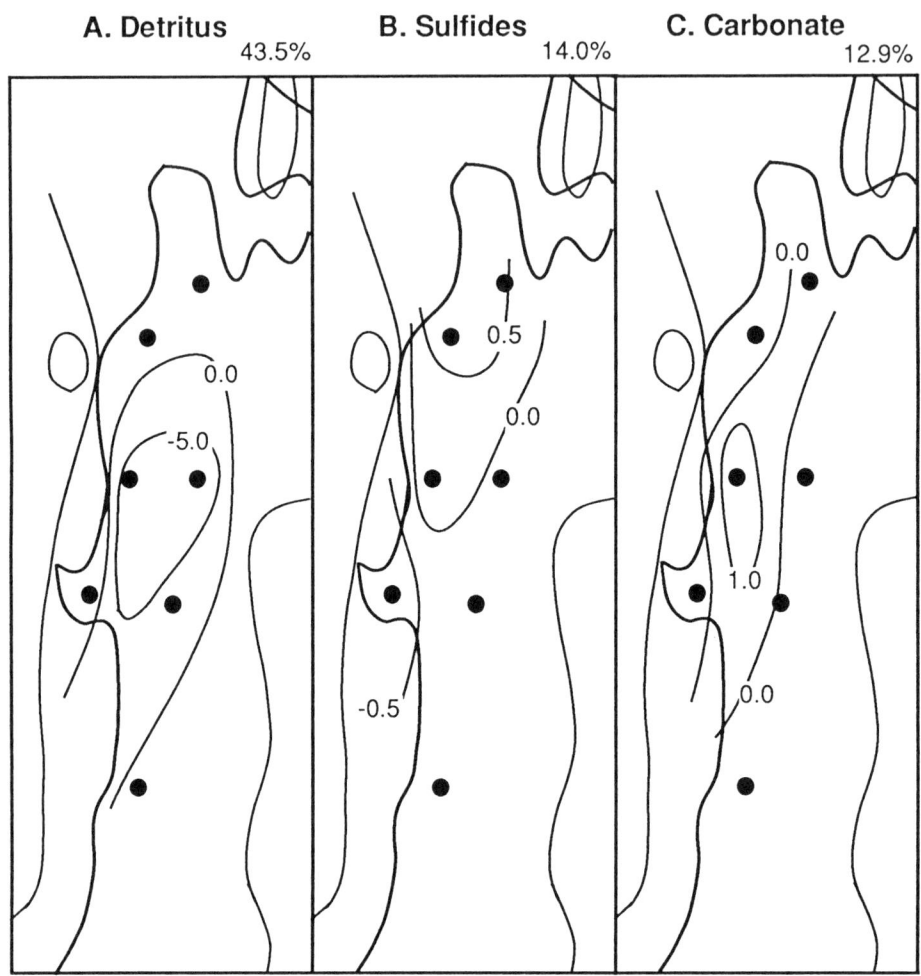

Figure 8. Contour maps of the distribution of average scores of the first three factors derived from the analysis based on the bulk chemical composition. Hair lines on the map indicate the margins of shallow sea areas based on Fig. 1.

(less than approximately 50 %) show negative regression (r = -0.67). This implies that the biogenic fraction controlled the degree of pyritization under oxygenated conditions where the flux of reactive organic material limited pyrite formation.

The degree of pyritization does not fluctuate from the low values in higher BIO regions (>50 %). This trend is due to the fact that available iron limited the formation of pyrite under euxinic conditions. This implies that the higher biogenic silica flux produced anoxic to euxinic bottom-water conditions in the Onnagawa Sea. Thus, high metal sulfide factors are coupled with high biogenic silica fraction under conditions of less detritus flux, and the bottom-water conditions are assumed to have been generally euxinic. This condition applies to the western side of the basin in the Noshiro - Gojonome areas, which coincides well with the area where laminae are well preserved.

High productivity of siliceous organism, bottom water stagnation and smaller detritus input occurred simultaneously in the deepest part of the basin (Figs. 8A, 8B, 10B and 10C). This

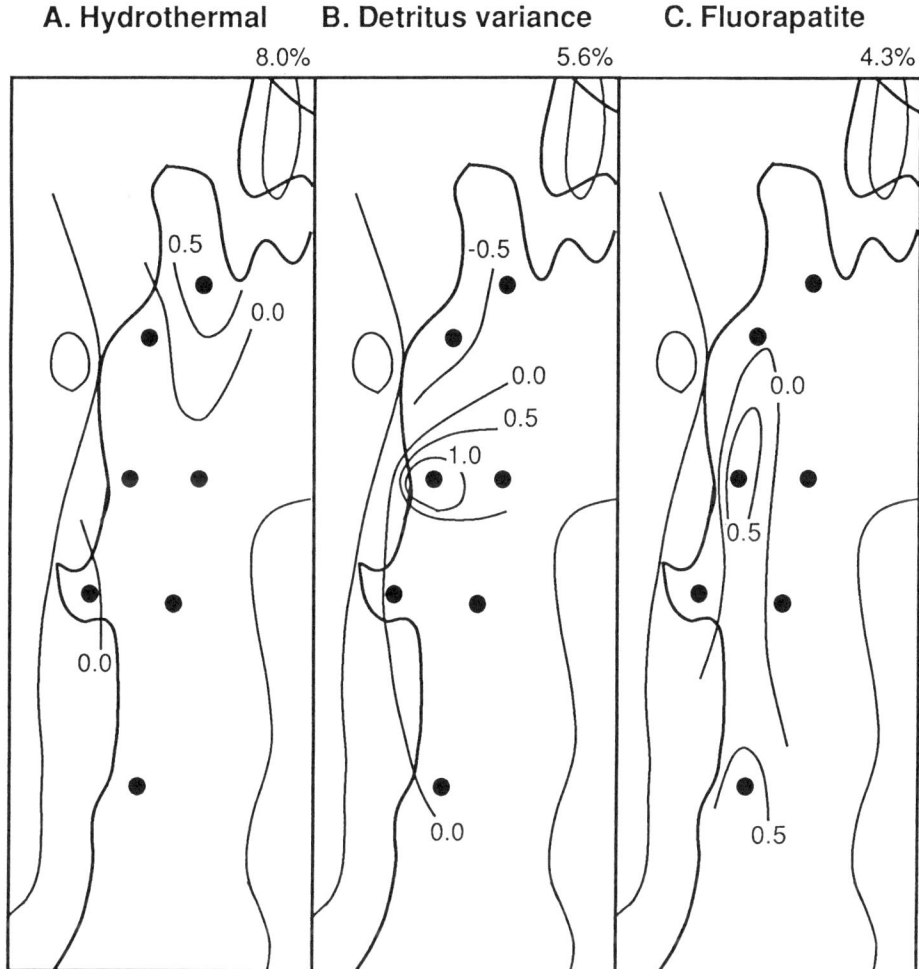

Figure 9. Contour maps of the distribution of average scores of the fourth to sixth factors derived from the analysis based on the bulk chemical composition. Hair lines on the map indicate the margins of shallow sea areas based on Fig. 1.

supports the stagnation model due to restricted deep-water circulation, rather than an occurrence of an oxygen minimum zone in a region of high organic productivity, because the stagnation occurred in the center of the basin in its deepest part.

CONCLUSION

The Miocene Onnagawa Sea in northern Honshu consisted of N-S elongated axial deeps where the typical laminated diatomaceous sediments were accumulated along the western side. Statistical analyses of the chemical composition of these sediments indicates that they are admixtures of terrigenous detritus, biogenic silicea, metal sulfides, diagenetic carbonate,

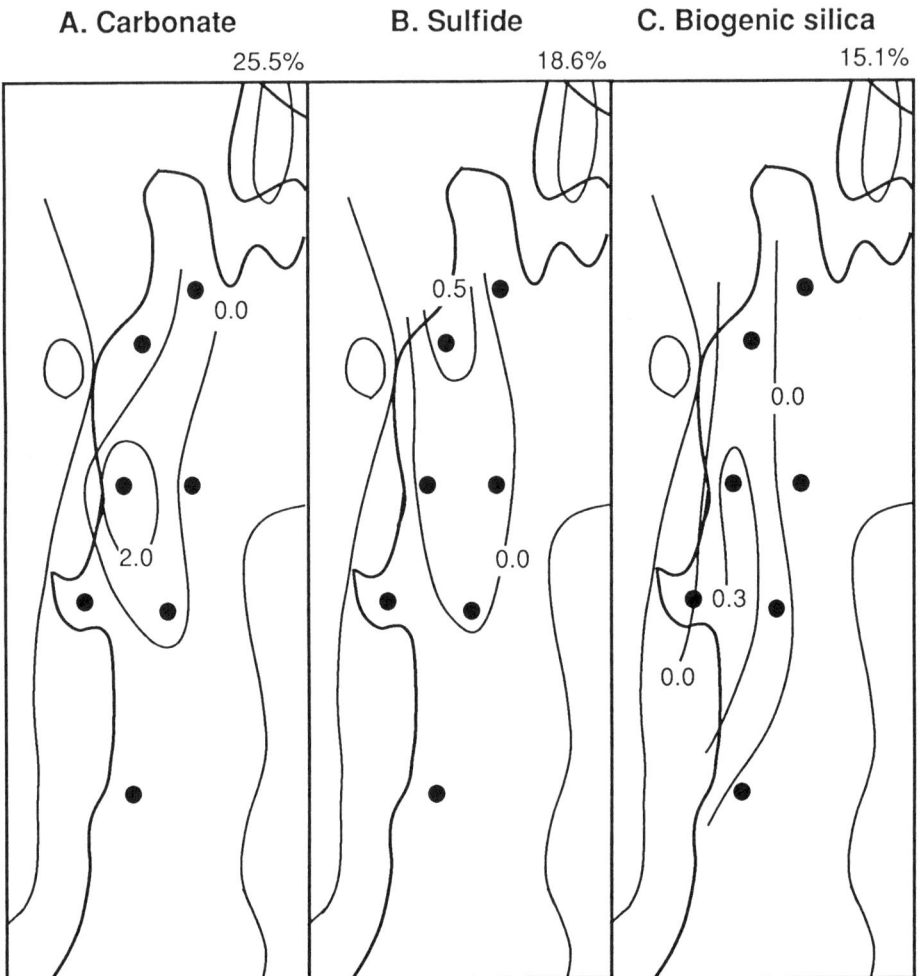

Figure 10. Contour maps of the distribution of average scores of the first three factors derived from the analysis based on the DET-normalized chemical composition. Hair lines on the map indicate the margins of shallow sea areas based on Fig. 1.

fluorapatite, and local hydrothermal precipitates. The paleogeographic distribution of the components was not uniform. However, the area with the least detrital dilution was along the western side of the basin, where the siliceous biogenic and metal sulfide fractions were enriched. This part of the basin was favorable for the accumulation of organic-rich siliceous sediments because the anoxic to euxinic bottom-water conditions probably occurred beneath an area of higher biogenic productivity.

Acknowledgments
This study was supported by the AIST Special Research Program "Research on three dimensional modeling for resources assessment". We are greatly indebted to Drs. K. Kodama and S. Tokuhashi of GSJ for the assistance during the program, Emeritus Professor A. Iijima, Drs. R. Tada and R. Matsumoto of the University of Tokyo and Dr. T. Tsuji of JAPEX Co. Ltd. for helpful discussions and

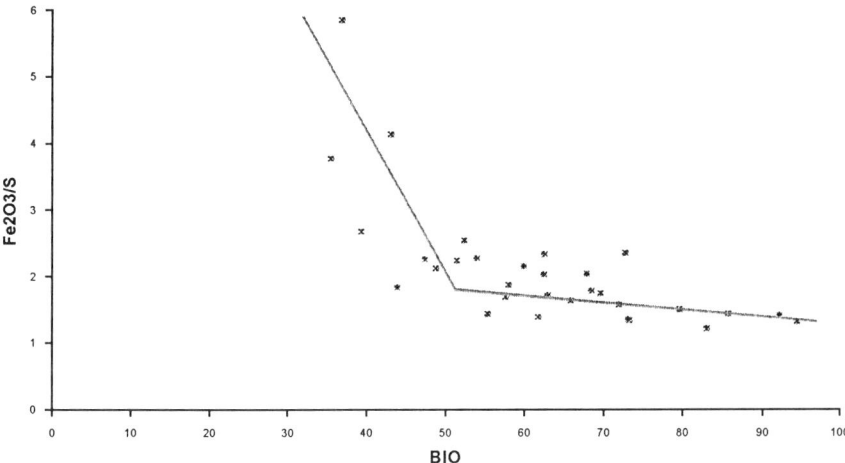

Figure 11. Relationship between Fe_2O_3/S and biogenic silica fraction (BIO). There are two trends in the scatter in the regions of lower and higher BIO values probably reflecting the mechanism of the pyrite formation.

kind permission to use the unpublished data, Dr. E. Honza of GSJ for critical reading of an earlier manuscript, Dr. S. Togashi and Mr. S. Terashima of GSJ for analytical facilities, and Messrs. M. Watanabe and T. Sumii of GSJ and T. Sakamoto of the Hokkaido University for help during the field survey. The final manuscript has been greatly benefited by anonimous editors. We also thank the convenors of the 29th IGC symposium on "Siliceous, phosphatic and glauconitic sediments of the Tertiary and Mesozoic" for organizing the session.

REFERENCES

1. J.C. Ingle, Jr. Origin of Neogene diatomites around the North Pacific Rim. In: *The Monterey Formation and Related Siliceous Rocks of California.* R.E. Garrison and R.G. Douglas (Eds). pp.159-179. Pasific Section, Soc. Econ. Paleontol. Mineral. (1981).
2. K. Taguchi and K. Sasaki. Organogeochemistry and its relation to the geology of petroleum accumulation in Japan, In: *Proceedings of Symposium on Hydrogeochemistry and Biogeochemistry.* II Biogeochemistry, pp.133. The Clark Company, Washington, D.C. (1973).
3. H. Fukusawa, T. Sakamoto and I. Koizumi. Rhythmical oceanographic changes recorded in laminated diatomaceous mudstones of the Neogene Nakayama Formation, Sado, *Monthly Tikyu* **13**, 467-469 (1991). (in Japanese).
4. M. Watanabe. Stratigraphy of the Neogene sequence in the Himi-Nadaura area, Toyama Prefecture, central Japan -with special reference to the hiatus between the Sugata Formation and overlying formations-, *Jour. Geol. Soc. Japan,* **96**, 11, 915-936 (1992). (in Japanese).
5. K. Taguchi, K. Sasaki and N. Ushijima. Geochemical significance of porphyrin pigments in the stratigraphic correlation of the Neogene Tertiary rocks 1. Yashima Oil Field, Akita Prefecture, Japan. *Sci. Rept. Tohoku Univ. III,* **10**, 333-348 (1969).
6. R. Tada. Origin of rhythmical bedding in middle Miocene siliceous rocks of the Onnagawa Formation, northern Japan, *Jour. Sed. Petrol.* **61**, 1123-1145 (1991).
7. H. Fukusawa. Preliminary report on the relationship between the sedimentation of Neogene biosiliceous shale and paleoceanographic environments, *Geol. Soc. Japan Mem.* no. 37, 219-226 (1992). (in Japanese).
8. A. Iijima, R. Tada and Y. Watanabe. Developments of Neogene sedimentary basins in the northeastern Honshu Arc with emphasis on Miocene siliceous deposits, *Jour. Fac. Sci. Univ. Tokyo. Sec II,* **21**, 417-

446 (1988).
9. K.A. Mertz, Jr. Origin of hemipelagic source rocks during Early and Middle Miocene, Monterey formation, Salinas Basin, California, *AAPG. Bull.* **73**, 510-524 (1989).
10. A. Iijima and R. Tada. Evolution of Tertiary sedimentary basins of Japan in reference to opening of the Japan Sea, *Jour. Fac. Sci. Univ. Tokyo, Sec II*, **22**, 121-171 (1990).
11. H. Sato and K. Amano. Relationship between tectonics, volcanism, sedimentation and basin development, Late Cenozoic, central part of Northern Honshu, Japan, *Sed. Geol.* **74**, 323-343 (1991).
12. M. Usuta. Geotectonic history of the southern part of Akita Prefecture, Northeast Japan, *Geol. Soc. Japan, Mem.*, no. 32, 57-80 (1989) (in Japanese).
13. T. Shiraishi and Y. Matoba. Neogene paleogeography and paleoenvironment in Akita and Yamagata Prefectures, Japan Sea side of northerast Honshu, Japan, *Geol. Soc. Japan, Mem.*, no. 37, 39-51 (1992) (in Japanese).
14. R. Tada, Y. Watanabe and A. Iijima. Accumulation of laminated and bioturbated Neogene siliceous deposits in Ajigasawa and Goshogawara areas, Aomori Prefecture, Northeast Japan, *Jour. Fac. Sci. Univ. Tokyo, Sec II*, 21, 139-167 (1986).
15. Y. Watanabe and R. Tada. Chemical composition of Neogene siliceous rocks in Aomori Prefecture, In: *Report of Comprehensive Research on Tertiary Siliceous Shales.* A. Iijima (Ed.), pp.129-148. University of Tokyo (1988) (in Japanese).
16. T. Tsuji, Y. Masui, A. Waseda, Y. Inoue, H. Kurita and K. Kai. The Onnagawa Formation in the vicinity of the Yashima town, Akita Prefecture, Northern Japan -with special reference to the lithologic units, the depositional environments and their relation to source rock characteristics-, *Research report of Japan Petroleum Exploration Co.Ltd.* 7, 45-99 (1991MS). (in Japanese).
17. I. Koizumi and Y. Matoba. On the top of the Nishikurosawa Stage, *Geol. Soc. Japan, Mem.*, no. 32, 187-195 (1989) (in Japanese).
18. F. Akiba and C. Hiramatu. Neogene diatom biostratigraphy in Ajigasawa, Goshogawara and Shimokita areas, Aomori Prefecture, In: *Report of Comprehensive Research on Tertiary Siliceous Shales.* A. Iijima (Ed.), pp.35-51. University of Tokyo (1988) (in Japanese).
19. Y. Aita and Y. Matoba. Neogene radiolarian fossils in the Ajigasawa and Goshogawara areas and Shimokita Peninsula, Aomori Prefecture, In: *Report of Comprehensive Research on Tertiary Siliceous Shales.* A. Iijima (Ed.), pp.63-80. University of Tokyo (1988) (in Japanese).
20. M. Sato. Diatom biostratigraphy of the Onnagawa formation and its equivalents in Akita oilfields. Master thesis at the Akita Univ. (1985MS).
21. Y. Matoba and I. Koizumi, On the top of the Nishikurosawa Stage, *Let. Northeast. Br. Geol. Soc. Japan*, no. 16, 15-16 (1986) (in Japanese).
22. M. Ogasawara. Trace element analysis of rock samples by x-ray fluorescence spectrometry, using Rh anode tube, *Bull. Geol. Surv. Japan* **38**, 57-68 (1987) (in Japanese).
23. N. Imai. Multielement analysis of stream sediment by ICP-AES, *Bunseki Kagaku* **36**, T41-T45, (1987) (in Japanese).
24. N. Imai. Multielement anlysis of rocks with the use of geological certified reference material by inductively coupled plasma mass spectrometry. *Anal. Sci.* **6**, 389-396 (1990).
25. S. Terashima, T. Yamashige and A. Ando. Determination of major and minor elements on the six GSJ rock reference samples, *Bull. Geol. Surv. Japan* **35**, 171-177 (1984).
26. M. Yamamoto and Y. Watanabe, Biomarker geochemistry and paleoceanography of Miocene Onnagawa diatomaceous sediments, northern Honshu, Japan. In: *Siliceous, phosphatic and glauconitic sediments in the Tertiary and Mesozoic.* A. Iijima, R.B. Garrison and A.M. Abed (Eds). Proc. 29th Intern. Geol. Congress. VSP Sci. Publ. (in press).
27. A. Y. Huc. Origin and formation of organic matter in recent sediments and its relation to kerogen. In: *Kerogen.* B. Durand (Ed.). pp.445-474. Technip, Paris (1980).
28. J.B. Maynard, M. Jansen and G. Schuette. Composition of modern deep-sea sands from arc-related basins, *Geol. Soc. Lond. Spec. Publ.*, **10**, 551 (1982).
29. H.-J. Brumsack and J. Thurow. The geochemical facies of black shales from the Cenomanian/Turonian boundary event (CTBE). *Mitt. Geol. -Palaont. Inst. Univ. Hamburg* **60**, 242-265 (1986).
30. J.H. Patterson, A.R. Ramsden and L.S. Dale. Geochemistry and mineralogical residences of trace elements in oil shales from the Condor deposit, Queensland, Australia, *Chemical Geology*, **67**, 327-340 (1988).
31. H.-J. Brumsack. Geochemistry of recent TOC-rich sediments from the Gulf of California and the Black Sea, *Geologische Rundschau* **78**, 851-882 (1989).
32. R. Matsumoto and H. Matsuda. Occurrence, chemical composition and stable isotope of the carbonate

concretions in the Neogene silicous sediments in Ajigasawa, Goshogawara and Shimokita areas, Aomori Prefecture. In: *Report of Comprehensive Research on Tertiary Siliceous Shales*. A. Iijima (Ed.), pp.163-176. University of Tokyo (1988) (in Japanese).
33. K. Bostrom. T. Kraemer and S. Gartner, Provenance and accumulation rates of biogenic silica, Al, Ti, Fe, Mn, Cu, Ni and Co in Pacific pelagic sediment. *Chem. Geol.* **11**, 123-148 (1973).
34. S. Calvert and N. Price. Geochemical variation in ferromanganese nodules and associated sediments from the Pacific Ocean, *Mar. Chem.* **5**, 43-74 (1977).
35. M. Lyle, D.M. Murray, B.P. Finney, J. Dymond, J.M. Robbins and K. Brooksforce. Copper-nickel-enriched ferromanganese nodules and associated crusts from the Bauer Basin, northwest Nazca Plate, *Earth Planet. Sci. Lett.* **35**, 55-64 (1977).
36. W.M. Landing and R.A. Feely. The chemistry and vertical flux of particles in the northeastern Gulf of Alaska, *Deep-Sea Res.* **28A**, 19-37 (1977).
37. A. Ozawa, Y. Ikebe, J. Hirayama, Y. Awata and T. Takayasu. *Geology of the Noshiro District. With geological sheet map at 1:50,000 Aomori*. Geol. Surv. Japan, Tsukuba (1984). (in Japanese).
38. A.P. Vinogradov. The elementary chemical composition of marine organisms, Yale University, New Heaven, (1953).
39. R.E. Garrison, M. Kaster and Y.Kolodny. Phosphorites and phosphatic rocks in the Monterey Formation and related Miocene units, coastal California, In: *Cenozoic basin development of coastal California, Rubey Volume VI*: R.V. Ingersoll and W.G. Ernst (Eds), pp. 348-381, Englewood Cliffs, New Jersey, Prentice Hall (1987).
40. A. Iijima. Clay and zeolitic alteration zones surrounding Kuroko deposits in the Hokuroku district, northern Akita, as submarine hydrothermal-diagenetic alteration product. *Doc. Min. Geol. Japan, Spec. Issue* no. 5, 267-289 (1974).
41. T. Sakamoto. Sedimentary rhythm of the Nakayama Formation (Middle Miocene-Early Pliocene) in the Sado Island, *Jour. Geol. Soc. Japan,* **98**, 7, 611-633 (1992). (in Japanese).
42. R. Railswell and R.A. Berner. Pyrite and organic matter in Phanerozoic normal marine shales, *Geochim. Cosmochim. Acta* **50**, 1967-1976 (1986).
43. R.A. Berner, Sedimentary pyrite formation: An update, *Geochim. Cosmochim. Acta.* **48**, 605-615 (1984).

Biomarker geochemistry and paleoceanography of Miocene Onnagawa diatomaceous sediments, northern Honshu, Japan

M. YAMAMOTO[1] and Y. WATANABE[2]
[1,2]*Geological Survey of Japan, 1-1-3 Higashi, Tsukuba, Ibaraki 305 Japan*

Abstract--Biomarker compounds and major and metal elements were analyzed for diatomaceous sediments from the Onnagawa, Funakawa and Tentokuji Formations (from middle Miocene to Pliocene) in the Akita oil-producing basin, northern Honshu, Japan.
Onnagawa diatomaceous sediments are characterized by high inputs of algal and bacterial organic matters and low inputs of terrigenous organic matter. The relative contribution of bacteria and algae reflects bacterial activity during early diagenesis controlled by both detrital input and bottom-water redox potential. Variations of preservation degree of lamination, homohopane index and gammacerane/hopane ratio in Yashima area indicate a gradual oxygenation of bottom-water through the deposition of the upper part of the Onnagawa Formation (ca. 10-6 Ma). Variations of C_{27}/C_{29} sterane ratio indicate high algal production through the deposition of the upper part of the Onnagawa Formation and lower part of the Funakawa Formation (ca. 9-5 Ma), suggesting regionally intensified upwelling currents in Yashima area.
High concentrations of 24-norcholestanes and 28,30-bisnorhopane are found in the horizons indicating high algal productivity and oxygenated depositional environments. This occurrence suggests that 24-norcholestanes originated from some algal species flourishing in high productivity areas, and that the 28,30-bisnorhopane precursor is a metabolic product discharged from some sulfur-oxydizing bacterium living near oxic-anoxic interface in the surface sediments.

Keywords: diatomaceous sediments, the Onnagawa Formation, biomarker, 24-norcholestanes, 28,30-bisnorhopane, bacterial activity, regional upwelling.

INTRODUCTION

A large volume of Neogene diatomaceous sediments were deposited along the northern Pacific rim [1,2]. Their synchronous sedimentation is attributed to the combination of (1) increasingly vigorous atmospheric-oceanic circulation and prolific diatom productivity at mid and high latitudes in response to mid-Miocene (16-15 Ma) buildup of the Antarctic ice cap and global change to glacial climate mode, and (2) a widespread late Oligocene - mid-Miocene episode of tectonism and synchronous marginal basin formation around the Pacific rim likely triggered by an increased rate of spreading on the East Pacific Rise beginning at 25-22 Ma [2]. However, local variations are found in their lithological association, timing of beginning deposition of diatomaceous sediments, and organic richness. This suggests the importance of revealing local oceanographic and tectonic settings for better understanding of the event deposition of these diatomaceous sediments.
These sediments have also attracted petroleum geologists and geochemists, because they are the main source rocks (Monterey Formation of California [3]; Kamchatka Peninsula [4]; Sakhalin Island [5]; Onnagawa Formation of northern Honshu Island [6]) and reservoir rocks (Monterey

Formation [7]; Onnagawa Formation [8,9]) in oil and gas fields around northern circum-Pacific area. Special attention of petroleum geochemists has been focused on immature generation of liquid oil likely from sulfur-rich kerogen (California [10,11,12,13]; Kamchtka and Sakhalin [4]). Since the formation of sulfur-rich kerogen needs highly-reducing depositional environment [12], it is important for petroleum exploration and evaluation in those areas to reveal the oceanographic controls on the formation of such environments.

The Onnagawa Formation (from 12 Ma till 10 to 6 Ma) and its equivalents consist mainly of diatomite, porcelanite and chert, and are widely distributed in northern Japan (e.g., Wakkanai Fm. of Tenpoku Basin, Odoji Fm of Ajigasawa Basin, Kusanagi Fm. of Yamagata Basin, Teradomari Fm. of Niigata Basin, Nakayama Fm. of Sado Island, Sugata Fm. of Noto Peninsula) and Sea of Japan (ODP Legs 127 and 128, Units 3 and 4, [14]). Recent sedimentological research has made the paleoceanographic setting of the Onnagawa Formation and its equivalents clearer. For examples, Iijima et al. [15] and Iijima and Tada [16] compiled a paleobathymetric map of the Neogene basin where the Onnagawa Formation and its equivalents were deposited, and showed that the basin was a silled basin whose sill depth was probably upper bathyal (150 to 500 m). Based on the cyclic changes in mass accumulation rates of biogenic silica and detrital matter related to orbital cycles in the Onnagawa Formation, Tada [17] proposed a model whereby the intrusion of OMZ water from the Pacific Ocean via the sill to the basin caused by a sea level rise resulted in bottom-water anoxia in the silled basin. Fukusawa [18] suggested a seasonal wind upwelling model similar to that found in the recent Gulf of California [19] as the oceanographic model for the Wakkanai Formation in the Tenpoku Basin, based on the distribution in the diversity of diatom assemblages, total organic carbon contents and sedimentary structures in the basin. Despite the progress as mentioned above, the sedimentological approach can provide only limited information on critical biological and hydrological aspects of the paleoceanographic setting. Therefore, more detailed paleoceanographic reconstruction needs some additional approaches, e.g., more likely geochemical approaches on the stable isotopes of biological elements and/or organic molecules derived from ancient living organisms.

Biomarker geochemistry may contribute to paleoceanographic reconstruction of the Onnagawa diatomaceous sediments. A biomarker, which is also called "biological marker", "molecular fossil" and "chemical fossil", is an organic compound having a molecular structure that can be linked to a specific biological source [20]. Although the application of biomarker geochemistry to paleoenvironmental reconstruction has some difficulties due to its unconfirmed premise on some precursor-product relationships, the biomarker is undoubtedly a powerful tool to provide valuable information on variable aspects, e.g., organic source (algal/higher plant [21]; angiosperm [22], dinoflagellates [23]), paleotemperature [24], paleosalinity [25,26] and bottom-water redox potential [27]. The application of biomarker geochemistry to paleoenvironmental analysis of Neogene sedimentary rocks in Japanese oil-producing basins was started by Taguchi and his co-workers' pioneering works [28]. They pointed out concurrent changes of nickel and vanadyl porphyrin contents related to bottom-water oxygenation in those basins, and local variation of perylene content likely related to detrital input [29]. However, most of the following biomarker studies have focused on diagenetic transformation of organic molecules (e.g., fatty acids [30,31]; steranes and triterpanes [32,33,34]; nucleic acids [35], while some studies have focused on the relationships to source organisms (e.g., amino acids [36]; steranes and triterpanes [37,38,39]).

This paper presents our case study on an application of biomarker geochemistry to paleoceanographic reconstruction of Japanese Neogene diatomaceous sediments. The authors report the stratigraphic variations of biomarker distributions in the Onnagawa Formation and the overlying Funakawa and Tentokuji Formations, and discuss the changes in source organisms, bottom-water redox condition and surface algal productivity through middle Miocene-Pliocene time. We also refer to the factors controlling bacterial activity during early diagenesis and report on the occurrence and significance of 24-norcholestanes and 28,30-bisnorhopane which show high abundances in some Onnagawa and Funakawa samples.

GEOLOGICAL BACKGROUND

The following four tectonic stages are recognized in the Neogene northern Honshu arc: (1) *Rifting Stage* (22?-15 Ma), (2) *Backarc Basin Opening Stage* (15-13 Ma), (3) *Transitional Stage* (13-2.4 Ma), (3a) *Thermal Subsidence Substage* (13-8 Ma), (3b) *Incipient Compressional Substage* (8-2.4 Ma), (4) *Shortening Deformation Stage* (2.4 Ma- present) [40]. The Onnagawa Formation and its equivalents were deposited mostly during the *Thermal Subsidence Substage*, a time of relatively little volcanisms, in a NNE-SSW trending silled basin, situated on the western side of the northern Honshu Arc. This basin was approximately 150 km wide and 500 km long with its eastern margin bounded by the Kitakami and Abukuma massifs and its western margin by off-shore banks with possible small islands [15] (Figure 1). The deposition of the Onnagawa Formation started at approximately 13 Ma and continued till 10 to 6 Ma, and the diatomaceous rock (diatomite, porcelanite and quartz chert) changes gradually upwards to more argillaceous diatomaceous mudstone of the Funakawa Formation and the overlying Tentokuji Formation.

The paleobathymetric map [15] (Figure 1) shows that the Yashima area was situated on the western slope of a glauconite-topped small bank during the deposition of the Onnagawa

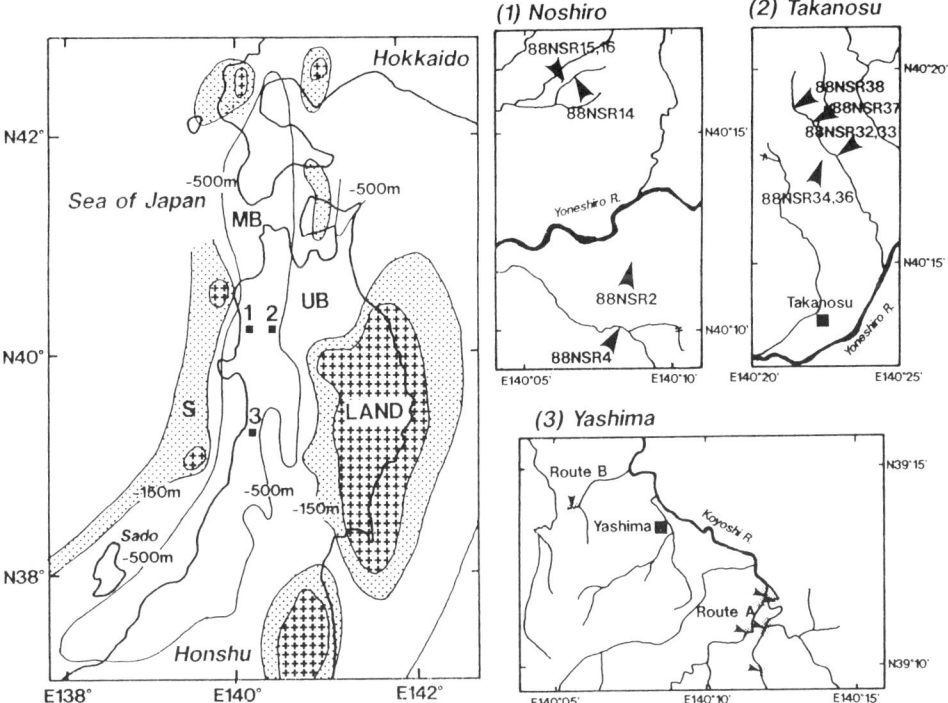

Figure 1. Maps showing locations of areas studied and the middle Miocene paleogeography of northern Honshu. The paleobathymetric information is from Iijima et al. [15]. S = shallow (0 to -150 meters), UB = upper bathyal (-150 to -500 meters) and MB = middle bathyal (-500 to -2000 meters).

Formation. Tsuji et al. (1991MS) point out that the paleodepth of the Onnagawa Formation in Yashima area was middle bathyal (-500 to -2000 meters) except for its uppermost horizon which contains upper bathyal (-150 to -500 meters) assemblages of benthic foraminifera. The paleobathymetric map also shows that Noshiro and Takanosu areas were situated in central part of the basin, and were middle bathyal.

SAMPLES AND METHODS

Totally twenty-nine outcrop rock samples were taken from the Onnagawa, Funakawa and Tentokuji Formations from middle Miocene to Pliocene ages, occurring in the Yashima, Noshiro and Takanosu areas in the Akita oil-producing Neogene basin, northern Honshu (Figures 1 and 2; Table 1). The rock samples were crushed and milled to a particle size of about 200 mesh.
Determination of total organic carbon was carried out using the Yanagimoto MT-2 CHN analyzer and the Kokusai Denki VK-111 carbon and sulfur analyzer after acid treatment of the sample to remove carbonate carbon with 3N hydrochloric acid. Determination of total nitrogen and total sulfur was carried out using Yanagimoto MT-2 and MT-5 CHN analyzers and a Horiba EMIA-520 carbon and sulfur analyzer.
Determination of content of the major and metallic elements was carried out using a Seiko SPS-1200 ICP system, a Yokogawa Electric PMS-200 ICP-MS system and a Nippon Jarrell-Ash AA-781 atomic absorption and flame emission spectrometer according the solution method [42,43].

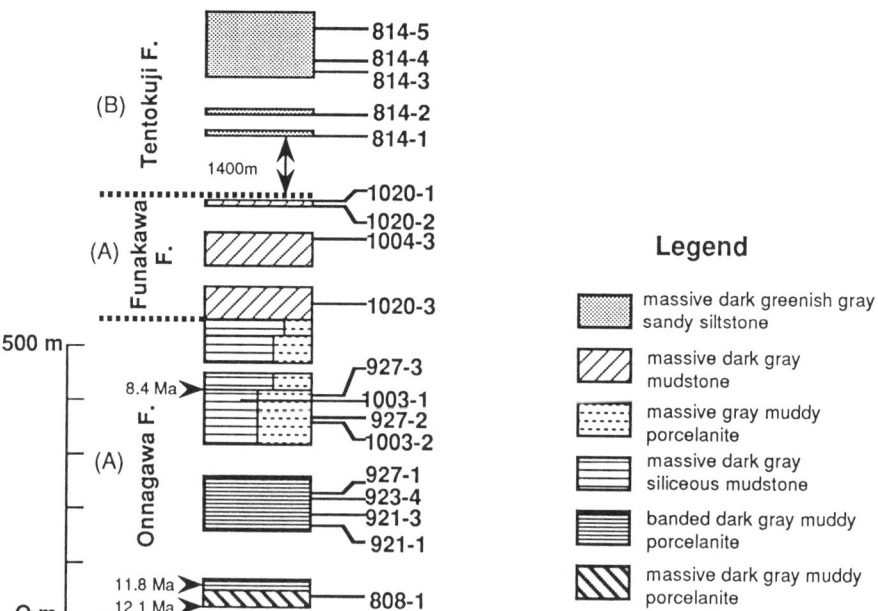

Figure 2. Schematic stratigraphic and lithologic sections of Yashima area. Numbers in the column indicate sampling points. Age data are from Tsuji et al. [41].

Table 1.
List of samples

Sample No.	Lithology	Lamination/ Bioturbation	Formation name	Geologic age
	Yashima area			
814-5	massive dark greenish gray sandy siltstone	bioturbated	Tentokuji	Pliocene
814-4	massive dark greenish gray sandy siltstone	bioturbated	Tentokuji	Pliocene
814-3	massive dark greenish gray sandy siltstone	bioturbated	Tentokuji	Pliocene
814-2	massive dark greenish gray sandy siltstone	bioturbated	Tentokuji	Pliocene
814-1	massive dark greenish gray sandy siltstone	bioturbated	Tentokuji	Pliocene
1020-1	massive dark gray mudstone	bioturbated	Funakawa	Late Miocene-Early Pliocene
1020-2	massive dark gray mudstone	bioturbated	Funakawa	Late Miocene-Early Pliocene
1004-3	massive dark gray mudstone	bioturbated	Funakawa	Late Miocene-Early Pliocene
1020-3	massive dark gray mudstone	bioturbated	Funakawa	Late Miocene-Early Pliocene
927-3	massive gray muddy porcelanite	slightly lam.	Onnagawa	Late Miocene
1003-1	massive dark gray siliceous mudstone	slightly lam.	Onnagawa	Late Miocene
927-2	massive gray muddy porcelanite	slightly lam.	Onnagawa	Late Miocene
1003-2	massive gray muddy porcelanite	slightly lam.	Onnagawa	Late Miocene
927-1	banded dark gray muddy porcelanite	laminated	Onnagawa	Middle Miocene
923-4	banded dark gray muddy porcelanite	laminated	Onnagawa	Middle Miocene
921-3	banded dark gray porcelanite	laminated	Onnagawa	Middle Miocene
921-1	banded dark gray muddy porcelanite	slightly lam.	Onnagawa	Middle Miocene
808-1	massive dark gray muddy porcelanite	bioturbated	Onnagawa	Middle Miocene
	Noshiro area			
88NSR2	banded black porcelanite	laminated	Onnagawa	Miocene
88NSR4	banded black porcelanite	slightly lam.	Onnagawa	Miocene
88NSR14	banded drak gray muddy porcelanite	well lam.	Onnagawa	Miocene
88NSR15	banded drak gray porcelanite	well lam.	Onnagawa	Miocene
88NSR16	banded drak gray porcelanite	well lam.	Onnagawa	Miocene
	Takanosu area			
88NSR32	banded black muddy porcelanite	slightly lam.	Onnagawa	Miocene
88NSR33	banded dark gray porcelanite	slightly lam.	Onnagawa	Miocene
88NSR35	banded black porcelanite	slightly lam.	Onnagawa	Miocene
88NSR36	massive dark greenish gray diatomaite	bioturbated	Onnagawa	Miocene
88NSR37	banded dark gray chert	bioturbated	Onnagawa	Miocene
88NSR38	banded dark gray tuffaceous siltstone	bioturbated	Onnagawa	Miocene

slightly lam. = slightly laminated; well lam. = well laminated

Extraction and isolation of biomarker compounds from rock samples were carried out according to the method of Sakata et al. [44,45]. The biomarker compounds thus isolated were analyzed on a Hewlett Packard 5890A+5970B gas chromatography-mass spectrometry (GC/MS) system. Assignment of peaks was carried out by comparison with standards and literature, e.g., [46,47,48]. Relative abundance of the compounds was measured in peak areas in the fragmentgrams of m/z=191 for triterpanes and m/z=217 for steranes.

Preservation degree of lamination was determined according to the method of Tada [17]. The preservation degree consists of four classes: *bioturbated* (Class E by [17]), *slightly laminated* (Class D), *laminated* (Classes B and C) and *well laminated* (Class A).

RESULTS AND DISCUSSION

An example of reconstructed ion chromatogram and fragmentgrams of hydrocarbon fraction from an Onnagawa sample (no. 88NSR16) and the list of biomarker compounds assigned on the fragmentgrams are shown in Figure 3 and Table 2, respectively. Contents of total organic carbon, total sulfur, total nitrogen and major inorganic and metal elements are shown in Table 3. The relative abundance of steranes and triterpanes are shown in Table 4.

Major and metal element distribution
Close positive correlations between Al_2O_3, TiO_2, Na_2O, K_2O, Fe_2O_3 and MgO exist in all samples from the Onnagawa, Funakawa and Tentokuji Formations, suggesting a detrital origin for those elements, while no correlation is found between Al_2O_3 and the other major components: TOC, TN, TS, MnO, P_2O_5 and CaO (Figure 4). The negative correlation between

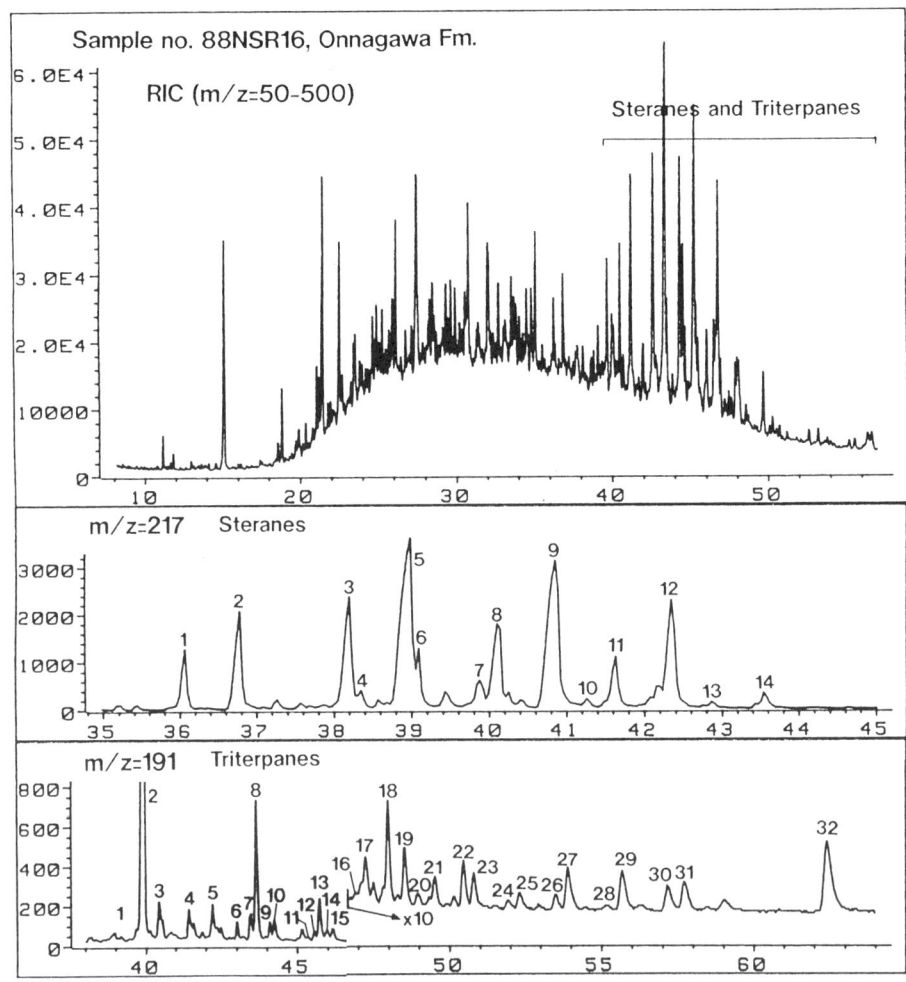

Figure 3. Reconstructed ion chromatogram and mass fragmentograms of steranes (m/z=217) and triterpanes (m/z=191) in the saturated hydrocarbon fraction from a Onnagawa diatomaceous sediment (no. 88NSR16). GC/MS conditions: Ultra-1 capillary column (25 m x 0.20 mm); the temperature was programmed from 60 °C to 160 °C at 20 °C/min, from 160 °C to 300 °C at 3 °C/min (RIC) or 4 °C/min (m/z=217 and 191), and then isothermal. See Table 2 for assignment of peaks.

Table 2.
Assignment of peaks on the chromatograms shown in Figure 3

Peak no. Steranes (m/z=217)

1. 5β,14α,17α-24-norcholestane 20R (C_{26})
2. 5α,14α,17α-24-norcholestane 20R (C_{26})
3. 5β,14α,17α-cholestane 20R (C_{27})
4. 13β,17α-diastigmastane 20S
5. 5α,14α,17α-cholestane 20R (C_{27})
6. 13β,17α-diastigmastane 20S
7. 5α,14α,17α-ergostane 20S (C_{28})
8. 5β,14α,17α-ergostane 20R (C_{28})
9. 5α,14α,17α-ergostane 20R (C_{28})
10. 5α,14α,17α-stigmastane 20S (C_{29})
11. 5β,14α,17α-stigmastane 20R (C_{29})
12. 5α,14α,17α-stigmastane 20R (C_{29})
13. 5β,14α,17α-propyl-cholestane 20R (C_{30})
14. 5α,14α,17α-propyl-cholestane 20R (C_{30})

Peak no. Triterpanes (m/z=191)

1. 18α-22,29,30-trisnorneohopane (C_{27})
2. 17α-22,29,30-trisnorhopane (C_{27})
3. 17β-22,29,30-trisnorhopane (C_{27})
4. 17α,18α,21β-28,30-bisnorhopane (C_{28})
5. 17α,21β-30-norhopane (C_{29})
6. 17β,21α-30-norhopane (C_{29})
7. 18α-oleanane
8. 17α,21β-hopane (C_{30})
9. 17β,21β-30-norhopane (C_{29})
10. 17β,21α-moretane(hopane) (C_{30})
11. 17α,21β-homohopane 22S (C_{31})
12. 17α,21β-homohopane 22R (C_{31})
13. gammacerane
14. 17β,21β-hopane (C_{30})
15. 17β,21α-homohopane (C_{31})
16. 17α,21β-bishomohopane 22S (C_{32})
17. 17α,21β-bishomohopane 22R (C_{32})
18. 17β,21α-bishomohopane (C_{32})
19. 17β,21β-homohopane (C_{31})
20. 17α,21β-trishomohopane 22S (C_{33})
21. 17α,21β-trishomohopane 22R (C_{33})
22. 17β,21α-trishomohopane (C_{33})
23. 17β,21β-bishomohopane (C_{32})
24. 17α,21β-tetrakishomohopane 22S (C_{34})
25. 17α,21β-tetrakishomohopane 22R (C_{34})
26. 17β,21α-tetrakishomohopane (C_{34})
27. 17β,21β-trishomohopane 22S (C_{33})
28. 17α,21β-pentakishomohopane 22S (C_{35})
29. 17α,21β-pentakishomohopane 22R (C_{35})
30. 17β,21α-pentakishomohopane (C_{35})
31. 17β,21β-tetrakishomohopane (C_{34})
32. 17β,21β-pentakishomohopane (C_{35})

Table 3.
Contents of TOC, TS, TN and major inorganic and metal components in the Onnagawa, Funakawa and Tentokuji Formations.

Sample no.	TOC (%)	TS (%)	TN (%)	MnO (%)	Fe$_2$O$_3$^{T} (%)	MgO (%)	CaO (%)	TiO$_2$ (%)	Al$_2$O$_3$ (%)	K$_2$O (%)	Na$_2$O (%)	P$_2$O$_5$ (%)	TOC/TS	TOC/TN	SiO$_2$* (%)	V (ppm)	Cu (ppm)	Zn (ppm)
Yashima area																		
814-5	0.97	1.08	0.10	0.04	4.68	1.70	2.17	0.44	12.57	2.00	1.34	0.075	0.90	9.70	72.83	72.37	27.93	92.23
814-4	0.93	0.86	0.09	0.04	3.73	1.73	1.53	0.50	13.09	2.35	1.60	0.087	1.08	10.00	73.47	82.87	28.05	96.44
814-3	0.98	0.98	0.10	0.04	4.31	1.74	1.92	0.52	13.16	1.95	1.37	0.096	1.00	9.80	72.85	81.71	21.99	88.78
814-2	0.98	1.12	0.09	0.03	4.46	1.75	1.36	0.50	12.78	1.81	1.26	0.087	0.88	10.89	73.78	71.79	23.73	94.67
814-1	0.95	0.86	0.09	0.04	4.00	1.78	1.53	0.50	13.80	2.18	1.56	0.093	1.10	10.56	72.62	79.23	26.07	95.36
1020-1	0.66	1.10	0.10	0.04	4.50	1.31	1.24	0.50	14.65	1.94	1.30	0.048	0.60	6.60	72.62	68.07	22.71	95.76
1020-2	0.76	0.93	0.07	0.01	2.70	0.81	0.90	0.41	13.03	1.95	0.81	0.068	0.82	10.41	77.55	47.63	17.73	108.62
1004-3	0.69	1.19	0.12	0.04	4.46	1.05	1.22	0.48	13.70	1.85	1.93	0.142	0.58	5.75	73.12	63.34	20.70	90.87
1020-3	0.66	0.86	0.16	0.04	2.63	0.88	0.54	0.30	7.34	1.58	1.38	0.071	0.77	4.13	83.56	44.05	20.69	91.48
927-3	1.53	1.33	0.18	0.04	2.68	0.62	3.68	0.32	7.08	1.26	0.67	0.139	1.15	8.50	80.48	66.94	36.12	104.77
1003-1	1.34	1.39	0.20	0.03	3.17	0.80	1.41	0.38	10.06	1.62	1.17	0.097	0.96	6.70	78.33	72.34	33.90	91.87
927-2	1.13	0.95	0.18	0.03	2.05	0.53	0.64	0.21	5.62	0.96	0.59	0.086	1.19	6.28	87.03	35.90	28.70	79.40
1003-2	1.23	1.40	0.19	0.03	2.58	0.80	0.81	0.30	8.59	1.37	0.80	0.085	0.88	6.47	81.81	48.69	35.95	125.11
927-1	1.59	1.02	0.16	0.02	1.75	0.54	0.53	0.26	6.01	1.33	0.52	0.070	1.56	9.94	86.21	61.74	45.97	153.74
923-4	1.38	0.91	0.13	0.02	1.54	0.60	1.34	0.22	6.14	1.31	0.62	0.563	1.52	10.62	85.23	70.37	40.22	183.55
921-3	1.05	0.71	0.10	0.01	1.24	0.42	0.38	0.17	4.23	0.76	0.42	0.071	1.48	10.50	90.42	40.31	37.44	121.64
921-1	1.09	1.39	0.12	0.02	3.17	1.37	0.89	0.35	7.70	1.32	0.67	0.088	0.78	9.08	81.82	56.42	41.31	104.46
808-1	1.16	1.64	0.10	0.02	3.07	1.36	0.69	0.36	7.91	1.50	0.88	0.086	0.71	12.08	81.23	71.85	43.54	130.89
Noshiro area																		
88NSR2	1.80	0.34	0.08	0.01	0.80	0.64	0.26	0.13	4.31	0.77	0.28	0.204	5.29	22.50	90.38	86.98	57.66	57.57
88NSR4	1.22	0.97	0.08	0.02	1.53	0.58	0.11	0.17	4.50	0.77	0.26	0.069	1.26	15.25	89.73	51.00	50.06	150.36
88NSR14	1.69	1.23	0.11	0.02	1.76	0.72	0.05	0.31	7.11	2.76	0.34	0.049	1.37	15.36	83.84	136.89	135.48	434.51
88NSR15	1.94	1.39	0.11	0.07	1.94	2.80	4.93	0.18	4.70	1.00	0.43	0.514	1.40	17.64	79.99	156.58	47.42	93.20
88NSR16	1.20	0.65	0.08	0.04	0.79	1.53	2.79	0.08	1.90	0.33	0.25	0.034	1.85	15.00	90.32	114.33	38.12	59.27
Takanosu area																		
88NSR32	0.96	0.84	0.07	0.02	1.50	0.81	0.58	0.18	5.02	0.91	0.38	0.084	1.14	13.71	88.64	62.05	42.72	217.47
88NSR33	1.37	0.97	0.09	0.03	1.98	0.99	0.23	0.21	4.94	1.17	0.33	0.066	1.41	15.22	87.61	72.16	51.16	251.69
88NSR35	1.31	0.53	0.08	0.01	0.76	0.22	0.22	0.09	2.14	0.33	0.13	0.020	2.47	16.38	94.16	32.11	25.17	81.88
88NSR36	2.18	1.01	0.14	0.01	1.36	0.50	0.20	0.17	4.13	0.53	0.34	0.055	2.16	15.57	89.38	69.63	49.36	144.31
88NSR37	1.04	0.75	0.07	0.31	1.13	0.47	0.44	0.12	3.10	0.63	0.33	0.061	1.39	14.86	91.85	33.47	33.07	79.82
88NSR38	1.22	1.44	0.07	0.02	3.26	1.67	1.49	0.35	8.06	1.60	1.00	0.116	0.85	17.43	79.70	111.15	57.08	125.21

TOC=total carbon content, TS=total sulfur, TN=total nitrogen, Fe$_2$O$_3$^{T}=FeO+Fe$_2$O$_3$.
SiO$_2$* (%) =100%−(TOC+TS+TN+MnO+Fe$_2$O$_3$^{T}+MgO+CaO+TiO$_2$+Al$_2$O$_3$+K$_2$O+Na$_2$O+P$_2$O$_5$)=SiO$_2$+H$_2$O+CO$_2$.

Table 4.1.
Bitumen content and biomarker distribution in the Onnagawa, Funakawa and Tentokuji Formations

Sample no.	Bitumen (%)	C_{27}/C_{29}	$C_{26}(\%)$	$C_{27}(\%)$	$C_{28}(\%)$	$C_{29}(\%)$	$C_{30}(\%)$	$S/S+R$ C_{28}	$S/S+R$ C_{29}	Oleanane I.	Gam./Hops(%)	Hop/St
					Steranes							
							Yashima area					
814-5	0.05	0.85	5.63	28.86	29.19	33.80	2.52	0.08	0.13	11.63	0.00	1.09
814-4	0.05	0.99	5.31	33.33	25.15	33.54	2.67	0.12	0.19	0.00	0.00	1.32
814-3	0.05	0.94	4.92	28.63	31.56	30.56	4.32	0.15	0.08	2.72	0.00	1.16
814-2	0.05	0.49	2.96	22.84	27.51	46.69	0.00	0.11	0.08	2.93	0.00	1.09
814-1	0.08	0.89	6.55	31.09	24.91	34.88	2.58	0.11	0.16	0.00	0.00	1.28
1020-1	0.08	0.85	5.82	34.06	15.83	40.23	4.07	0.00	0.00	0.00	0.00	1.65
1020-2	0.09	0.82	3.43	34.26	19.51	41.80	1.00	0.16	0.04	0.52	0.00	1.88
1004-3	0.04	1.15	9.47	34.87	22.57	30.28	2.82	0.00	0.00	0.00	0.00	0.83
1020-3	0.05	2.15	35.23	29.62	17.09	13.80	4.27	0.16	0.00	6.15	0.00	0.50
927-3	0.15	1.76	19.95	33.87	22.34	19.20	4.64	0.11	0.03	3.74	1.74	0.07
1003-1	0.16	1.30	10.58	31.94	28.65	24.50	4.33	0.09	0.04	8.52	1.71	0.05
927-2	0.20	1.02	5.80	27.48	35.14	26.88	4.69	0.06	0.04	9.59	4.08	0.14
1003-2	0.15	0.91	4.87	25.50	36.92	28.03	4.68	0.05	0.04	0.00	0.00	0.11
927-1	0.20	1.02	5.37	23.37	44.15	22.90	4.22	0.05	0.04	10.35	2.40	0.12
923-4	n.d.	1.03	6.33	22.92	43.91	22.29	4.55	0.07	0.06	12.81	2.43	0.09
921-3	0.17	1.32	6.19	26.54	42.85	20.08	4.34	0.05	0.05	15.56	2.57	0.13
921-1	0.09	1.30	5.62	28.38	40.96	21.79	3.25	0.08	0.04	6.75	2.61	0.15
808-1	0.13	1.28	5.89	30.70	35.87	23.90	3.64	0.10	0.06	19.24	5.31	0.21
							Noshiro area					
88NSR2	0.23	1.68	6.13	35.50	31.58	21.18	2.99	0.11	0.07	8.65	2.53	0.66
88NSR4	0.11	1.13	2.88	36.13	24.73	31.93	2.74	0.08	0.03	1.87	0.00	0.78
88NSR14	0.22	2.59	7.95	41.71	28.47	16.12	2.63	0.19	0.09	5.22	6.58	0.18
88NSR15	0.25	2.39	7.62	40.33	30.32	16.86	2.12	0.13	0.07	15.98	3.89	0.18
88NSR16	0.19	1.95	9.46	33.92	32.94	17.42	2.89	0.14	0.08	16.55	5.39	0.21
							Takanosu area					
88NSR32	0.11	1.17	2.75	40.11	20.00	34.15	1.86	0.10	0.07	1.36	0.11	3.40
88NSR33	0.17	0.90	6.91	28.93	19.13	32.22	7.58	0.27	0.28	6.35	1.48	5.70
88NSR35	0.19	2.02	20.17	35.59	11.57	17.66	2.91	0.40	0.24	6.69	0.00	2.82
88NSR36	0.22	0.73	2.80	27.15	22.70	37.30	7.93	0.08	0.03	0.51	0.00	6.27
88NSR37	0.22	1.82	10.75	36.93	23.78	20.29	1.76	0.32	0.23	6.38	0.72	0.77
88NSR38	0.11	2.20	3.64	46.10	23.65	20.98	3.51	0.18	0.18	2.16	0.60	0.98

Steranes (based on the relative intensity on the fragmentogram of m/z=217); $C_n(\%) = \alpha\alpha\alpha R\text{-}C_n/\Sigma(\alpha\alpha\alpha R\text{-}C_{26} - \alpha\alpha\alpha R\text{-}C_{30}) \times 100$; n=26,27,28,29,30.
Oleanane I. = 18α-oleanane/C_{30} hopanes x100. Gam./Hop(%)=gammacerane/$\Sigma(C_{27}\text{-}C_{35}$ hopanes)x100 (m/z=191).
Hop/St=$\Sigma(C_{27}\text{-}C_{35}$ hopanes:m/z=191)/$\Sigma(C_{26}\text{-}C_{30}$ steranes:m/z=217). n.d.=not determined.

Table 4.2.
Bitumen content and biomarker distribution in the Onnagawa, Funakawa and Tentokuji Formations

Sample no.	C_{27}(%)	C_{28}(%)	C_{29}(%)	C_{30}(%)	C_{31}(%)	C_{32}(%)	C_{33}(%)	C_{34}(%)	C_{35}(%)	Homohop.I.	ββ/total C_{30}	S/S+R C_{31}
Yashima area												
814-5	18.61	1.34	25.79	39.23	15.02	0.00	0.00	0.00	0.00	0.00	0.39	0.00
814-4	15.75	0.00	29.39	30.23	24.63	0.00	0.00	0.00	0.00	0.00	0.31	0.21
814-3	13.29	0.41	21.04	32.06	28.08	3.83	1.29	0.00	0.00	0.00	0.51	0.34
814-2	19.34	1.37	26.70	37.59	15.00	0.00	0.00	0.00	0.00	0.00	0.38	0.32
814-1	16.01	0.00	29.92	28.00	26.06	0.00	0.00	0.00	0.00	0.00	0.33	0.23
1020-1	9.53	12.52	26.10	29.18	22.68	0.00	0.00	0.00	0.00	0.00	0.41	0.00
1020-2	10.75	7.18	14.81	33.40	32.01	1.83	0.00	0.00	0.00	0.00	0.62	0.10
1004-3	7.30	19.11	18.69	26.06	24.98	3.86	0.00	0.00	0.00	0.00	0.58	0.27
1020-3	5.64	28.00	29.91	19.60	13.44	3.41	0.00	0.00	0.00	0.00	0.41	0.35
927-3	8.94	32.10	31.02	17.68	7.53	2.72	0.00	0.00	0.00	0.00	0.41	0.44
1003-1	5.59	12.05	31.29	28.91	9.97	6.09	3.01	1.24	1.84	8.32	0.34	0.36
927-2	6.11	1.27	26.80	43.81	15.54	3.62	1.22	0.00	1.62	7.34	0.31	0.30
1003-2	11.77	12.07	32.27	31.39	10.60	1.90	0.00	0.00	0.00	0.00	0.21	0.18
927-1	12.43	2.52	24.11	46.60	10.24	3.14	0.95	0.00	0.00	0.00	0.09	0.20
923-4	9.20	2.77	23.26	48.00	8.42	5.24	2.13	0.00	0.98	5.85	0.08	0.18
921-3	16.13	0.64	15.53	26.00	10.85	9.88	4.98	3.21	12.77	30.63	0.15	0.28
921-1	25.66	3.59	25.10	22.49	13.19	8.80	1.17	0.00	0.00	0.00	0.18	0.19
808-1	7.60	6.76	28.06	27.69	16.64	7.72	3.33	0.77	1.43	4.78	0.16	0.19
Noshiro area												
88NSR2	44.57	1.05	19.78	22.45	6.00	1.67	1.79	0.62	2.09	17.18	0.30	0.18
88NSR4	20.00	1.83	26.20	39.01	10.73	1.52	0.32	0.00	0.37	2.89	0.29	0.23
88NSR14	25.01	0.00	24.45	23.00	13.34	5.28	1.91	1.48	5.53	20.07	0.30	0.16
88NSR15	55.84	0.50	11.61	18.18	4.05	2.93	1.83	1.35	3.72	26.82	0.10	0.29
88NSR16	60.04	3.54	8.05	15.56	3.24	2.71	1.98	1.29	3.58	27.98	0.08	0.00
Takanosu area												
88NSR32	18.31	4.24	23.90	35.76	15.44	1.80	0.55	0.00	0.00	0.00	0.40	0.44
88NSR33	11.91	1.97	23.68	34.41	18.91	6.57	1.37	0.59	0.58	2.08	0.36	0.43
88NSR35	27.74	1.66	34.67	24.50	8.30	2.59	0.44	0.00	0.10	0.86	0.18	0.44
88NSR36	8.13	0.00	10.34	73.39	7.79	0.33	0.02	0.00	0.00	0.00	0.11	0.14
88NSR37	32.88	0.75	28.25	22.42	7.25	3.20	2.37	0.82	2.07	13.16	0.11	0.50
88NSR38	17.89	0.00	31.52	36.08	7.55	6.59	0.37	0.00	0.00	0.00	0.38	0.16

Hopanes(based on the relative intensity on the fragmengram of m/z=191); C_n(%)=$C_n/\Sigma(C_{27}-C_{35})\times 100$; n=27,28,29,30,31,32,33,34,35.
Homohop.I.=Homohopane Index=C_{35} hopanes/(C_{31}-C_{35} hopanes)×100 (m/z=191).

Al$_2$O$_3$ and SiO$_2$* means that the samples consist of two major end components, i.e. detrital matter and biogenic silica components. As is shown in Table 3, the Onnagawa Formation is characterized by relatively low contents of detrital components and high silica content, while the Funakawa and Tentokuji Formations by relatively high contents of detrital components and low silica content. Factors controlling the distribution of these major elements in the Onnagawa Formation are discussed in Watanabe et al.[49].

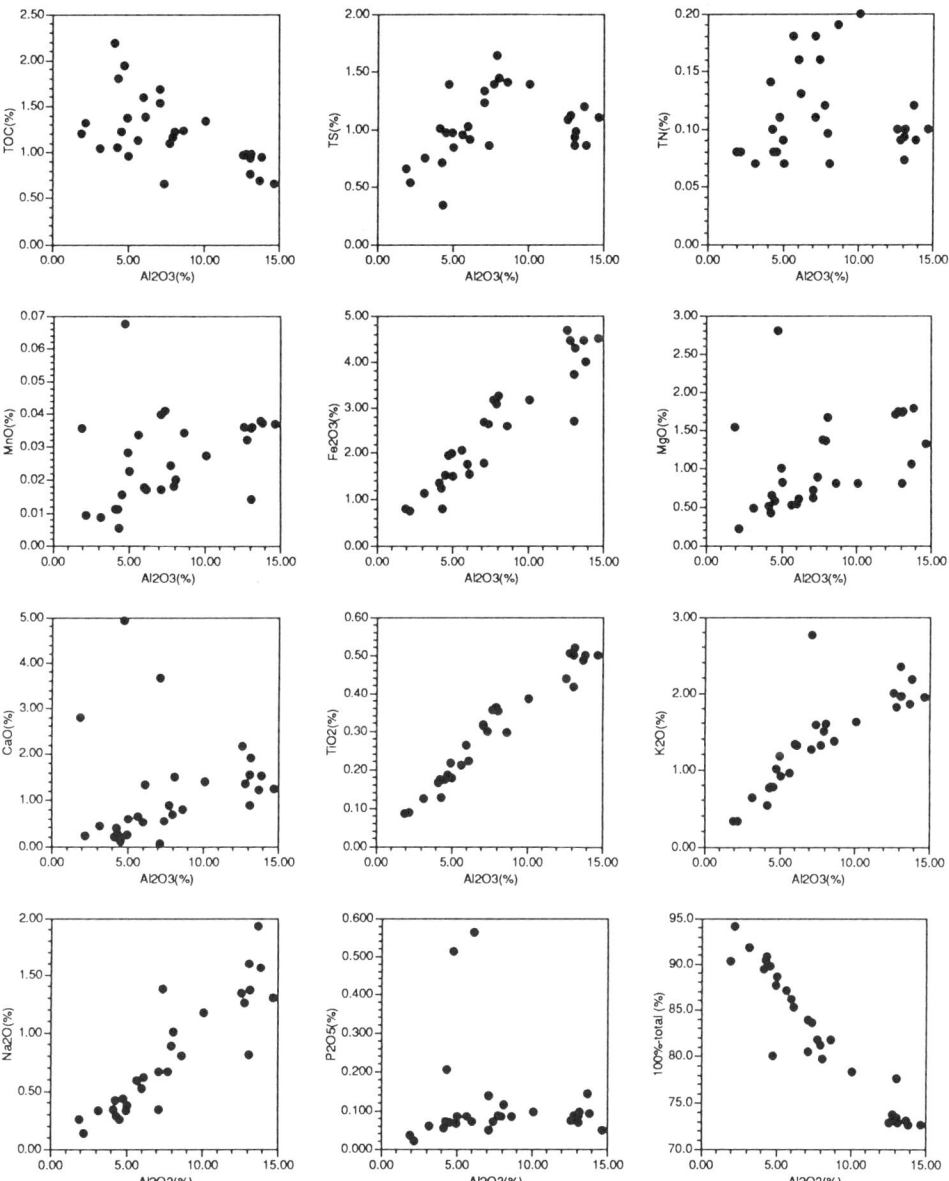

Figure 4. Scattergraphs of several major elements against alumina content. 100 - total = SiO$_2$*.

Maturity of organic matter
The presence of thermally unstable molecules, i.e. 5β, 14α, 17α-steranes and 17β, 21β-hopane, in all samples analyzed here (Figure 3) means that those samples are in an immature stage below the oil generation window.
A consistent decrease occur in ββ/total hopane ratio, which is a maturity parameter based on the isomerization (ββ,βα to αβ isomers), from younger to older horizons in Yashima area (Table 4). Since the degrees of isomerization of ββ/total hopane ratio = 0.24, 0.11 and 0.04 correlate to vitrinite reflectances Ro = 0.3, 0.4 and 0.5% respectively in a Neogene oil-producing basin of northern Honshu, Japan [44], the Tentokuji and Funakawa Formations have the maturity below Ro = 0.3 %, and the Onnagawa Formation has the maturity of Ro = 0.3-0.4 % in the Yashima area. Similarly, the Onnagawa samples from the Noshiro and Takanosu areas have maturities below Ro = 0.4 %.
It should be noted that unusual higher values of S/S+R ratios of ergostanes, stigmastanes and homohopanes, which are also used as maturity parameters based on their epimerization (R to S isomers), than those expected from the stages of silica diagenesis and their ββ/total hopane ratios are observed in all Tentokuji samples from the Yashima area and two Onnagawa samples (nos. 88NSR33 and 88NSR35) from the Takanosu area (Table 4). Some researchers report the unusual characteristics of these maturity parameters in Japanese Cenozoic sediments [38,50,51]. Although there is no palynological evidence, the high aluminum contents in the Tentokuji Formation suggest that the unusual higher values of S/S+R ratio in the Tentokuji samples are due to the increased input of terrigenous matured organic matter. However, this reason fails to explain those in Onnagawa samples from Takanosu area, because their low aluminum contents (Table 3) suggest that terrigenous organic matter is poor in those samples. ten Haven et al. [52] report that immature bitumens from hypersaline rocks can appear mature due to special diagenetic pathways of steranes and hopanes. Both Onnagawa samples showing unusual S/S+R ratios have significant amounts of tetrakishomohopanes (C_{34} hopanes) and pentakishomohopanes (C_{35} hopanes), suggesting a highly reducing depositional environment, as is discussed in following section, which is similar to a hypersaline environment. Thus it is assumed that some common factor in both highly reducing and hypersaline depositional environments, most likely the reaction of hydrogen sulfide to the diagenetic evolution of steroids and hopanoids, causes the unusual S/S+R ratios of the two Onnagawa samples.

Source of organic matter
The samples from the Onnagawa Formation and lower part of the Funakawa Formation show high cholestane/stigmastane (C_{27}/C_{29} sterane) ratios, while those from the upper part of the Funakawa Formation and the Tentokuji Formation show low C_{27}/C_{29} sterane ratios (Table 4). Since the C_{27}/C_{29} sterane ratio is a source parameter indicating the relative inputs of algae and higher plants based on the dominance of steroids in most algae and C_{27}/C_{29} steroids in higher plants [21], the relatively higher C_{27}/C_{29} sterane ratios in the Onnagawa Formation and the lower part of the Funakawa Formation indicate high algal input and low higher plant input. It is also noted that extremely high C_{27}/C_{29} sterane ratios are shown in the upper part of the Onnagawa Formation and the lower part of the Funakawa Formation. Since a gradual increase of aluminum content upwards in the Onnagawa Formation suggests a gradual increase of terrigenous organic matter, it is suggested that those high C_{27}/C_{29} sterane ratios do not reflect decreased terrigenous organic matter, but rather increased algal productivity.
Oleanane index, which is defined as the percentage of 18α-oleanane to C_{30} hopanes and is often used as an indicator of angiosperm input [22], is generally higher in the Onnagawa samples than in Funakawa and Tentokuji samples. This may be attributed to the increase of relative abundance of C_{30} hopanes over the increase of oleanane contents upwards.
Hopane/sterane ratios, which reflect relative inputs of each source organism, i.e., bacteria and

algae, show relatively high values in the Funakawa and Tentokuji samples, and very wide variations in Onnagawa samples (Table 4). It is also pointed out that bioturbated samples from the Onnagawa Formation have relatively higher hopane/sterane ratios than laminated samples. Figure 5 shows that the hopane/sterane ratios of Onnagawa samples appear to reflect both alumina content (detrital input) and preservation degree of lamination (bottom-water redox potential). Mackenzie [53] stated that the hopane concentration reflects not only the bacterial input to the sediment, but also the input of bacterially reworked terrigenous organic matter. If the input of bacterially reworked organic matter is reflected by hopane/sterane ratio, there should be a positive correlation between this ratio and alumina content. However, a negative correlation is found between them (Figure 5), suggesting that hopane/sterane ratio in the Onnagawa Formation reflects the relative inputs of authigenic bacteria and algae, i.e., bacterial activity during early diagenesis. It is known that clay minerals tend to protect organic matter against bacterial degradation by the formation of clay-organic matter complexes [54]. It is also known that bacterial activity is higher in oxygenated environments than in oxygen-deficient environments. These two factors probably affected the bacterial activity in Onnagawa diatomaceous sediments.

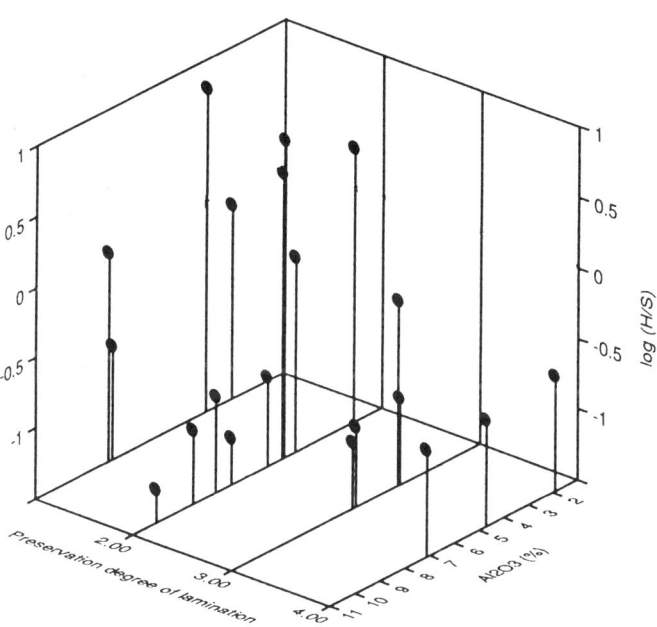

Figure 5. Three-dimensional diagram of hopane/sterane ratio against alumina content and preservation degree of lamination. Preservation degree of lamination; 1 = bioturbated, 2 = slightly laminated, 3 = laminated, 4 = well laminated.

Bottom-water redox condition

The samples from the lower part of the Onnagawa Formation in Yashima area preserve laminations and show a high homohopane index ($C_{35}/(C_{31}-C_{35})$ hopane ratio [27]), while those from the upper part of the Onnagawa Formation and the Funakawa and Tentokuji

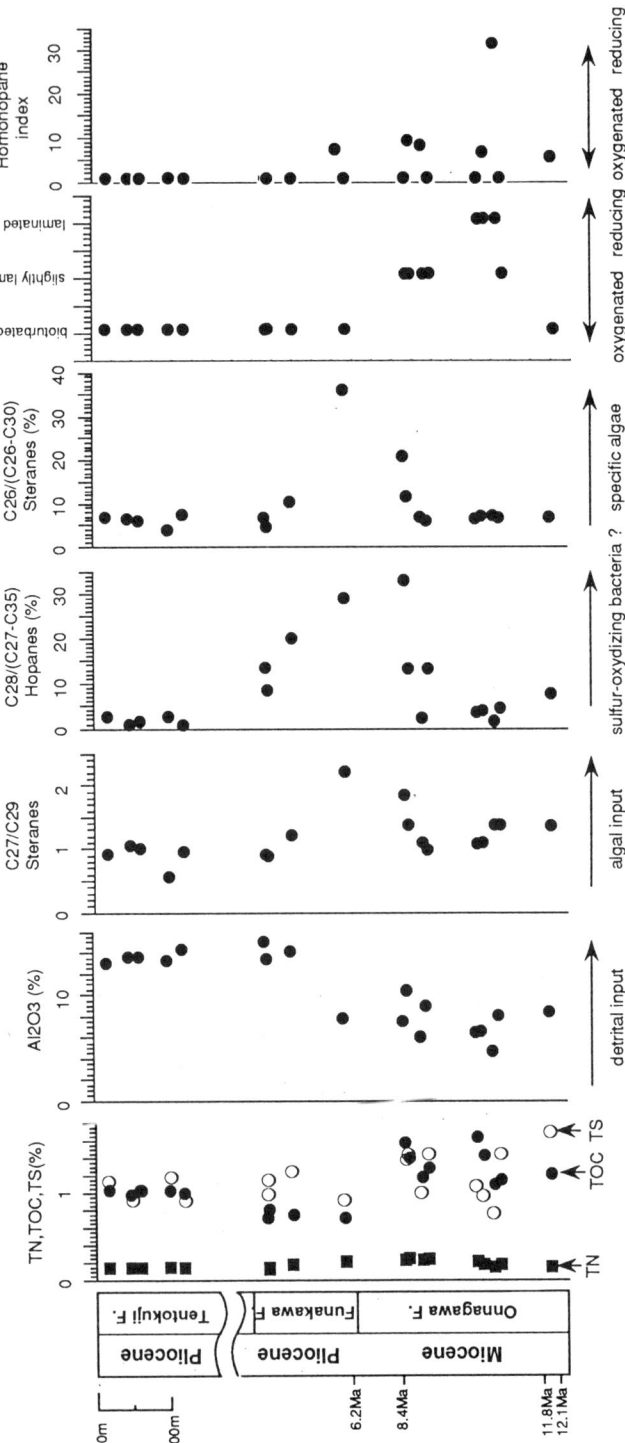

Figure 6. Stratigraphic variation of contents of total organic carbon, total nitrogen, total sulfur and alumina, C_{27}/C_{29} sterane ratio, C_{28} hopane/hopane ratio, C_{26} sterane/sterane ratio, preservation degree of lamination and homohopane index in Yashima area.

Formations show less preserved laminated structure and a lesser amount or absence of C_{35} hopanes (Figure 6). Laminated structure means absence of benthic organisms, indicating a oxygen-deficient depositional environment [55]. The homohopane index is an environmental parameter indicating a reducing depositional environment [27]. Consequently the changes in those parameters from the lower part of the Onnagawa Formation to the overlying formations means that bottom-water redox conditions changed gradually from reducing to oxygenated during the deposition of the upper part of the Onnagawa Formation (10-6 Ma).

Noshiro samples show well laminated structure and a high homohopane index, indicating the deposition in an anoxic environment, while most Takanosu samples show bioturbated structure and are poor in C_{35} hopanes, indicating an oxic depositional environment (Table 4). Whether this difference between the Noshiro and Takanosu samples reflects differences in the horizons sampled in those two areas or whether it reflects different topographic settings cannot be determined, because of lack of information on the biostratigraphy of those two areas at present It is noted that some Onnagawa samples show the presence of gammacerane. A positive correlation (r=0.53) between gammacerane/hopane ratio and homohopane index supports the hypothesis that gammacerane originates from some Protozoa living in highly reducing environment [56].

Occurrence and significance of 24-norcholestanes and 28,30-bisnorhopane

Two unusual compounds, 24-norcholestanes and 28,30-bisnorhopane, are found in some Onnagawa and Funakawa samples (Figure 7).

24-norcholestanes (C_{26} steranes) are detected in some crude oils from the Monterey Formation of California, the Upper Cretaceous of Angola and the Lower Cretaceous Mowry Formation of Central USA [57] and bitumen from the Onnagawa Formation [58]. Traces of the probable precursor, 24-norcholesterols occur in living marine algae and invertebrates, suggesting an origin in eukaryotes with or without prokaryotic symbionts [59]. Moldowan et al. [57] and Suzuki and Sampei [58] suggest the specific algal origin of 24-norcholestanes. In this study, a positive correlation (r=0.52) is found between the 24-norcholestanes/steranes ratio (C_{26}/C_{26}-C_{30} sterane ratio) and the cholestane/stigmastane ratio (C_{27}/C_{29} sterane ratio). Since the high C_{27}/C_{29} sterane ratio means increased algal input due to increased algal productivity as discussed before, this positive correlation suggests that 24-norcholestanes originated from some algal species flourishing in high productivity areas.

28,30-bisnorhopane (C_{28} hopane) is detected in some petroleum source rocks (e.g., Monterey Formation of California [13,60,61,62]; Kimmeridge shale of the North Sea [63]; Middle Ordovician Guttenberg Oil Rock of Wisconsin [64]). Abundant fossilized bacterial mats in the Monterey Formation led Katz and Elrod [61] to suggest the sulfur-oxidizing bacterium *Thioploca* as a possible source for the diagenetic precursor of 28,30-bisnorhopane [65]. In this study, a significant amount of 28,30-bisnorhopane is observed in restricted horizons, i.e., the upper part of the Onnagawa Formation and the lower part of the Funakawa Formation of Yashima area (Figure 6), characterized by high algal productivity and suboxic to oxic depositional environments. In such depositional environments, it is supposed that the combination of an increased input of organic matter from surface waters to the sediment and the high availability of oxygen and sulfate ion in bottom water accelerates the biogeochemical sulfur cycle, reduction of sulfate ion to reduced sulfur species and reoxidation of reduced sulfur species to sulfate ion, in the surface sediments resulting in the production of the 28,30-bisnorhopane precursor which originated from sulfur oxidizing bacteria. The state of 28,30-bisnorhopane in sedimentary rocks, which does not appear to be generated from kerogen, but rather occurs as free molecules in the original bitumen [66], and the lack of hopanoids in lipids extracted from *Thioploca* [67] suggest that 28,30-bisnorhopane precursor is a metabolic product discharged from some sulfur oxidizing bacterium.

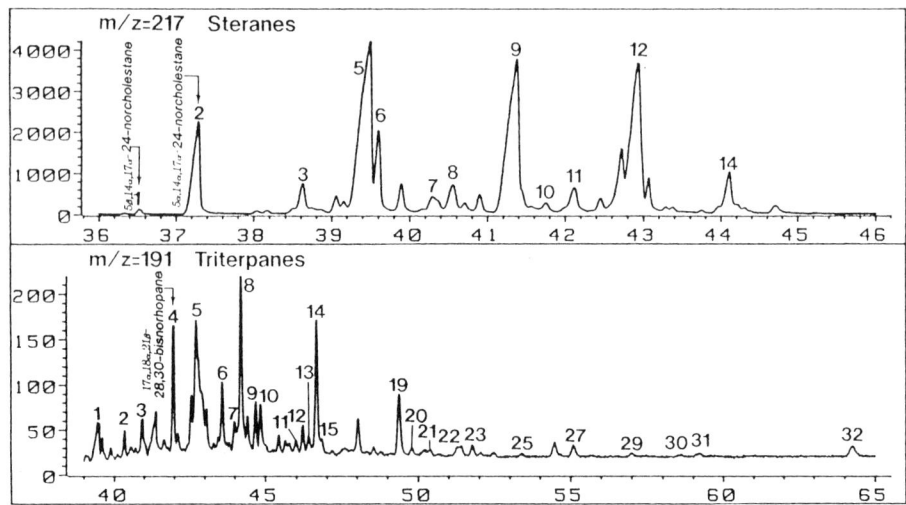

Figure 7. Mass fragmentgrams of steranes (m/z=217) and triterpanes (m/z=191) in an Onnagawa sample (no. 1003-1). Note unusual high abundance of 24-norcholestanes and 28,30-bisnorhopane. GC/MS conditions: Ultra-1 capillary column (25 m x 0.20 mm); the temperature was programmed from 60 °C to 160 °C at 20 °C/min, from 160 °C to 300 °C at 4 °C/min (m/z=217 and 191), and then isothermal. See Table 3 for assignment of peaks.

Paleoceanography of Onnagawa diatomaceous sediments

A gradual increase of aluminum content upwards in the Funakawa Formation (ca. 6 - 4 Ma) of Yashima area (Figure 6) means an increased proportion of detrital matter in the sediments. Tada et al. [68] concluded that the increase in proportion of detrital matter upwards in the Ajigasawa area was caused by an increase of detrital input rather than a decrease of biogenic input. Since the sedimentation rates of the Funakawa and Tentokuji Formations are higher than that of the Onnagawa Formation in Yashima area [69], it is supposed that this increase resulted rates of sedimentation of detrital matter. It is known that the tectonic stress regime of the northern Honshu Arc changed from the *Thermal subsidence substage* with restricted volcanism to the *Incipient Compressional substage* with formation of large caldera structures associated with subaerial felsic volcanism along the present Backbone ranges at ca. 8 Ma; the Backbone Range region began to uplift and expanded its land area [40]. Also, the detrital matter in the upper part of the Onnagawa Formation of Yashima area is composed mainly of volcanic detritus [41]. Considering these aspects, the increase of detrital matter contents in the Funakawa Formation is undoubtedly attributed to the increasing input of volcanic detritus as the Backbone volcanic arc was uplifted.

Variations of C_{27}/C_{29} sterane ratio in the Yashima area (Figure 6) indicate that high algal productivity occurred during deposition of the upper part of the Onnagawa Formation and the lower part of the Funakawa Formation (ca. 9 - 5 Ma). Several diatomaceous sequences deposited in the paleo-Japan Sea during this period show the following unique characteristics: extremely heavy carbon isotopic composition of kerogen in the Yurihara Oil Field in the Akita Basin [9], presence of phosphorite nodules in the Nangai area of the Akita basin [70,71], increased sedimentation rates of diatomite on Sado Island [72,73], unusual presence of magnesite in the Kita-Yamato trough (Unit 3, ODP Leg 127 Site 794) [74] and relatively high silica contents in northern Japan basin (Unit 3, ODP Leg 127 Site 795) [14]. The authors believe that these phenomena can be attributed to the concurrent increased algal production in those areas during ca. 9-5 Ma. However, it should be noted that the differences in the period of high productivity are found between these areas and the other sites. Results of ODP Leg 127 in the Japan Sea show that the productivity was low during 10-6 Ma, while it was relatively high during 6-2.5 Ma, in the Yamato Basin (Sites 794 and 797) and the Japan Basin (Site 796)[14]. This means that the concurrent increase in algal productivity occurred only in some restricted areas, suggesting regional intensification of upwelling currents in the paleo-Japan Sea during this period. It is known that late Miocene was globally a time of coolest climate and lowest sea level [75]. Submarine hiatuses during the interval 8.5-6.6 Ma in the Yamato Basin [76] and the Noto Peninsula [77] suggests the presence of strong bottom currents in the paleo-Japan Sea [76]. Also, the low sea level should have resulted in an undulating sea bottom in the "Green Tuff Basin" [15], which influenced its ocean circulation. Thus the combination of the strong bottom-currents and the undulating sea bottom relief possibly served to intensify the regional upwelling which supported regional high algal productivity.

Variations of preservation degree of lamination and homohopane index in the Yashima area (Figure 6) indicate that bottom-water redox conditions changed gradually from reducing to oxygenated during the deposition of the upper part of the Onnagawa Formation (ca. 10 - 6 Ma). Recently a gradual oxygenation during same period has been reported in Ajigasawa [68,78], Gojonome [17], Nangai [71], Sado Island [79], the Japan Basin [14,80], the Yamato Basin [14] and the Kita-Yamato basin [74]. This means synchronous oxygenation of bottom water during ca. 10-6 Ma in a wide area of the paleo-Japan Sea. Tada [17] proposed a model in which the intrusion of OMZ water from the Pacific Ocean via the sill into the Onnagawa silled basin caused bottom-water anoxia. His model suggests that the decrease of OMZ water input due to falling sea level through late Miocene decreased the total algal production in surface water masses of the paleo-Japan Sea due to the decrease of nutrient supply into the basin, resulting in concurrent oxygenation of its bottom water. Alternatively, it is also possible that the turbulence caused by intensified bottom-water current contributed to the bottom-water oxygenation during this period; however we can find no positive evidence to support this idea at present.

Comparison of the Onnagawa Formation with the Monterey Formation on their organic and petroleum geochemical characteristics

Both the Onnagawa and Monterey Formations are characterized by the presence of sulfur-rich kerogen (type II-S) [12,81], extremely heavy carbon isotopic compositions of kerogen and oil [9,12], high C_{27}/C_{29} sterane ratio [62] and homohopane index [48], the presence of 24-norcholestanes [57,58], 28,30-bisnorhopane [60] and gammacerane [48] (Table 5). This similarity probably reflects the abundance of some common algal species and the presence of a common bacterial fauna.

In contrast, the total organic carbon contents of the Onnagawa Formation are remarkably lower than those of the Monterey Formation. Kano [82] suggested that low organic matter contents of the Onnagawa Formation can be attributed to turbulence of oxygen minimum zone due to the inflow of turbidity currents and volcanic detritus. Iijima et al. [15] point out two possibilities: (1) lack of significant upwelling and (2) dilution effects of organic matter by biogenic silica. This study shows that the horizons showing high C_{27}/C_{29} sterane ratio and 28,30-bisnorhopane, which indicate intensified upwelling, are restricted in time and space in the Onnagawa Formation; also these same horizons show oxygenated depositional environments. This suggests that lack of combination of intensified upwelling and oxygen-deficient bottom waters during deposition of the Onnagawa Formation resulted in the relatively low contents of organic matter.

It is widely accepted that sulfur-rich kerogen of the Monterey Formation formed in highly reducing environments has generated a significant amount of immature oil [12]. Thus the presence of pentakishomohopanes and sulfur-rich kerogen [78] in the Onnagawa Formation may suggest an immature oil generation from such sulfur-rich kerogen. However, crude oils from the main oil fields in the Akita oil-producing basin do not show immature characteristics

Table 5.
Comparison of the Onnagawa Formation with the Montery Formation on their organic geochemical characteristics

	Onnagawa Formation, Akita, northern Honshu, Japan	Monterey Formation, California
Total organic carbon content	1.93% [87]	6.79% (Siliceous Facies) [62]
	1.32% [88]	7.90% (Phosphate Facies) [62]
	1.58% [9]	
	1.37% [this sudy]	
Kerogen type	type II, type II-III, type II-S [81,89]	type II-S, II [12]
$\delta 13C$ of kerogen	-22.44 per mil [9]	-21.8 per mil [12]
$\delta 13C$ of crude oil	-21.97 per mil [9]	-22.6 per mil [12]
C_{27}/C_{29} Sterane ratio	1.5 [this study]	2.2 [62]
C_{26} Steranes(24-norcholestanes)	present [58]	present [57]
C_{28} Hopanes(28,30-bisnorhopanes)	present [this study]	present [13]
C_{35} Hopanes(Pentakishopanes)	present [this sudy]	present [27]
Gammacerane	present [this study]	present [27]

but early mature characteristics (S/S+R ratio of stigmastane = 0.29-0.50; Ro=0.6-0.8%) [45,83,84]. During last decade, it has come to be accepted that the expulsion of hydrocarbons from source rocks is triggered by the increase of capillary pressure caused by cracking of high-molecular-weight organic matter in the source rock, which is controlled mainly by the organic matter content of the source rock [85,86]. Considering this, the absence of immature oil in the Akita oil-producing basin may be attributed to the low organic matter contents of the Onnagawa diatomaceous sediments rather than the quality of organic matter.

CONCLUSIONS

1. The Onnagawa diatomaceous sediments are characterized by high inputs of algal and bacterial organic matter and low inputs of terrigenous organic matter. The relative contribution of bacteria and algae reflects bacterial activity during early diagenesis controlled by both detrital input and bottom-water redox potential.
2. A gradual oxygenation of bottom-waters during the deposition of the upper part of the Onnagawa Formation (ca. 10-6 Ma) in the Yashima area is shown in the variations of preservation degree of lamination, homohopane index and gammacerane/hopane ratio.
3. High algal production through the deposition of the upper part of the Onnagawa Formation and lower part of the Funakawa Formation (ca. 9-5 Ma) indicated by variations of C_{27}/C_{29} sterane ratio during the period suggests regionally intensified upwelling in the Yashima area.
4. High concentrations of 24-norcholestanes and 28,30-bisnorhopane found in the horizons showing high algal productivity and oxygenated depositional environments suggests that 24-norcholestanes originated from some algal species flourishing in high productivity areas, and that the 28,30-bisnorhopane precursor is a metabolic product discharged from some sulfur-oxydizing bacterium living near oxic-anoxic interfaces in the surface sediments.

*Acknowledgment*ts
We sincerely thank the following persons for their kind help with this manuscript. This work includes one of the authors' (M. Y.) undergraduate research project supervised by Prof. Kazuo Taguchi, Tohoku University. This work was supported by financial aid from the Agency of Industrial Science and Technology (AIST special research "Research on three dimensional modeling for resources assessment" supervised by Drs. Teruki Miyazaki, Kisaburo Kodama and Shuichi Tokuhashi). Dr. Eiichi Honza and Mr. Mahito Watanabe, Geological Survey of

Japan (GSJ), read the original manuscript and gave us critical comments. Dr. Takashi Tsuji, JAPEX Co.Ltd., allowed us to refer to his MS manuscript [41]. Dr. Noriyuki Suzuki, Shimane University, and Dr. Yukio Yanagisawa, GSJ, gave us an important suggestion. Drs. Noboru Imai, Minako Terashima, Susumu Sakata, Shigeru Terashima and Mr. Nobuyuki Kaneko, GSJ, gave us kind support on our experiments and field survey. We also thank Prof. Robert E. Garrison, University of California, Santa Cruz, and Emeritus Prof. Azuma Iijima, University of Tokyo, for critical reading of the manuscript.

REFERENCES

1. R.E. Garrison. Neogene Diatomaceous Sedimentation in East Asia: A review with Recommendations for Further Study. In: *United Nations Economic and Social Commission for Asia and the Pacific, Committee for Co-ordination of Joint Prospecting for Mineral Resources in Asian Offshore Areas, Technical Bulltin* 9, pp.57-69 (1975).
2. J.C. Ingle, Jr. Origin of Neogene diatomites around the North Pacific Rim. In: *The Monterey Formation and Related Siliceous Rocks of California.* R.E. Garrison et al. (Eds). pp.159-179. SEPM Special Publication. Soc.Econ.Paleontol.Mineral. (1981).
3. J.C. Taylor. Geologic appraisal of the petroleum potential of offshore southern California: The Borderland compared to onshore coastal basins, *U.S. Geological Survey Circular* **730**, 1-14 (1976).
4. O.K. Bazhenova and O.A. Arefiev. Immature oils as the products of early catagenetic transformation of bacterial-algal organic matter, *Org. Geochem.* **16**, 307-311 (1990).
5. V.V. Ivanov and O.V. Shcherban. Mineral matrix influence on the dynamics and products of organic matter catagenetic transformation, *Org.Geochem.* **4**, 185-194 (1983).
6. K. Taguchi and K. Sasaki. Organogeochemistry and its relation to the geology of petroleum accumulation in Japan, In: *Proceeding of Symposium on Hydrogeochemistry and Biogeochemistry.* II Biogeochemistry, p.133. The Clark Company, Washington, D.C. (1973).
7. S.A. Graham and L.A. Williams. Tectonic, depositional, and diagenetic history of Monterey Formation (Miocene), central San Joaquin basin, California, *Am. Assoc. Petrol. Geol. Bull.* **69**, 385-411 (1985).
8. K. Aoyagi and A. Iijima. Petroleum occurrence, generation, and accumulation in the Miocene siliceous deposits of Japan. In: *Siliceous Sedimentary Rock-Hosted Ores and Petroleum.* J. R. Hein (Ed.). pp.117-137. Van Nostrand Reinhold, New York (1987).
9. A. Waseda and M. Omokawa. Generation, migration and accumulation of hydrocarbons in the Yurihara oil and gas field, *J. Jap. Assoc. Petrol. Technol.* **55**, 233-244 (1990). (in Japanese).
10. C.W.D. Milner, M.A. Rogers and C.R. Evans. Petroleum transformations in reservoirs, *J. Geochem. Explor.* **7**, 101-153 (1977).
11. N.F. Petersen and P.J. Hickey. Evidence of early generation of oil from Miocene source rocks, California coastal basins (abstract). In: *Petroleum Generation and Occurrence in the Miocene Monterey Formation.* C. M. Isaacs and R. E. Garrison (Eds). Pacific Section, SEPM. p.226. Los Angeles, California (1983).
12. W.L. Orr. Kerogen/asphaltene/sulfur relationships in sulfur-rich Monterey oils, *Org. Geochem.* **10**, 499-516 (1986).
13. J.A. Curial, D. Cameron and D.V. Davis. Biological marker distribution and significance in oils and rocks of the Monterey Formation, California, *Geochim. Cosmochim. Acta.* **49**, 271-288 (1984).
14. R. Tada and A. Iijima. Lithostratigraphy and compositional variation of Neogene hemipelagic sediments in the Japan Sea. In: *Proceedings of the Ocean Drilling Program, Scientific Results.* Vol. 127/128, pp.1229-1260. (1992).
15. A. Iijima, R. Tada and Y. Watanabe. Developments of Neogene sedimentary basins in the northeastern Honshu Arc with emphasis on Miocene siliceous deposits, *Fac. Sci. Univ. Tokyo.* Sec. II, **21**, 417-446 (1988).
16. A. Iijima and R. Tada. Evolution of Tertiary sedimentary basins of Japan in reference to opening of the Japan Sea, *Fac.Sci.Univ.Tokyo,* Sec II, **22**, 121-171 (1990).
17. R. Tada. Origin of rhythmical bedding in middle Miocene siliceous rocks of the Onnagawa Formation, northern Japan, *Jour. Sedim. Petrol.* **61**, 1123-1145 (1991).
18. H. Fukusawa. Preliminary report on the relationship between the sedimentation of Neogene biosiliceous shale and paleoceanographic environments, *Mem. Geol. Soc. Japan* no. 37, 219-226 (1992). (in Japanese).
19. T.H. van Andel. Gulf of California. In: *The Encyclopedia of Oceanography.* R. W. Fairbridge (Ed.). pp.312-315. Dowden Hutchingson & Ross Inc, Strousburg, Pennsylvania. (1966).
20. G. Eglington and M. Calvin. Chemical fossils, *Sci. Am.* **216**, 32-43 (1967).

21. W.-Y. Huang and W.G. Meinschein. Sterols as ecological indicators. *Geochim. Cosmochim. Acta* **43**, 739-745 (1979).
22. C.M. Ekwoezor, J.I. Okogun, D.E.U. Ekong and J.M. Maxwell. Preliminary organic geochemical studies of samples from the Nigel Delta (Nigeria). *Chem. Geol.* **27**, 29-37 (1979).
23. G.A. Wolff, N.A. Lamb and J.R. Maxwell. The origin and fate of 4-methyl steroids hydrocarbons I. 4-methyl sterenes, *Geochim. Cosmochim. Acta* **50**, 335-342 (1986).
24. S.C. Brassell, R.G. Brereton, G. Eglinton, J. Grimalt, G. Liebezeit, I.T. Marlowe, U. Pflaumann and M. Sarnthein. Paleoclimatic signals recognized by chemometric treatment of molecular stratigraphic data, *Org. Geochem.* **10**, 649-660 (1986).
25. H.L. ten Haven, J.W. de Leeuw and P.A. Schenck. Organic geochemical studies of a Messinian evaporitic basin, northern Apennines (Italy) I: Hydrocarbon biological markers for a hypersaline environment. *Geochim. Cosmochim. Acta* **46**, 2181-2191 (1985).
26. J.W. de Leeuw and J.S. Sinninghe Damste. Organic sulfur compounds and other biomarkers as indicators of paleosalinity. In: *Geochemistry of sulfur in fossil fuels.* W.L. Orr and C.M. White (Eds.). ACS symposium series 429, pp.417-443. The American Chemical Society, Washington, D.C. (1990).
27. K.E. Peters and J.M. Moldowan. Effects of source, thermal maturity, and biodegradation on the distribution and isomerization of homohopanes in petroleum, *Org. Geochem.* **17**, 47-61 (1991).
28. K. Taguchi, K. Sasaki and N. Ushijima. Geochemical significance of porphyrin pigments in the stratigraphic correlation of the Neogene Tertiary rocks 1. Yashima Oil Field, Akita Prefecture, Japan. *Sci.Rept.Tohoku Univ., Sec. III*, **3**, 333-348 (1969).
29. K. Taguchi, K. Sasaki and N. Ushijima. Porphyrin pigments in the Neogene Tertiary rocks of the Shinjo oil field, Yamagata Prefecture. -Sedimentological and stratigraphical study of the porphyrin pigments in Neogene Tertiary rocks 3-, *Geol.Soc.Japan.* **76**, 559-566 (1970). (in Japanese).
30. N. Suzuki and K. Taguchi. Diagenesis of extractable and bound fatty acids in possible source rocks in Japan, *Org. Geochem.* **6**, 125-133 (1983).
31. R. Ishiwatari and M. Shioya. Diagenetic change of organic compounds in relation to generation of petroleum hydrocarbons, In: *Contributions to Petroleum Geoscience, dedicated to Prof. K. Taguchi on the occasion of his retirement.* J. Aiba et al. (Eds.). pp.357-377. Sendai (1986). (in Japanese).
32. M. Shioya and R. Ishiwatari. Diagenetic alteration of organic matter (hydrocarbons) in Shinjo sedimentary rocks of Miocene age, *Sed.Soc.Japan* **17-19**, 33-39 (1983). (in Japanese).
33. N. Suzuki and I. Shimada. Considerations in epimerization of sterane and triterpane as indicators of thermal history of sedimentary rocks, *Sed.Soc.Japan* **17-19**, 47-55 (1983). (in Japanese).
34. K. Taguchi, S. Shimoyama, Y. Itihara, N. Imoto, R. Ishiwatari, A. Shimoyama, M. Akiyama and N. Suzuki. Relationship of organic and inorganic diagenesis of Neogene Tertiary rocks, northeastern Japan. In: *Roles of organic matter in sediment diagenesis.* D.L. Cautier (Ed.). pp.47-64. SEPM Special Publication 38. Soc.Econ.Paleontol.Mineral., Tulsa, Oklahoma (1986).
35. A. Shimoyama, S. Hagishita and K. Harada. Purines and pyrimidines in Neogene sediments of the Shinjo basin, *Geochem. J.* **22**, 143 (1988).
36. K. Sasaki. Organo-sedimentological study on amino acids in the Neogene Tertiary rocks of the Imokawa area, Niigata Prefecture, Japan, *Geol. Soc. Japan* **79**, 427-439 (1973). (in Japanese).
37. N. Suzuki and I. Shimada. Steroid hydrocarbons (5α-C_{27},C_{28},C_{29} steranes) in sedimentary rocks -relation between their compositions and paleoenvironments-, *Mem. Fac. Sci. Shimane Univ.* **16**, 125-142 (1982). (in Japanese).
38. T. Machihara. Steranes and triterpanes in sediments from an offshore well, Sea of Japan -Relation between source indicators and depositional environments. In: *Contributions to Petroleum Geoscience, dedicated to Prof. K. Taguchi on the occasion of his retirement.* J. Aiba et al. (Eds.). pp.255-268. Sendai (1986). (in Japanese).
39. M. Akiyama, H. Fukusawa and S. Sakata. Biomarkers of the Wakkanai Formation in northern Hokkaido with reference to the sedimentary environment, *Mem. Geol. Soc. Japan* no.37, 201-206 (1992). (in Japanese).
40. H. Sato and K. Amano. Relationship between tectonics, volcanism, sedimentation and basin development, Late Cenozoic, central part of Northern Honshu, Japan, *Sed. Geol.* **74**, 323-343 (1991).
41. T. Tsuji, Y. Masui, A. Waseda, Y. Inoue, H. Kurita and K. Kai. The Onnagawa Formation in the vicinity of the Yashima town, Akita Prefecture, Northern Japan -with special reference to the lithologic units, the depositional environments and their relation to source rock characteristics-, *Res. Rep. JAPEX* no.7, 45-99 (1991MS). (in Japanese).
42. N. Imai. Multielement analysis of stream sediment by ICP-AES, *Bunseki Kagaku* **36**, T41-45 (1987). (in Japanese).

43. N. Imai. Multielement analysis of rocks with the use of geological certified reference material by inductively coupled plasma mass spectrometry. *Anal. Sci.* 6, 389-396 (1990).
44. S. Sakata, Y. Suzuki and N. Kaneko. Biological markers in the Neogene Gas Field around Nagaoka, *J.Jap. Assoc.Petrol.Technol.* 52, 221-230 (1987). (in Japanese).
45. S.Sakata, N. Suzuki and N. Kaneko. A biomarker study of petroleum from the Neogene Tertiary sedimentary basins in Northeast Japan, *Geochem. J.* 22, 89-105 (1988).
46. W.K. Seifert and J.M. Moldowan. Applications of steranes, terpanes and monoaromatics to the maturation, migration and source of crude oils, *Geochim.Cosmochim.Acta* 42, 77-95 (1978).
47. R. P. Philp. *Fossil Fuel Biomarkers, Applications and Spectra.* Methods in Geochemistry and Geophysics 23. Elsevier, Amsterdam (1985).
48. K. E. Peters and J. M. Moldowan. *The Biomarker Guide, Interpreting Molecular Fossils in Petroleum and Ancient Sediments.* Prentice Hall, Englewood Cliffs, New Jersey (1993).
49. Y. Watanabe, M. Yamamoto and N. Imai. Sedimentary environment in the Onnagawa Sea: middle Miocene Japanese backarc trough. In: *Siliceous, phosphatic and glauconitic sediments of the Tertiary and Mesozoic.* A. Iijima, R.E. Garrison and A.M. Abed (Eds). VSP, Zeist (in press).
50. N. Kaneko, S. Sakata and T. Machihara. Biological markers in bitumen extracts and a crude oil from the Mitsuke oil field, *Bull. Geol. Surv. Japan* 41, 383-394 (1990). (in Japanese).
51. N. Kaneko and S. Sakata. Biomarker compositions in immature sedimentary samples from the Niigata Basin -Geological and geochemical explanations for the origin of themodynamically stable isomers and the mechanism of reversal maturation trend-, *J.Jap. Assoc. Petrol. Technol.* 57, 243-252 (1992). (in Japanese).
52. H.L. ten Haven, J.W. de Leeuw, T.M. Peakman and J.R. Maxwell. Anomalies in steroid and hopanoid maturity indices, *Geochim. Cosomochim. Acta* 50, 858-855 (1986).
53. A. S. Mackenzie. Applications of biological markers in petroleum geochemistry. In: *Advances in Petroleum Geochemistry* J. Brooks and D. H. Welte (Eds.). vol.1, pp.115-214. Academic Press, London (1984).
54. A. Y. Huc. Origin and formation of organic matter in recent sediments and its relation to kerogen. In: *Kerogen.* B. Durand (Ed.). pp.445-474. Technip, Paris (1980).
55. D. C. Rhoads and J. W. Morse. Evolutionary and ecologic significance of oxygen-deficient basins, *Lethaia* 4, 413-428 (1971).
56. E. Caspi, J.M. Zander, J.B. Greig, F.B. Mallory, R.L. Conner and J.R. Landrey. Evidence for nonoxidative cyclization of squalene in the biosynthesis of tetrahymanol, *J. Amer. Chem. Soc.* 90, 3563-3564 (1968).
57. J.M. Moldowan, C.Y. Lee, D.S. Watt, A. Jeganathan, N.-E. Slougui and E.J. Gallegos. Analysis and occurrence of C_{26}-steranes in petroleum and source rocks, *Geochim. Cosmochim. Acta* 55, 1065-1081 (1991).
58. N. Suzuki and Y. Sampei. C_{26}-Steroid hydrocarbon ($5\alpha(H)$-norcholestane) in Miocene Onnagawa shale (abstract), In: *Abstract of 1991 Annual Meeting of The Geochemical Society of Japan.* p.261. (1991). (in Japanese).
59. L.J. Goad and N. Withers. Identification of 27-nor-(24R)24-methylcholesta-5,22-dien-3β-ol and brassicasterol as the major sterols of the marine dinoflagellate Gemnodinium simplex, *Lipids* 17, 8583-858 (1982).
60. W. Seifert, J.M. Moldowan, G.W. Smith and E.V. Whitehead. First proof of structure of a C_{28}-pentacyclic triterpane in petroleum, *Nature* 271, 436-437 (1978).
61. B.J. Katz and L.W. Elrod. Organic geochemistry of DSDP Site 467, offshore California, Middle Miocene to Lower Pliocene strata. *Geochim. Cosmochim. Acta* 47, 389-396 (1983).
62. J. A. Curial and J. R. Odermatt. Short-term biomarker variability in the Monterey Formation, Santa Maria Basin, *Org. Geochem.* 14, 1-13 (1989).
63. P. J. Grantham, J. Posthuma and K. DeGroot. Variation and significance of and C_{28} triterpane content of a North Sea core and various North Sea crude oils. In: *Advances in Organic Geochemistry 1979.* A. G. Douglas and J. R. Maxell (Eds.). pp.29-38. Pergamon Press, New York (1980).
64. M. G. Fowler and A. G. Douglas. Distribution and structure of hydrocarbons in four organic-rich Ordovician rocks, *Org. Geochem.* 6, 105-114. (1984).
65. L. A. Williams. Subtidal stromatolites in Monterey Formation and other organic-rich rocks as suggested source contributors to petroleum formation, *Am. Assoc. Petrol. Geol. Bull.* 68, 1879-1893 (1984).
66. R. Noble, R. Alexander, and R. I. Kagi. The occurrence of bisnorhopane, trisnorhopane and 25-norhopane as free hydrocarbons in some Australian shales, *Org. Geochem.* 8, 171-176 (1985).
67. M. A. McCafferey, M. A. Farrington and D. J. Repeta. Geochemical implications of the lipid composition of *Thioploca spp.* from the Peru upwelling region -15 S, *Org. Geochem.* 14, 61-68 (1989)
68. R. Tada, Y. Watanabe and A. Iijima. Accumulation of laminated and bioturbated Neogene siliceous deposits in Ajigasawa and Goshogawara areas, Aomori Prefecture, Northeast Japan, *Fac. Sci. Univ. Tokyo, sec. II,* 21, 139-167 (1986).

69. U. Suzuki. Geology of Neogene Basins in the eastern part of the Sea of Japan. *Mem. Geol. Soc. Japan* **32**, 143-183 (1989). (in Japanese).
70. S. Ogihara and K. Taguchi. On Japanese phosphate nodules -A discovery of phosphate nodules from Neogene rocks, Northeast Japan (abstract). *Jap. Miner. Petrol. Econ. Geol.* **81**, 4, 155-156 (1986). (in Japanese).
71. S. Ogihara. Phosphate nodule from northern Dewa hill, northwest Yokote basin, Akita Prefecture. *Jap. Miner. Petrol. Econ. Geol.* (in press). (in Japanese).
72. Y. Watanabe, S. Kato, H. Kosaka and I. Kobayashi. Some analysis on peritic rocks of the Tsurushi and Nakayama Formations in Sado Island, Niigata Prefecture, Japan -especially on the relation between rock facies, organic carbon content, nitrogen content and diatom fossils. *Rep. Sado Museum.* **7**,103-111 (1977). (In Japanese).
73. H. Fukusawa, T. Sakamoto and I. Koizumi. Rhythmical oceanographic changes recorded in laminated diatomaceous mudstones of the Neogene Nakayama Formation, Sado, *Monthly Tikyu* **13**, 467-469 (1991). (in Japanese).
74. R. Matsumoto. Diagenetic dolomite, calcite, rhodochrosite, magnesite, and lansfordite from Site 799, Japan Sea -implications for depositional environments and the diagenesis of organic-rich sediments. In: *Proceedings of the Ocean Drilling Program, Scientific Results,* Vol. 127/128, pp.75-98 (1992).
75. H.B. Haq, J. Hardenbol and P.R. Vail. Chronology of fluctuating sea levels since the Triassic. *Science* **235**, 1156-1167 (1987).
76. F. Akiba. Middle Miocene to Quaternary diatom biostratigraphy in the Nankai Trough and Japan Trench, and modified Lower Miocene through Quaternary diatom zones for middle-to-high latitudes of the North Pacific. In: *Initial Report, Deep Sea Drilling Project,* vol. 87, pp.393-481. U.S.Goverment.Printing Office, Washington,D.C. (1986).
77. M. Watanabe. Stratigraphy of the Neogene sequence in the Himi-Nadaura area, Toyama Prefecture, central Japan -with special reference to the hiatus between the Sugata Formation and overlying formations-, *Geol.Soc.Japan* **96**, 915-936 (1992). (in Japanese).
78. K. Sasaki and M. Yamamoto. Geochemical study of the Odoji mudrocks of Nakamuragawa area, Aomori Prefecture, Japan, with special reference to the significance of n-alkane distribution maximizing at C_{25}. In: *Contributions to Petroleum Geoscience, dedicated to Prof. K. Taguchi on the occasion of his retirement.* Aiba et al. (Eds.). pp.223-239. Sendai, (1986). (in Japanese).
79. T. Sakamoto. Sedimentary rhythm of the Nakayama Formation (Middle Miocene-Early Pliocene) in the Sado Island, *Geol.Soc.Japan* **98**, 611-633 (1992). (in Japanese).
80. T. Masuzawa, J. Takada and R. Matsushita. Trace-element geochemistry of sediments and sulfur isotope geochemistry of framboidal pyrite from Site 795, Leg 127, Japan Sea. In: *Proceedings of the Ocean Drilling Program, Scientific Results.* vol. 127/128, pp.705-717 (1992).
81. Y. Sampei, T. Ichiba, N. Suzuki and K. Sekiguchi. Organic sulfur concentration of kerogens in Neogene rocks from Akita, Yamagata and Niigata areas (abstract). *J. Jap. Assoc. Petrol. Technol.* **57**, 362-363 (1992). (in Japanese).
82. K. Kano. Depositional environment of petroleum source rocks -a case study on the Miocene Onnagawa Formation-. In: *Contributions to Petroleum Geoscience, dedicated to Prof. K. Taguchi on the occasion of his retirement.* J. Aiba et al. (Eds.). pp.161-169. Sendai (1986). (in Japanese).
83. A. Hirai, T. Sato and T. Takashima. Geochemical study on the Yabase oil field, Akita, *J. Jap. Assoc. Petrol. Technol.* **55**, 37-47. (1990). (in Japanese).
84. M. Yamamoto. Fractionation of azaarenes during oil migration, *Org. Geochem.* **19**, 389-402 (1992).
85. P. Ungerer, F. Bessis, P.Y. Chenet, B. Durand, E. Nogaret, A. Chiarelli, J.L. Oudin and J. F. Perrin. Geological and geochemical models in oil exploration: principles and practical examples. In: *Petroleum Geochemistry and Basin Evaluation.* G. Demaison and R.J. Murris (Eds). AAPG Memoir 35, pp.53-77 (1984).
86. B. Durand. Understanding of HC migration in sedimentary basins (present state of knowledge), *Org. Geochem.* **13**, 445-459 (1988).
87. S. Sato, K. Sasaki and K. Taguchi. Distribution of organic carbon and extractable organic matter in Neogene Tertiary rocks of Akita and Niigata Districts, with particular reference to the removal of carbonate carbons in sediments, *Geol.Soc.Japan* **78**, 643-651 (1972). (in Japanese).
88. N. Hayashida and K. Taguchi. Evolution of non-hydrocarbon constituents in extractable organic matter under the effect of burial in some Japanese sediments. *Mem. Geol. Soc. Japan* no.15, 191-204 (1978). (in Japanese).
89. S. Sato. Organo-geochemical study on kerogen of sedimentary rocks in Japan. *Sci. Rept. Tohoku Univ., 3rd Ser.* **8**, 85-113 (1976).

Proc. 29th Int'l. Geol. Congr., Part C, pp. 75-87
A. Iijima *et al.* (Eds.)
© VSP 1994

Occurrence and properties of glauconite in Miocene biosiliceous sediments of the Noto Peninsula, Hokuriku District, Japan

T. NISHIMURA

Geoscience Institute, Hyogo University of Teacher Education, 942-1 Shimokume, Yashiro-cho, Kato-gun, Hyogo 673-14, JAPAN

Abstract-- Glauconite beds associated with Miocene biosiliceous sediments of the Noto Peninsula are composed of 11 to 60 vol.% glauconite grains of authigenic origin. The Noto glauconites show rather low crystallinity. Total iron content of the Noto glauconite pellets ranges from 16 to 27 wt.%, and the K_2O content from 5 to 8 wt.%. There is a proportional relationship between total iron and the K_2O contents, implying that parent materials were transformed to glauconite by substitution of iron for aluminium and fixation of potassium. A weak correlation between the amount of detrital grains and tetrahedral Al can be recognized, suggesting that the clastic materials reacted with sea water to release R^{3+} which was fixed by glauconite. The glauconite-bearing siliceous sediments of the Noto Peninsula were deposited with a sedimentation rate of 1.3 m/m.y. on an offshore bank which existed in the elongated, middle Middle Miocene sedimentary basin.

Keywords : glauconite, Miocene siliceous sediments, chemical composition, Noto Peninsula

INTRODUCTION

Neogene siliceous sediments are widely distributed in the Northeastern Honshu Arc and the Noto Peninsula, Japan [1,2]. In many areas, glauconite-bearing beds are associated with these siliceous sediments, exclusively with middle Middle Miocene beds [3]. It is said that authigenic glauconite grains are a good marker of rate and site of sedimentation [4,5]. At present, authigenic glauconite pellets are known to concentrate on offshore banks with a water-depth of 50-500 m off southern California [4] and on the shelf off the west coast of Vancouver Island [6]. Glauconite grains in siliceous sediments of the Northeastern Honshu Arc and Noto Peninsula thus should offer significant information relating to the opening of the Sea of Japan.
In this article, the occurrence and properties of glauconite grains from the Noto Peninsula are described, and their origin and geological meaning are discussed.

GEOLOGIC SETTING

The Noto Peninsula belongs to the San'in-Hokuriku Province, a distinct geologic province formed during the Miocene. Neogene strata distributed in this Province have similar lithofacies, stratigraphy, and fossil content to those of the Sea of Japan side of Northeast Japan. Structural deformation is weaker in the San'in-Hokuriku Province than in Northeast Japan. Synclinal and anticlinal structures with E-W trending axes are recognized in Neogene

strata, especially in siliceous deposits in the northeastern part of the Noto Peninsula [7].
The stratigraphy of the northeastern part of the Noto Peninsula is summarized as follows after Funayama [7], Kaseno [8] and Fujii et al.[9]. Miocene strata are divided into three formations : the Anamizu Formation, the Yanaida Formation, and the Suzu Formation in ascending order. The Anamizu Formation, more than 50 m thick, is mainly composed of andesitic lava and pyroclastics with a small amount of clastic rocks of Early Miocene age. The Yanaida Formation, 500 m thick, comprises dacitic pyroclastics and basaltic lava of Early Miocene age. It is considered that a part of these two Formations was deposited on land. The Early to Middle Miocene marine Suzu Formation has a thickness of about 1000 m, and is subdivided into four members: i.e., in ascending order, the Higashi-Innai alternated beds, the Akagami mudstone, the Awagura tuff, and the Najimi siliceous mudstone members. The Higashi-Innai alternated beds member, 200 m thick, comprises alternating beds of conglomerate, sandstone, and mudstone which were deposited in a shallow marine to upper bathyal basin. The Akagami mudstone member, 200 m thick, is composed of dark gray mudstones associated with upper bathyal foraminifers. The Awagura tuff member consists of white to pale yellow massive tuff. The thickness of this member varies from 200 m or more in the western area to 1 m in the eastern area. The Najimi siliceous mudstone member, 450 m thick, is composed of dark gray-colored massive siliceous mudstones associated with benthic foraminifers indicating an upper bathyal environment.

SAMPLING AND METHODS OF STUDY

Glauconite rocks were sampled from eight localities in the Suzu Area (S1 to S8) and three localities in the Nanao Area (N1 to N3). Sampling localities are shown in Figure 1.

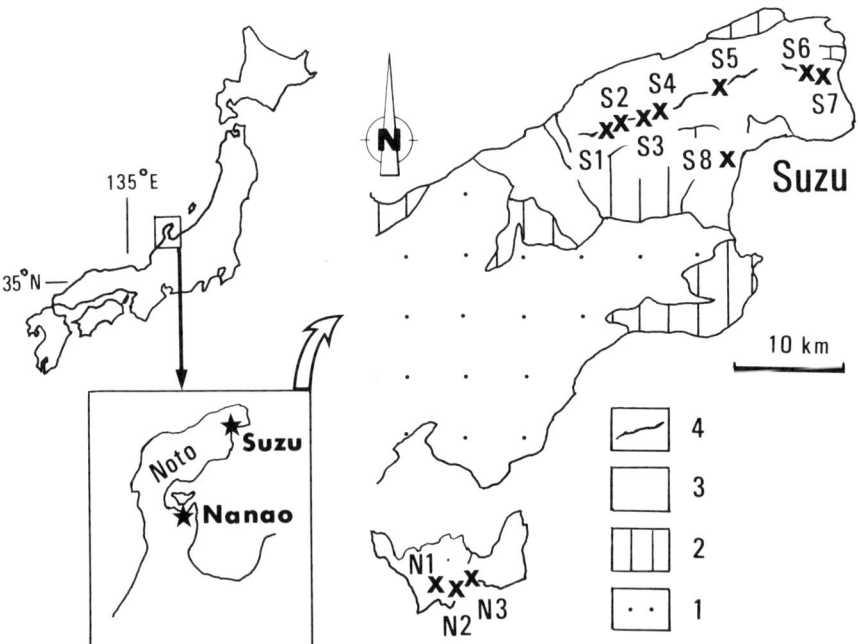

Figure 1. Index map and sampling localities. 1: Anamizu Formation, 2: Yanaida Formation, 3: Suzu Formation and younger deposits, 4: glauconite bed.

Modal composition of glauconite rocks, and X-ray diffraction, grain-size distribution, and chemical composition of glauconite pellets were investigated.

Modal compositions of glauconite rocks from eleven localities were obtained by counting 500 points per each thin section.

The X-ray diffractograms of the oriented slide of powdered glauconite pellets picked up from each sample were obtained in order to investigate the crystallinity of glauconite pellets of the Noto Peninsula. Analytical conditions are as follows : 30 kV, 20 mA, Cu target, Ni filter, time constant 4 sec., full scale 500 cps.

Grain-size distribution of glauconite pellets in glauconite rocks from eleven localities was analysed through the measurement of 100 spherical or spheroidal grains per sample. Size of glauconite pellets on the fracture plane of each sample was measured under a binocular microscope. Grains less than 5 phi are not counted.

Chemical compositions of glauconite grains free from pyrite were obtained by the use of the energy dispersive X-ray microanalyser for 9 major elements. Experimental conditions are 15 kV of the accelerating voltage, 0.2 nA of the emission current, 52.5 degrees of the take-off angle, and 100 seconds of the exposure time. The correction was made by the Bence and Albee method [10]. Total iron is expressed as ferric iron oxide ($Fe_2O_3^*$). Water content was not measured. Two or three dark green pellets and pale green pellets for each sample were selected and analysed. Chemical composition of a given pellet is the average value of those of two or three points. Total number of analysed glauconite pellets is 61.

MODE OF OCCURRENCE

A glauconite bed having a thickness of 0.5 - 2 m can be readily traced, from S1 to S7, in the basal part of the Najimi siliceous mudstone member in the Suzu Area. In the western part of this area, the glauconite bed is 1.3 - 2 m thick and directly overlies the Awagura tuff member, whereas in the eastern part of the area it occurs in the basal part of the Najimi siliceous mudstone member, and its thickness ranges from 0.5 to 1 m (Figure 2). At S8, a glauconite bed, 1 m thick, rests directly on the Awagura tuff member and is overlain by the Najimi siliceous mudstone member. In the Nanao Area, a glauconite bed, 0.7 - 1 m thick, occurs in the basal part of the Wakura diatomaceous mudstone member which is correlated to the Najimi siliceous mudstone member [7]. Glauconite beds of the Suzu and Nanao Areas are of middle Middle Miocene age [8].

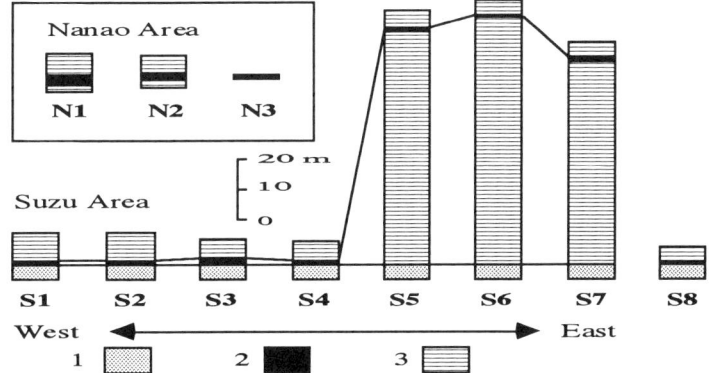

Figure 2. Columnar sections. 1: Awagura tuff member, 2: glauconite bed, 3: Najimi siliceous mudstone member.

In thin sections, glauconite rocks are composed of glauconite grains, quartz, plagioclase, potassium feldspar, volcanic rock fragments, volcanic glass shards, mica flakes, and framboidal pyrite with clay matrix and carbonate cement.
Glauconite grains occur commonly in the spherical or spheroidal form of dark to pale green pellets. Glauconite pellets occasionally have several radial cracks (Figure 3). There are two types of these radial cracks : one, deep cracks extending from the rim to the core, and the other, shallow cracks extending about a half or one third depth to the core. Radial cracks have a wedge-like shape, and the crack width is larger at the rim than at the core. Framboidal pyrites very often fill these radial cracks, but some cracks are not filled. Deep and/or shallow radial cracks can be more often recognized in coarse-grained and dark green pellets than in fine-grained and pale green pellets. Almost all glauconite pellets have "fibroradiated rims" as defined by Triplehorn [11]. Fibroradiated rims are peripheral zones of minute elongate crystals with radial orientation. Concentric layering is also present. Fibroradiated rims are continuous around the periphery of pellets free from radial cracks, but discontinuous at the aperture of cracks (Figure 3). Not uncommonly glauconite fills pore spaces of sponge spicules and radiolarians. In addition, glauconite frequently entirely or partly replaces angular to subangular volcanic rock fragments, volcanic glass shards, and mica flakes. All glauconite pellets and grains are composed of aggregates of microcrystalline glauconite crystals.
Detrital grains such as quartz, feldspars, and volcanic rock fragments are fine-grained, and are angular to subangular. Authigenic framboidal pyrites occur frequently in the core and/or radial cracks of glauconite grains and in the matrix. The matrix is composed of minute, light brown glass shards and clay minerals. Carbonate cements, determined as ankerite by X-ray diffractometry, occur in samples S3 and S6.
Modal compositions of glauconite rocks are shown in Table 1. Figure 4 shows the triangular diagram of glauconite - detrital grains - matrix . The content of glauconite pellets ranges from 10.8 to 51.6%, angular glauconite 0 to 8.6%, detrital grains 0 to 9.8%, and matrix 20.6 to 67.4%. Detrital grains such as quartz, feldspars, and rock fragments are abundant in sample S3, little in S7 and N1, and rare to lacking in others.

Figure 3. Microphotograph showing radial cracks of glauconite pellets from N2. Scale bar represents 1 mm.

Table 1.
Modal compositions of glauconite rocks of the Noto Peninsula.

	S1	S2	S3	S4	S5	S6
Glauconite pellets						
dark-colored	19.4 %	39.6 %	3.4 %	3.2 %	19.0 %	26.6 %
dark-colored, including pyrite	21.0	1.0	0.6	14.6	0.2	2.4
pale-colored	4.4	11.8	6.8	6.6	12.2	5.0
pale-colored, including pyrite	6.6	0.2	0.0	5.0	0.0	0.0
Glauconite replacing clastics						
dark-colored	5.0	3.0	0.0	0.6	0.6	4.4
pale-colored	2.2	3.8	0.0	1.0	5.8	1.2
Detrital grains						
Quartz	0.0	0.0	2.6	0.2	0.4	0.0
Feldspar	0.2	0.2	7.2	1.4	0.2	0.0
Rock fragment	0.0	0.0	0.0	0.0	0.0	0.0
Matrix	41.2	40.4	29.4	67.4	61.6	20.6
Cement	0.0	0.0	50.0	0.0	0.0	39.8
Total	100.0	100.0	100.0	100.0	100.0	100.0
Glauconite	58.6 %	59.4 %	21.6 %	31.0 %	37.8 %	65.8 %
Detrital grains	0.2	0.2	19.6	1.6	0.6	0.0
Matrix	41.2	40.4	58.8	67.4	61.6	34.2

	S7	S8	N1	N2	N3
Glauconite pellets					
dark-colored	46.8 %	3.4 %	25.0 %	24.0 %	28.6 %
dark-colored, including pyrite	0.2	25.2	6.2	13.6	2.6
pale-colored	4.6	9.0	6.0	9.6	5.0
pale-colored, including pyrite	0.0	7.4	0.4	0.8	0.2
Glauconite replacing clastics					
dark-colored	7.2	1.2	2.4	2.0	4.4
pale-colored	1.4	1.6	1.8	1.6	0.6
Detrital grains					
Quartz	0.4	0.0	1.0	0.4	0.4
Feldspar	1.0	0.0	2.6	0.2	0.0
Rock fragment	1.2	0.0	0.2	0.0	0.0
Matrix	37.2	52.2	54.4	47.8	58.2
Cement	0.0	0.0	0.0	0.0	0.0
Total	100.0	100.0	100.0	100.0	100.0
Glauconite	60.2 %	47.8 %	41.8 %	51.6 %	41.4 %
Detrital grains	2.6	0.0	3.8	0.6	0.4
Matrix	37.2	52.2	54.4	47.8	58.2

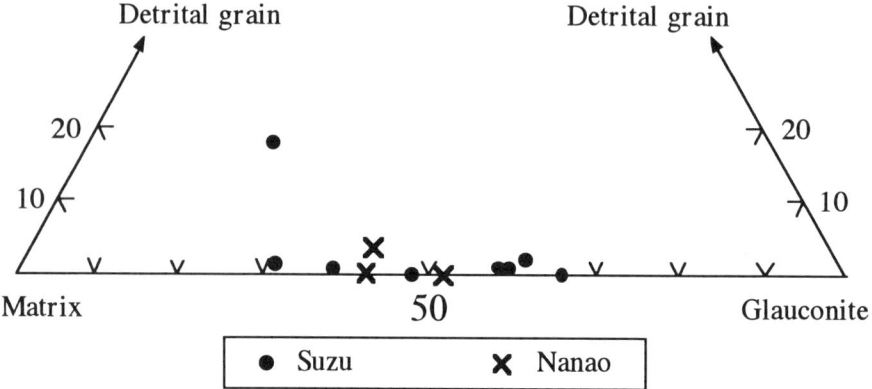

Figure 4. Glauconite - detrital grain - matrix diagram of glauconite rocks of the Noto Peninsula.

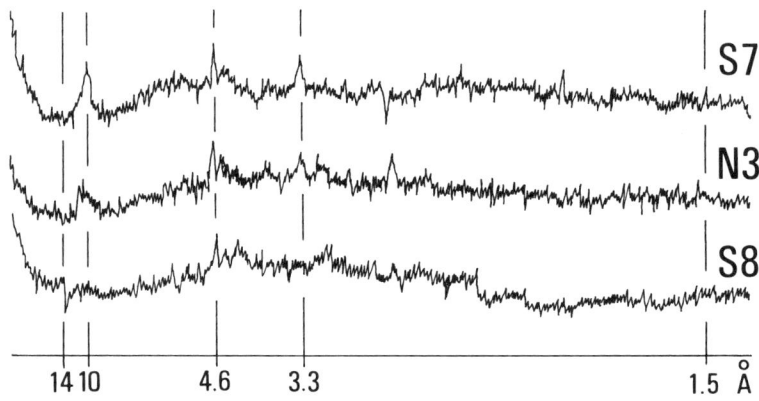

Figure 5. X-ray diffractograms of glauconites of Sample S7, N3, and S8.

PROPERTIES OF GLAUCONITE

X-Ray diffraction
Glauconites from localities S2, S6, S7, N1, and N3 show only a very broad pattern or have very weak peaks, suggesting that the crystallinity of glauconites is low, in other word, less evolved. Glauconites from S7 show a rather broad peak at 10.6 Å, medium peaks at 4.6 Å, 3.3 Å, and a very weak peak at 1.5 Å (Figure 5). Glauconites from N3 have a broad peak at 11.5 Å. The broad peak at 11 Å or so did not shift to lower angle side, when glycolated. Glauconites from other localities show patterns that are almost amorphous, but an extremely weak peak at 14 Å may be recognized for glauconites from S8 (Figure 5).

Grain-size distribution
Figure 6 shows the grain-size distribution of glauconite pellets in glauconite rocks from eleven localities. Diameter of glauconite pellets ranges from -1 to 5 phi. However, most pellets fall into the 0 - 2 phi fractions or the 1 - 3 phi fractions, and they are well sorted.

Suzu Area

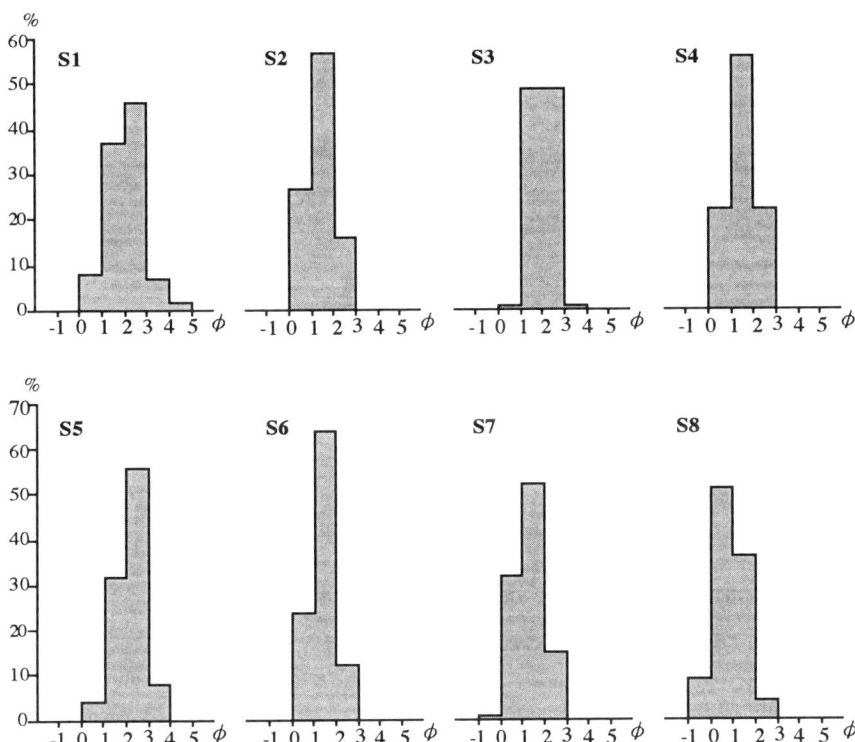

Nanao Area

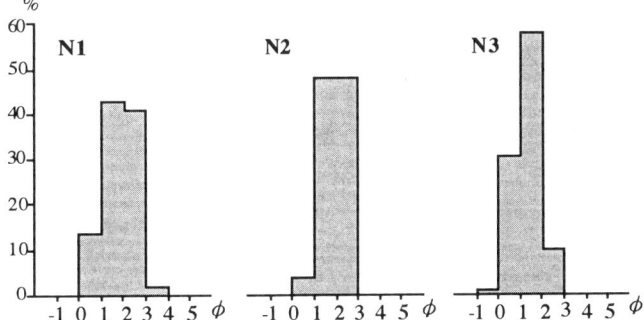

Figure 6. Histograms of size distribution of glauconite pellets.

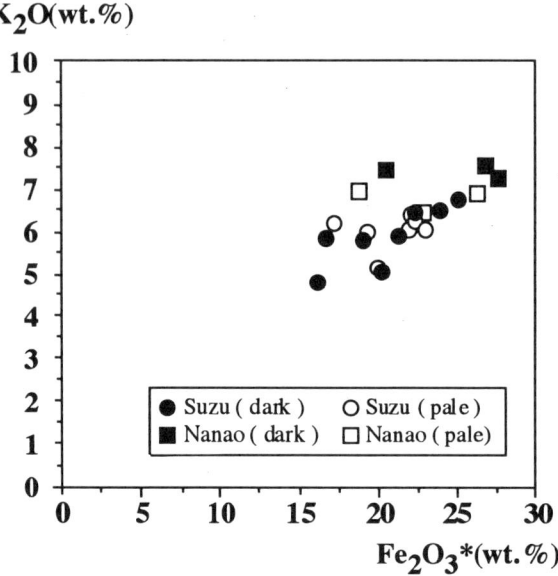

Figure 7. K_2O - $Fe_2O_3^*$ diagram of glauconites of the Noto Peninsula.

Chemical composition

The average chemical compositions of dark green and pale green glauconite pellets at each locality are shown in Table 2.

The silica content varies from 42.57 to 56.62 wt.%, the alumina content from 2.40 to 7.60 wt.%, the total iron content from 16.09 to 27.21 wt.%, and the K_2O content from 4.85 to 7.57 wt.%. As a whole, there is a proportional relationship between the K_2O and $Fe_2O_3^*$ contents in the glauconite pellets of the Noto Peninsula (Figure 7). However, it is difficult to find significant differences in composition between dark and pale green pellets.

Numbers of ions were calculated on the basis of 11 oxygens with the assumption that the glauconite is a mica (Table 2). The average values of cations of the interlayer and octahedral sites are nearly constant at all localities except for S3 and S4 : glauconites from S3 and S4 are poorer in K^+, and richer in octahedral Al^{3+} and Mg^{2+}. The glauconite pellets of the Noto Peninsula are richer in octahedral trivalent ions and poorer in octahedral divalent ions than ideal glauconite. This may be due to the expression of total iron as ferric iron oxide.

DISCUSSION

Glauconite grains of the Noto Peninsula, as already stated, occur most frequently in the spherical or spheroidal form of dark to pale green pellets. These glauconite pellets occasionally have radial cracks extending from the rim to the core. Radial cracks, often filled with framboidal pyrites, have a wedge-like shape, and the crack width is larger at the rim than at the core. This type of crack may be formed by shrinkage of the pellets due to desiccation of pellets during glauconitization. Fibroradiated rims are continuous around the periphery of pellets free from radial cracks, but discontinuous at the aperture of cracks. These characteristics indicate that glauconite pellets of the Noto Peninsula are of authigenic origin.

Table 2.
Chemical compositions of glauconite pelles of the Noto Peninsula. The subscripts d and p of each locality number indicate the dark green and pale green pellets, respectively.

(wt.%)	$S1_d$	$S1_p$	$S2_d$	$S2_p$	$S3_d$	$S3_p$	$S4_d$	$S4_p$	$S5_d$	$S5_p$	$S6_d$	$S6_p$
SiO_2	50.17	50.52	46.65	47.72	50.32	50.71	56.62	55.64	51.93	53.38	53.01	49.94
TiO_2	0.33	0.36	0.18	0.22	0.35	0.35	0.45	0.41	0.13	0.40	0.18	0.12
Al_2O_3	6.90	7.60	2.40	2.72	5.93	6.07	7.26	7.31	3.52	4.45	2.89	2.57
Fe_2O_3*	16.60	17.29	21.15	22.10	20.11	20.01	19.09	19.29	22.25	22.39	23.71	21.90
MnO	0.10	0.06	0.12	0.11	0.17	0.05	0.17	0.06	0.03	0.06	0.05	0.07
MgO	4.16	4.41	4.11	4.09	5.24	4.94	5.88	5.57	5.04	5.26	5.08	4.81
CaO	0.02	0.00	0.00	0.01	0.07	0.04	0.01	0.01	0.06	0.15	0.00	0.00
Na_2O	0.20	0.34	0.20	0.20	0.05	0.09	0.15	0.16	0.07	0.18	0.07	0.04
K_2O	5.94	6.23	5.93	6.46	5.08	5.23	5.83	6.02	6.52	6.29	6.50	6.10
TOTAL	84.42	86.80	80.73	83.62	87.32	87.48	95.46	94.47	89.55	92.54	91.49	85.56
					Numbers of ions on the basis of 11 (O)							
Si	3.88	3.82	3.89	3.86	3.80	3.82	3.87	3.85	3.88	3.84	3.89	3.91
Ti	0.02	0.02	0.01	0.01	0.02	0.02	0.02	0.02	0.01	0.02	0.01	0.01
Al^{IV}	0.12	0.18	0.11	0.14	0.20	0.18	0.13	0.15	0.12	0.16	0.11	0.09
Al^{VIII}	0.51	0.50	0.13	0.12	0.33	0.36	0.45	0.45	0.19	0.22	0.14	0.15
Fe^{3+}	0.97	0.98	1.33	1.34	1.14	1.13	0.98	1.00	1.25	1.21	1.31	1.29
Mn	0.01	0.00	0.01	0.01	0.01	0.00	0.01	0.00	0.00	0.00	0.00	0.00
Mg	0.48	0.50	0.51	0.49	0.59	0.55	0.60	0.57	0.56	0.56	0.56	0.56
Ca	0.00	0.00	0.00	0.00	0.01	0.00	0.00	0.00	0.00	0.01	0.00	0.00
Na	0.03	0.05	0.03	0.03	0.01	0.01	0.02	0.02	0.01	0.02	0.01	0.01
K	0.59	0.60	0.63	0.67	0.49	0.50	0.51	0.53	0.62	0.58	0.61	0.61

(wt.%)	$S7_d$	$S7_p$	$S8_d$	$N1_d$	$N1_p$	$N2_d$	$N2_p$	$N3_d$	$N3_p$
SiO_2	49.42	45.73	42.57	53.79	51.95	49.35	48.87	53.32	49.22
TiO_2	0.39	0.23	0.35	0.21	0.20	0.11	0.22	0.10	0.21
Al_2O_3	3.38	3.22	4.36	3.85	3.85	4.30	5.05	3.24	4.00
Fe_2O_3*	24.98	22.89	16.09	27.21	26.10	20.28	18.78	26.58	22.84
MnO	0.20	0.17	0.02	0.08	0.06	0.17	0.08	0.11	0.16
MgO	3.74	3.44	3.93	4.86	4.74	5.14	4.94	4.83	4.66
CaO	0.00	0.00	0.00	0.00	0.00	0.00	0.00	0.00	0.00
Na_2O	0.14	0.22	0.17	0.24	0.11	0.12	0.10	0.07	0.14
K_2O	6.82	6.14	4.85	7.35	7.01	7.50	7.04	7.57	6.52
TOTAL	89.07	82.03	72.34	97.60	94.03	86.96	85.06	95.80	87.75
				Numbers of ions on the basis of 11 (O)					
Si	3.78	3.79	3.88	3.82	3.83	3.79	3.79		
Ti	0.02	0.01	0.02	0.01	0.01	0.01	0.01	0.01	0.01
Al^{IV}	0.22	0.21	0.12	0.25	0.24	0.18	0.17	0.21	0.21
Al^{VIII}	0.08	0.11	0.35	0.07	0.08	0.21	0.30	0.06	0.15
Fe^{3+}	1.44	1.43	1.10	1.43	1.42	1.18	1.11	1.42	1.32
Mn	0.01	0.01	0.00	0.00	0.00	0.01	0.01	0.01	0.01
Mg	0.43	0.42	0.53	0.51	0.51	0.59	0.58	0.51	0.53
Ca	0.00	0.00	0.00	0.00	0.00	0.00	0.00	0.00	0.00
Na	0.02	0.04	0.03	0.03	0.02	0.02	0.01	0.01	0.02
K	0.67	0.65	0.56	0.65	0.65	0.74	0.70	0.69	0.64

* Fe_2O_3 is represented as total iron.

With respect to the parent material for glauconite formation, Galliher [12] proposed that glauconitization is caused by the alteration of biotite, and Takahashi [13] suggested that volcanic glass, mica, clays in fecal pellets, and feldspar were dehydrated and gelatinized. Burst [14] and Hower [15] considered that any degraded (expandable) layer lattice mineral can be the starting material in the formation of glauconite. Glauconite grains of the Noto Peninsula occur most frequently in the spherical or spheroidal form of dark to pale green pellets. And not uncommonly, glauconite fills pore spaces of sponge spicules and radiolarians. In addition, glauconite frequently partly or entirely replaces angular to subangular volcanic rock fragments, volcanic glass shards, and mica flakes. Therefore, it can be concluded that the parent materials of glauconite pellets of the Noto Peninsula are fecal pellets of benthic animals, microfossil remains, volcanic rock fragments, volcanic glass shards, and mica flakes. And adsorption of potassium and iron from sea water by these parent materials would occur. As already mentioned, a proportional relationship between the K_2O and Fe_2O_3* contents in the glauconite pellets of the Noto Peninsula can be recognized (Figure 7). This relationship implies that starting materials were transformed to glauconite by substitution of iron for aluminium and fixation of potassium. In order to investigate chemical characteristics in more detail, values of cations of glauconite pellets of the Noto Peninsula were plotted on the celadonite - muscovite - pyrophyllite diagram (Figure 8). The asterisk represents the ideal glauconite, and the broken lines delimit the fields of dioctahedral micas and related minerals after Yoder and Eugster [16]. The Noto glauconites occupy an intermediate field between well-crystallized glauconite and smectite. This is consistent with the fact that glauconites of the Noto Peninsula have low crystallinity. Glauconite pellets at localities S3, S7, N1, and N3 are richer in tetrahedral R^{3+} than those of the other localities. The glauconite rocks at N1, S7, and S3 contain more detrital grains such as quartz, feldspars, and volcanic rock fragments (Table 1). A weak correlation between the amount of detrital grains and tetrahedral Al can be recognized as shown in Figure 9. It is likely that the clastic materials reacted with sea water to release R^{3+} which was fixed by glauconite.

Because the process of glauconitization consists of adsorption of potassium and iron from sea water by the parent materials, it is necessary for them to have free access to sea water for a long time during glauconitization [15]. A favorable site for glauconite formation is immediately on the sea floor or in the interface zone of sea water and sediments. Moreover, if the rate of sedimentation is high, the glauconitization process would not proceed. For this reason, sedimentation rate of the glauconite bed of the Noto Peninsula was estimated.

Micropaleontological study of the siliceous sediments of the Suzu Area has revealed that the glauconite bed, 2 m thick, of this area encompasses four diatom zones : the uppermost part of *Denticulopsis lauta* zone, the *D. hyalina* zone, the *D. hustedtii* zone, and the lowermost part of *D. nicobarica* zone [7]. The total duration of these zones is about 1.5 million years. Consequently, the sedimentation rate of the glauconite bed is calculated as follows :

2 m/1.5 m.y. = ca. 1.3 m/m.y.

The most favorable place which satisfies the conditions that few clastic materials are transported from land, the sedimentation rate is low, and that benthic animals can survive, is an offshore bank or a shelf of an isolated island without large rivers.

According to Iijima and Tada [17], in the early to middle Middle Miocene, a deep-sea trough was formed by rapid subsidence along the Sea of Japan coast from the south Noto Peninsula to the Tsushima Strait. This trough was bounded on the northwest by a shallow-sea zone extending southwestwards from the north Noto Peninsula. In the middle Middle Miocene, the trough widened and developed rather complicated submarine topography. This progress of rifting correlates with the rapid opening of the Sea of Japan. Southwest Japan became extensively uplifted to form a very broad landmass. In this elongated trough, glauconite-topped offshore banks existed in the north Noto Peninsula and along the Sea of Japan coast of the

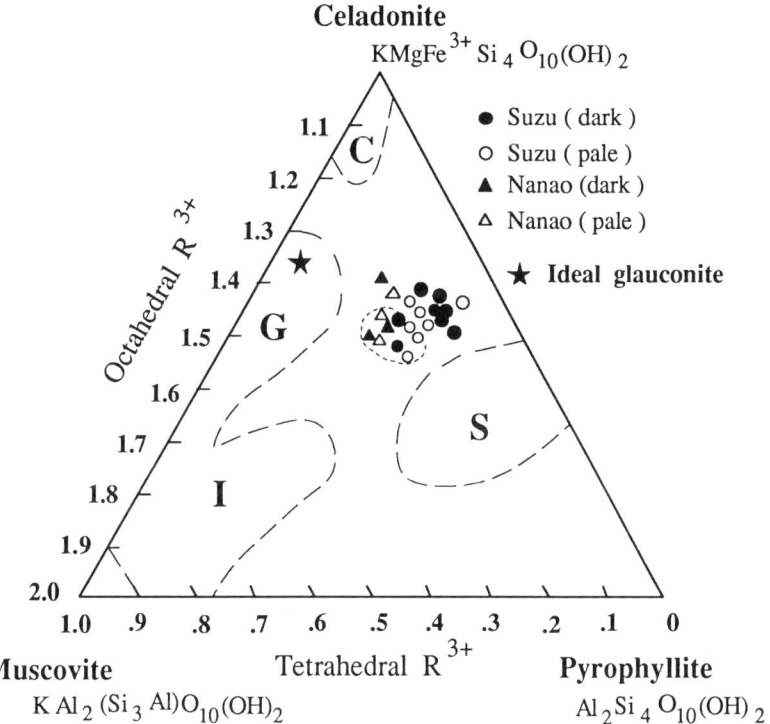

Figure 8. Celadonite - muscovite - pyrophyllite diagram. Asterisk represents the ideal glauconite, and broken lines show the fields of glauconite(G), celadonite(C), illite(I), and smectite(S) after Yoder and Eugster [16]. S3, S7, N1, and N3 are enclosed by dotted line.

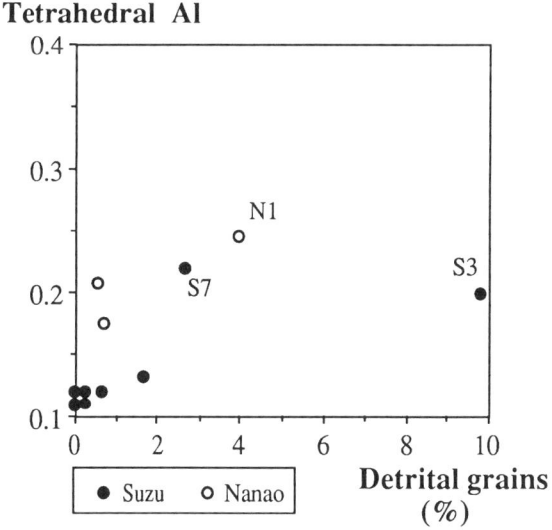

Figure 9. Detrital grains content - tetrahedral Al diagram of glauconites of the Noto Peninsula.

Northeastern Honshu Arc.

Miocene siliceous sediments are also widely distributed in California [18], the Monterey Formation, whose lithofacies resemble those distributed along the Sea of Japan coast of the Northeastern Honshu Arc. However, there exists a conspicuous difference between the Monterey Formation and Miocene siliceous sediments along the Sea of Japan coast : phosphatic facies characterizing the Monterey Formation [19] are not recognized in Japan, though phosphatic and glauconitic facies are often found together in marine formations [20]. Development of phosphatic beds needs large-scale upwelling which supplies nutrients from intermediate and bottom waters [19]. California, Peru and the western coast of Africa, where Recent to Neogene phosphatic facies develop [19, 20], are situated in the eastern rims of the Pacific Ocean and the Atlantic Ocean respectively. In these eastern rims, large-scale coastal upwelling persistently occurs. On the contrary, not abyssal but bathyal environment is recognized in Middle Miocene sedimentary basins of the Northeastern Honshu Arc [17]. A possibility of coastal upwelling occurring in Middle to Late Miocene bathyal basins of northern Japan is pointed out [21]. This coastal upwelling in the northwestern Pacific Rim, however, might not be persistent and might not be of so large scale as in the eastern Pacific Rim. Instead of phosphatic facies, glauconitic facies develop in offshore bank environments in the northwestern Pacific Rim.

SUMMARY AND CONCLUSIONS

1) Glauconite beds associated with Miocene siliceous sediments of the Noto Peninsula are composed of 11 to 60 vol.% glauconite grains of authigenic origin.

2) Glauconites of the Noto Peninsula show rather low crystallinity, in other words, are less evolved.

3) Grain size of the Noto glauconite pellets ranges from -1 to 5 phi. However, they are well sorted.

4) Total iron content of the Noto glauconite pellets ranges from 16 to 27wt.%, and the K_2O content from 5 to 8 wt.%. There is a proportional relationship between the K_2O and Fe_2O_3* contents, implying that parent materials such as fecal pellets were transformed to glauconite by substitution of iron for aluminium and fixation of potassium.

5) A weak correlation between the amount of detrital grains and tetrahedral Al can be recognized, suggesting that the clastic materials reacted with sea water to release R^{3+} which was fixed by glauconite.

6) The glauconite-bearing siliceous sediments of the Noto Peninsula were deposited at a sedimentation rate of 1.3 m/m.y. on an offshore bank which existed in an elongated sedimentary basin of middle Middle Miocene age.

Acknowledgements
I wish to express my sincere gratitude to Professors R.E. Garrison and A. Iijima for their critical reading of the manuscript and invaluable advice.

REFERENCES

1. A. Iijima and R. Tada. Silica diagenesis of Neogene diatomaceous and volcaniclastic sediments in northern Japan, *Sedimentology* **28**, 185-200 (1981).

2. A. Iijima and M. Utada. Recent developments in the sedimentology of siliceous deposits in Japan. In: *Siliceous Deposits in the Pacific Region*. A. Iijima, J.R. Hein and R. Siever (Eds). Developments in Sedimentology. vol. 36, pp. 45-64. Elsevier, Amsterdam (1983).
3. A. Iijima, R. Tada and Y. Watanabe. Developments of Neogene sedimentary basins in the Northeastern Honshu Arc with emphasis on Miocene siliceous deposits, *J. Fac. Sci. Univ. Tokyo, Sec. II* **21**, 417-446 (1988).
4. S.G. McRae. Glauconite, *Earth Sci. Rev.* **8**, 397-440 (1972).
5. G.S. Odin and A. Matter. De glauconiarum origine, *Sedimentology* **28**, 611-641 (1981).
6. B.D. Bornhold and P. Giresse. Glauconitic sediments on the continental shelf of Vancouver Island, British Columbia, *J. Sedim. Petrol.* **55**, 653-664 (1985).
7. M. Funayama. Miocene radiolarian stratigraphy of the Suzu area, northeastern part of the Noto Peninsula, Japan, *Tohoku Univ. Inst. Geol. Pal. Contr.* no. 91, 15-41 (1988). (in Japanese).
8. Y. Kaseno (Ed.). *Geological map of Ishikawa Prefecture and its explanatory text*. Ishikawa Prefectural Office, Kanazawa (1977). (in Japanese).
9. S. Fujii, Y. Kaseno and T. Nakagawa. Neogene paleogeography in the Hokuriku region, central Japan, based on the revised stratigraphic correlation, *Mem. Geol. Soc. Japan* no. 37, 85-95 (1992). (in Japanese).
10. A.E. Bence and A.L. Albee. Empirical correction factors for the electron microanalysis of silicates and oxides, *J. Geol.* **76**, 382-403 (1968).
11. D.M. Triplehorn. Morphology, internal structure, and origin of glauconite pellets, *Sedimentology* **6**, 247-266 (1966).
12. E.W. Galliher. Glauconite genesis, *Bull. Geol. Soc. Amer.* **46**, 1351-1366 (1935).
13. J. Takahashi. Synopsis of glauconitization, In: *Recent Marine Sediments*. P.D. Trask (Ed.). pp. 503-512. American Association of Petroleum Geologists, Tulsa (1939).
14. J.F. Burst. "Glauconite" pellets : their mineral nature and applications to stratigraphic interpretations, *A. A. P. G. Bull.* **42**, 310-327 (1958).
15. J. Hower. Some factors concerning the nature and origin of glauconite, *Am. Mineral.* **46**, 313-334 (1961).
16. H.S. Yoder and H.P. Eugster. Synthetic and natural muscovites, *Geochim. Cosmochim. Acta* **8**, 225-280 (1955).
17. A. Iijima and R. Tada. Evolution of Tertiary sedimentary basins of Japan in reference to opening of the Japan Sea, *J. Fac. Sci. Univ. Tokyo, Sec. II* **22**, 121-171 (1990).
18. C. Isaacs. Compositional variation and sequence in the Miocene Monterey Formation, Santa Barbara coastal area, California. In: *Cenozoic Marine Sedimentation, Pacific Margin, U.S.A.* D.K. Larue and R.J. Steel (Eds). pp. 117-132. S.E.P.M. Pacific Section, Los Angeles (1983).
19. R.E. Garrison, M. Kastner and Y. Kolodny. Phosphorites and phosphatic rocks in the Monterey Formation and related Miocene units, coastal California. In: *Cenozoic basin development of Coastal California*. R.V. Ingersol and W.G. Ernst (Eds). Rubey Volume VI, pp. 349-381. Prentice-Hall, Englewood Cliffs (1987).
20. G.S. Odin and R. Letolle. Glauconitization and phosphatization environments : A tentative comparison. In: *Marine Phosphorites*. Y.K. Bentor (Ed.). S.E.P.M. Spec. Publ. vol. 29, pp. 227-237. S.E.P.M., Tulsa (1980).
21. H. Fukusawa. Sedimentary mechanism of Neogene bedded siliceous rocks - on Late Miocene Wakkanai Formation of northern Hokkaido, Japan - , *J. Geol. Soc. Japan* **94**, 669-688 (1988).

Genesis of siliceous deposits in back-arc shelf basins of Kamchatka

A.R. GEPTNER
Geological Institute, Russian Academy of Sciences, Pyzhevsky per., 7, 109017 Moscow, Russia

Abstract. The rhythmic-bedded siliceous deposits (containing up to 60 - 70% SiO2) are characteristic elements of Cenozoic shelf sedimentary basins of West Kamchatka. They are considered to be terrigenous sediments enriched in fine fraction of tephra. The pyroclastics have been washed down from the drainage areas and mingled with other sedimentary products. In the course of this process, some granulometric and mineralogical differentiation took place, and all fine-grain sediments, as a result, come to be enriched in acid volcanic glass.

Key words: siliceous deposits, shelf basins, Kamchatka.

INTRODUCTION

Kamchatka Peninsula is located in the zone of transition from the Asian continent to the Pacific Ocean. The position may be the reason for major contrasts in the geodynamic settings involving high rates and amplitudes of tectonic movements and active volcanism. Following strong orogeny late in the Cretaceous and early in the Paleogene, the Cenozoic record showed the emergence and evolution of the West, East and Central Kamchatka troughs, with the Central Kamchatka volcanic belt developing there from the Oligocene on. These troughs accommodated sequences of terrigenous volcanogenic and volcano-terrigenous sedimentary rocks (from 1 - 2 to 10 km). The thickest, mostly flysch-rock, sequences formed in the East Kamchatka trough. In the vast West Kamchatka trough, there are distinct lateral and vertical rows of continental and largely shallow-water shelf units. Paleogene and Neogene strata of Western Kamchatka break up in 4 series: Tigil, Kovachina, Voyampolka and Kavran. Lithologically, the most remarkable are the Voyampolka (suites: Amana, Gakkha (Oligocene), Utkholok, Viventek (Lower Miocene), comprising fine-grain siliceous deposits in thick sequences (1 km and more), with a pronounced flyschoid cyclicity.

Oligocene - Miocene rhythmically bedded deposits of the West Kamchatka paleo-shelf
In the Paleocene and Eocene, Western Kamchatka was dominated by continental and shallow-water sediments. Subaerial and rarely underwater volcanics in some areas seem to be suggestive of land volcanism, the first such in the Cenozoic history, that existed there at that time in the form of a chain of volcanic islands [1].
In the Oligocene and Early Miocene, the area of sedimentation expanded to add most of Western Kamchatka. At that time, an abruptly reduced inflow of terrigenous material was due to extensive lava sheets. At the same time, the volcanic belt was arising as a rich source of loose volcanics (various andesitic and dacitic tephras). Discontinuous eruptions, cyclic patterns of stratovolcanoes and accumulation of loose sediments at their feet, may be some but not the only factors

responsible for the formation on the shelf of a peculiar flysch-rock complex of the Voyampolka series, in which tephra was very important.

Oligocene-Miocene sequences of Western Kamchatka are dislocated, folded and faulted into blocks. The original composition of sediments has been altered katagenically. Katagenic alterations affected largely fine, pelitic and partly silty components of the sediment.

Flyschoid sequences were being deposited on a vast shelf, apparently far away from explosive centres (Fig. 1). Over a 40-km-distance, northwest to southeast from the coastal outcrops, the grade of the sediments and tuffs in the enclosed layers does not vary, staying mostly fine-grain. The flyschoid nature of the bedding remains intact.

The Voyampolka series shows thick sequences of rhythmically alternating mudstones, siltstones and fine-grain tuffs. The stratigraphic range of these sequences may vary. In most coastal outcrops, rhythmically bedded sequences form two suites: Gakkh and Viventek. Their thickness varies significantly. In the Tochilo coastal section (Fig.1) the Gakkh unit is 760 m and the Viventek unit is 200 m thick, while in the Maynach coastal section they are 300 and 500 m respectively. But in the Tochilo coastal section, where, as most researchers believe, the most deep-water sequences are exposed, the stratigraphic range of rhythmic beds is somewhat expanded. On the whole, in the sequence of the Voyampolka series, two rhythmic sequences are separated by horizontally bedded but more uniform sandy mudstones and siltstones of the Utkholok suite, also rich in tuff horizons. The flyschoid patterns of these sediments are graphic in the Gakkh and Viventek suites where they are due to frequent and regular alternation of rocks of the four main types.

One type is represented in the Gakkh suite by cherty siltstones and opoka-like siliceous mudstones (siliceous mudstones hereafter), either grey, bleached, very hard, massive, or with a vague subparallel banding camouflaged by cleavage jointing. Fractured at a right angle to bedding, they often show that the coarse-grain silt is concentrated in minute (1-2 mm) elongated lenses, emphasizing the horizontal bedding of uniform fine-grain sediments. The thin horizontal bedding is particularly apparent under microscope in thin sections oriented perpendicular to the bedding plane. Some beds may abound in fine fragments of fossilized plant tissues. Partly replaced by micro-globular pyrite, the plant detritus and elongated aggregates of pyrite are subparallel. The micro-layers are fractions of a millimetre, or rarely 1-2 mm thick.

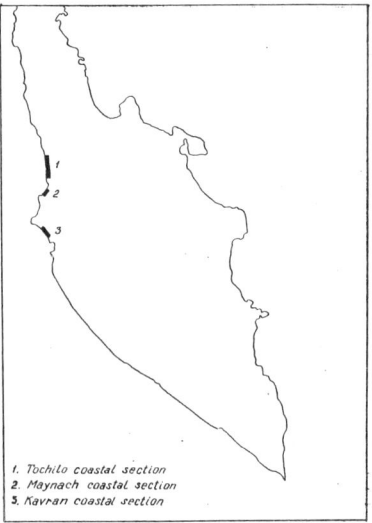

Figure 1. Location map.

More frequently, however, rock of the first type has a homogeneous structure, within a single bed, with rare, irregularly distributed coarse sand particles. There are sometimes numerous worm trails identified by a darker, light brown colour at the background of the grey rock. Thin sections often demonstrate portions of rock fully or partly reworked by the ooze-feeding fauna (worms). It is thus highly probable that the thin-bedded patterns have disappeared due to the workings of the ooze worms.

The beds of this type are from 0.01 to 0.8 m thick, with an average of 0.1-0.5 m. The bedding surface and vertical sections demonstrate (especially well on the weathered surfaces) nodular lens-like formations due to katagenic redistribution of siliceous substance (Fig. 2). The fact of this process at work is supported by chemical analyses of rocks sampled from the zone of lenticular silicification of the beds, showing high concentrations of SiO_2, maximum for these strata (Table 1).

In the stratigraphically higher (after Utkholok) Viventek suite, this type is represented by equally high-siliceous, fine-grain rocks, which contain more tephra. They are grey and blue grey, bleached on the walls of natural fractures. The beds are massive, with a vague subparallel banding occasionally emphasized by the lenticular arrangement of the sandy or silty fractions of volcaniclastics. Worm trails are abundant and particularly graphic when siliceous mudstones and siltstones are covered by a tuff layer, with a different colour and structure of fragments. The ash was mixed up by the worms, carried over into the underlying sediments, thus emphasizing the affected portions of the rock.

The second type of rhythmically bedded rocks includes clayey varieties of siltstones and siliceous mudstones with abundant clay minerals. It is a grey, often finely and parallelly bedded, foliated and thin-scree rock. The maximum thickness is 0.01-0.03 m, although routinely it is much less. There are also scattered sandy particles, sometimes distinctly streamlined by a fine-grain material.

The third type of rocks is argillated and carbonatizied vitroclastics or rarely vitrocrystalloclastic tuffs, grey, green grey, blue green and reddish. Tuffs sometimes show a granulometric differentiation, when lower particles are sandy, and the upper ones silty tephra. Tuff beds are usually 0.1-0.15 m thick, although there are some up to 0.5 - 0.8 m thick. The marker tuff bed at the top of the Viventek suite is almost 3 m thick. In the sequence, tuff beds occur every 1 to 5 m, or sometimes 10 m. They must have been laid down by strong explosions and ash fallouts (non-redeposited tephra). Of significance are clear-cut and even boundaries at the base

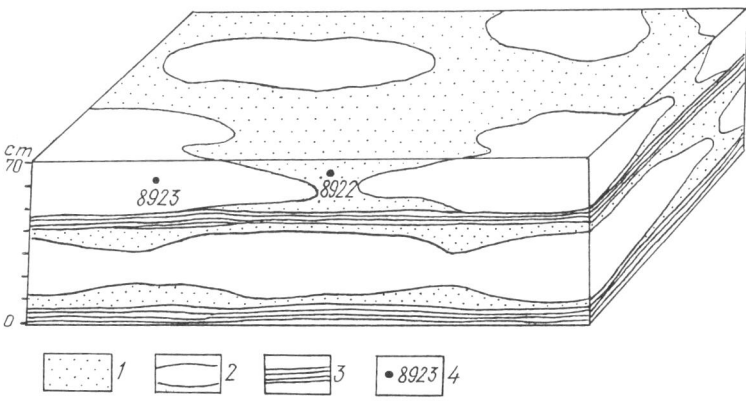

Figure 2. Intra-layer siliceous nodules, Gakkh suite, Tochilo section. 1 - silica-poor mudstones; 2 - silica-rich mudstones; 3 - clay siltstones; 4 - sampling sites (see Table 1).

Table 1.
Chemical composition of volcano-terrigenous rocks with distinct signs of katagenic silicification. Components, total. 1, 2 - Gakkha suite, Tochilino sequence: 1 - central part of the lens-shaped siliceous concretion (8923), 2 - a marginal part of the same concretion (8922). 3, 5, 7 - Viventek suite, Kavran sequence, stratiform siliceous lens-like swells in mudstones, 4, 6 - enclosing mudstones.

	1	2	3	4	5	6	7
SiO_2	82.54	78.87	79.58	64.76	85.24	68.60	86.64
TiO_2	0.63	0.80	0.32	0.62	0.26	0.62	0.20
Al_2O_3	5.66	6.50	8.13	14.88	5.44	14.56	5.32
Fe_2O_3	1.88	2.17	2.20	5.20	1.03	3.53	1.30
FeO	0.58	0.63	0.70	0.64	0.58	0.93	0.47
MnO	0.01	0.03	0.02	0.03	0.01	0.02	0.01
CaO	1.06	1.17	0.74	1.19	0.74	0.74	0.55
MgO	1.02	1.45	1.06	1.78	0.73	1.72	0.53
Na_2O	1.06	1.10	0.74	0.36	0.79	0.32	0.61
K_2O	0.99	1.17	1.71	0.64	1.77	0.55	0.82
H_2O^-	0.99	1.20	3.86	5.06	3.63	5.14	3.45
H_2O^+	1.61	2.15	n.d.	n.d.	n.d.	n.d.	n.d.
CO_2	0.45	-	<0.01	<0.01	0.07	<0.01	<0.01
C	0.27	0.27	n.d.	n.d.	n.d.	n.d.	n.d.
SO_3	n.d.	n.d.	0.80	2.44	0.40	1.29	0.48
P_2O_5	0.04	0.06	0.04	0.16	0.03	0.07	0.05
I.L.	1.26	2.34	n.d.	n.d.	n.d.	n.d.	n.d.
Total	100.05	99.91	99.91	97.77	100.72	98.10	100.44

and at the top of tuff beds, the missing or extremely vague traces of tephra rewashing or redeposition. Signs of slight redeposition of the sandy and silty tephras have been observed under microscope in oriented thin sections from overlying tuff beds. This, as well as the distinct granulometric differentiation of volcaniclastics in tuffs, is suggestive of a calm hydrodynamic environment of sedimentation beyond the effects of a strong wave action or currents. The fallen tephra accumulated slowly, sinking in the water and differentiating depending on its specific weight. Weak currents disturbed and rewashed only the upper thin layer of fine-grain volcaniclastics.

Less common are pure diatomites, or diatomaceous mudstones, siltstones and a tuff with a large amount of diatom frustules, which compose *the fourth, composite type* of rhythmically bedded rocks. However, some of the rare and poorly preserved diatom fossils can be observed in various types of rock throughout the sequence of the rhythmically bedded Gakkh and Viventek suites.

The sequence of highly siliceous rhythmically bedded strata shows three patterns of natural combinations of the above rocks that correspond to the definite sedimentary environments and strengths of explosive eruptions: (1) volcano-terrigenous, composed of intercalated highly siliceous mudstones and siltstones, and less so siliceous, clayey siltstones and mudstones with dispersed sandy material; (2) volcano-terrigenous-tephrogenic, composed of intercalated siliceous mudstones, siltstones, clay siltstones, mudstones, and tuffs. This association seems to reflect periods of strong explosive activities of the volcanoes. The sequences consisting of this association must have been synchronous with catastrophic caldera-producing eruptions marked by heavy tephra fallouts. Huge piles of tephra accumulated at the time of ash falls not only in the sea but also all over the catchment area, to become subsequently rewashed and repeatedly ensnared in sedimentation on the shelf (synchronous-redeposited tephra). At the time of waning volcanic activities, the basin of sedimentation received only the synchronous-redeposited tephra. The presence of tuff beds is not the only difference between these two associations. It shall be shown below that explosive volcanism affected not only the composition of volcano-terrigenous rocks but also the thickness of beds of intercalated siliceous mudstones and tuffs; (3) volcano-terrigenous-tephro-biogenic, composed of intercalated siliceous mudstones, siltstones, clayey siltstones, mudstones, tuffs and diatomites, frequently heavily

recrystallized and often devoid of the organogenic textures. Beds of diatomites, or mudstones rich in diatoms are believed to associate with periods of the abruptly reduced supplies, in some basinal zones, of volcano-terrigenous and tephrogenic materials alike (zones of chalistase). The tephrogenic component in this association may be missing sometimes.
Voyampolka strata, including the flysch-like Gakkh and Viventek sequences, contain glacial sediments which consist of gravels, pebbles and boulders. Genetically, the glacial sediments must be incorporating separate coarse sand particles and nestle-like aggregates of irregularly shaped sand.
Repeatedly alternating, the identified natural associations may occur at different levels in the Gakkh and Viventek suites. Because of the many hundreds of metres of these strata, it is difficult to establish the relations of those natural associations in the entire section of the Voyampolka series. Note only that the volcano-terrigenous natural association is the most common in the lower Gakkh sequences. The volcano-terrigenous-tephrogenic associations composes most of the Gakkh and Viventek suites, while the volcano-terrigenous-tephro-biogenic is more frequent in the Viventek suite.

Chemical composition of the rhythmically bedded deposits
All of the above strata are fairly high in silica: 60-85% SiO_2 (Table 2). They are now katagenically altered, particularly their fine-grain material. As a result, the original affinity of the uniform fine-grain rocks cannot be inferred unambiguously. Some suggest the biogenic origin of silica in these rocks, resulting from the accumulation of large masses of diatoms and their subsequent alteration, which removed the organogenic textures [2]. The author, however, failed to give reasons for the rhythmic patterns of bedding. A different opinion [3] holds that Voyampolka sequences must be a huge fan (or a system of fans) deposited by gravity flows triggered by frequent eruptions. Kuralenko [4] argues that intercalation of volcano-terrigenous and volcano-terrigenous-tephrogenic associations may be due to intermittent actions of autokinetic flows on the slopes of insular volcanoes, since there are reasons to believe that the complex of the above sequences was laid down in the peripheral portions of underwater aprons aligning the island arc. Kuralenko also thinks that an alternative reason for the rhythmic bedding may be intermittent effects of weak contour currents.
As a result of detailed studies of the material and structures, Geptner [5] collected additional facts that appear to account for an environment conducive to rhythmically bedded siliceous sequences of the Voyampolka series, outwardly very similar to flysch.
Mostly silty and fine-sandy tuffs in these strata suggest that the flysch-like siliceous sequences of the Voyampolka series must have formed far away from the eruptive zones of mostly andesitic volcanoes. Thick accumulations may have something to do with synsedimentary downwarping of some shelf zones. With the transportation and multiple rewashing of loose tephra (synchronous-redeposited material), accompanied by granulometric and mineralogical differentiations, the shelf received mostly silts and pelites dominated by acid volcanic glass. This is believed to be responsible for the silica-high composition of the Voyampolka sequences.
The above sequences appear to be rather high silica (60-85%). Now the rocks are altered, and the original nature of the fine silts and particularly pelites cannot be unambiguously interpreted under an optic or electron microscope.
If we assume that high silica content in the studied rocks was the result of high inflow of siliceous diatom shells, given slow or fully suppressed sedimentation of volcano-terrigenous and tephrogenic materials, dilution with biogenic silica must have produced a negative SiO_2/Ti correlation, due to the low-mobile and invariable Ti in various volcaniclastics. But this is not the case. Virtually, most of the studied rocks from the rhythmic Gakkh and Viventec sequences, irrespective of SiO_2 amounts (from 56 to 88%), fall in terms of their TiO_2 (recalculated to the silica-free sediment) in the 1 to 2.5% range (Fig. 3, Table 2).

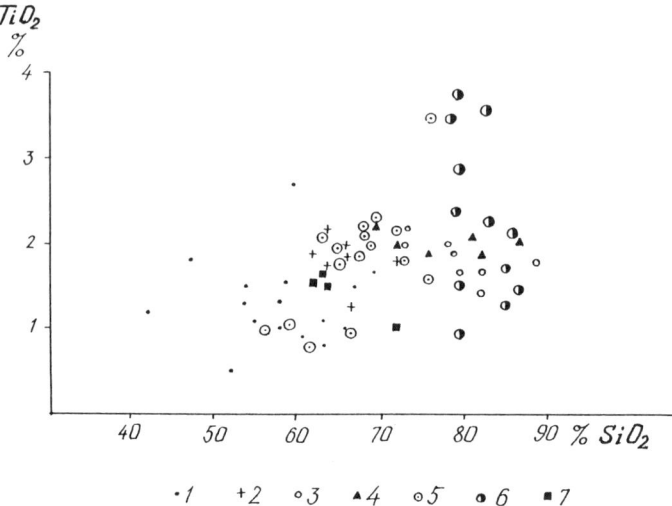

Figure 3. TiO_2/SiO_2 ratio in Oligocene and modern volcano-terrigenous silt-pelitic rocks, and Oligocene-Miocene tuffs. Data and interpretation by V.I.Grechin [4]: 1 - tuffs from various horizons of the Kavran and Voyampolka series; 2 - tuff diatomites of Kakert series, Kavran suite; 3 - opokas of Viventek suites, Voyampolka series; 4 - recrystallized opokas and siliceous tuff mudstones of the Gakkh suite, Voyampolka series. Data and interpretation by the authors: 5, 6 - volcano-terrigenous and volcano-terrigenous-tephrogenic unstratified rock associations, 6 - samples from horizons with distinct siliceous lens-shaped concretions (see Tables 1, 2); 7 - modern silt-pelitic sediment over the periphery of active andesitic volcanoes.

Estimates of TiO_2 concentrations recalculated to the silica-free sediment have been made for the following reasons. Local zeolitization, argillization and carbonatization might have led to locally increased or decreased concentrations of some petrogenic elements. A similar picture might have emerged back at the sedimentogenesis stage, when the volcano-terrigenics were diluted by biogenic silica. In our rocks, variations in SiO_2 and CaO are especially apparent. Epigenetically, the carbonatized portions may be easily identified and excluded from the chemical analyses of the rock. It is the genesis of silica in the studied rocks that is mostly the problem. Variations in other elements are largely insignificant and can be ignored. Exclusion of SiO_2 from the calculations of TiO_2 concentrations allows to avoid a possible effect of the added biogenic or chemogenic silica.

The TiO_2/SiO_2 plot in Fig. 3 shows that most of the studied (by us and by V.I. Grechin) volcano-sedimentary rocks have concentrations of TiO_2 similar to those of tuffs. There is more TiO_2 in some siliceous (75-82% SiO_2) rocks of the volcano-terrigenous association from the lower Gakkh suite (Tochifo sequence). Hence, the biogenic share of the silica, if present, cannot be important. It is likely that most of SiO_2 was supplied simultaneously with TiO_2 within the vitro-, crystallo- and lithoclastics with different origins. Local increases of SiO_2 by 10-15% may also be due to the katagenic alteration of rocks.

The substantial effect of katagenic alteration for Cenozoic rocks can be inferred from the mode of changes not only in Voyampolka sequences, but in some younger, Kavran strata [6]. Variably aged strata of the Kavran series in different regions of Western Kamchatka and Sredinny Ridge show the same assemblage of secondary minerals composed of (in the sequence of their formation) clay minerals, opal, zeolites and calcite. Clay minerals are dominated by smectites (Fe-montmorillonites), with celadonite only locally important. Development of secondary minerals depends

Table 2.
Chemical composition of the Gakkha (1-10) and Viventek (11-18) rocks, Kavran sequence.

	1	2	3	4	5	6	7	8	9
SiO_2	67.96	56.06	58.82	68.04	85.70	67.50	82.68	72.50	62.88
TiO_2	0.68	0.44	0.44	0.70	0.30	0.60	0.40	0.50	0.78
Al_2O_3	14.15	16.14	14.86	15.03	5.03	12.08	6.76	10.86	15.20
Fe_2O_3	2.67	8.58	6.23	2.27	1.17	2.62	1.50	2.42	3.79
FeO	0.99	0.93	0.82	0.93	0.70	1.05	0.76	0.87	0.70
MnO	0.02	0.02	0.02	0.03	0.02	0.02	0.01	0.03	0.02
CaO	2.39	1.93	2.57	2.39	0.83	2.57	0.74	2.02	2.76
MgO	1.32	4.29	3.70	1.59	0.73	1.98	0.79	1.19	1.85
Na_2O	1.11	0.20	0.28	1.43	0.32	0.83	0.40	0.57	0.70
K_2O	1.72	0.28	0.56	1.60	0.50	1.38	0.66	0.98	1.44
H_2O^-	3.42	13.60	10.08	2.94	2.95	4.60	3.77	5.97	5.81
CO_2	0.14	<0.01	<0.01	0.04	<0.01	<0.01	0.07	<0.01	<0.01
SO_3	0.68	0.05	0.04	0.72	0.65	1.80	0.75	2.54	0.24
P_2O_5	0.13	0.04	0.14	0.10	0.24	1.13	0.13	0.12	0.27
Total	97.38	102.56	91.42	104.95	99.14	98.16	99.42	100.57	96.44

	10	11	12	13	14	15	16	17	18
SiO_2	71.90	65.20	79.04	69.54	75.76	61.58	66.72	79.30	85.12
TiO_2	0.61	0.68	0.52	0.68	0.40	0.30	0.32	0.20	0.20
Al_2O_3	12.64	13.02	9.29	11.79	9.59	17.57	13.98	8.52	5.51
Fe_2O_3	1.76	7.94	1.94	4.30	2.87	4.82	4.69	1.57	1.50
FeO	0.82	1.17	0.58	1.63	0.70	0.58	0.58	0.58	0.53
MnO	0.02	0.02	0.01	0.04	0.03	0.02	0.03	0.02	0.01
CaO	2.02	1.29	1.10	0.92	1.84	1.47	1.29	1.29	0.55
MgO	1.45	1.92	1.06	1.52	1.06	3.30	2.25	1.32	1.26
Na_2O	0.66	1.00	0.57	0.66	0.36	0.40	0.38	0.38	0.28
K_2O	1.26	2.29	0.93	1.32	0.66	0.93	0.85	0.82	0.64
H_2O^-	4.15	4.27	2.82	4.31	4.34	10.22	6.68	4.50	3.78
CO_2	<0.01	0.07	<0.01	0.48	<0.01	0.04	0.07	<0.01	0.14
SO_3	0.15	0.34	1.25	0.74	1.50	0.13	0.37	0.34	1.28
P_2O_5	0.07	0.10	0.10	0.10	0.28	0.09	0.07	0.15	0.10
Total	97.51	99.31	99.21	98.03	99.39	101.45	98.28	98.99	100.90

largely on the permeability and grade of the rocks. The limits of the zones of development of secondary minerals were found to cross the sedimentary bedding. There are veins and concretion-cementing aggregates, sometimes with clear-cut concentric patterns and inliers of weakly altered rocks at the centre of the aggregates. This may be evidence of the superposed nature of the secondary mineralization, natural both in Kavran and in older strata. In the fine-grain Voyampolka sequences that are now tectonically at the same hypsometric level with Kavran, the cement is basically zeolites and siliceous minerals. Clay minerals which are not abundant but ubiquitous, are extensive only in permeable tuff layers, and often in heavily carbonatized ones too.

The zeolite-siliceous cementation has been identified in all Voyampolka horizons, irrespective of the facies or genesis of the strata. Zeolites, amorphous silica and fine-grain aggregates of chalcedony fill in the hollows of diatom frustules and other voids, but the biggest amounts of zeolites and siliceous material (mostly opal) fill in the cracks across the rock.

Such alteration of the Kavran and Voyampolka strata must definitely be associated with the effect of underground waters. At earlier stages, it must have been formation of the smectite cement which locally was replaced by celadonite. The data suggest that celadonite mineralization is either imposed on or replaces smectites. The subsequent alteration was due to tectonic deformations, folding, faulting and jointing. Instead of the clay cement, at that stage it was mostly zeolite-siliceous material [6]. The formation of zeolite-siliceous cement suggests that redistribution of silica was a critical factor of katagenic silicification of rocks. Silica in underground waters was derived from the country rock due to the solution not only of siliceous fossils but also of various volcanics, of which acid volcanic glass was the most remarkable.

Note also that modern, unaltered silt-pelites accumulating at the periphery of andesitic stratovolcanoes, composed mostly of vitro- and crystalloclastic parts of the synchronous-redeposited tephra, fit in the TiO_2/SiO_2 plot in the area corresponding to the Oligocene rocks of the Voyampolka series, i.e., the original silica content of these rocks might have been high solely due to the large contribution of synchronous-redeposited acid tephra.

Voyampolka opokas: tephra or biogenic silica ?

Undoubtedly diatom productivity at the time of deposition of shelf strata was fairly high. Biogenic opal in the sediments must be the result of the simultaneous effect of such agents as biological productivity, dilution by the terrigenous (volcano-terrigenous and tephrogenic, in this case) material, and chemical solution of the biogenic opal. The latter factor is particularly strong at the sediment-water interface and in the upper few centimetres of the sediments due to the effect of silica-undersaturated pore waters [7].

Given the low sedimentation rates, most of diatoms in the upper sedimentary layers are dissolved, while silica is released into bottom waters. Strong bioturbation may also be contributing to rapid solution of opal components [8].

Rhythmic sediments of the Voyampolka series accumulated slowly and were heavily disturbed by worms. As a result, the precipitated biogenic opal might have partly or wholly disappeared through solution back at the time of sedimentogenesis. In the Gakkh siliceous mudstones, diatom fossils, variably preserved, are fairly rare. But in these strata, in ash-rich layers, they are more abundant and better preserved. It may be due to the higher sedimentation rates there, and rapid burial and expulsion of opal sediments from the upper zones, where pore waters are more aggressive to SiO_2.

In the Viventek suite, the role of tephra is more important, suggesting that sedimentation rates must have also been high. As a result, diatoms are more frequent. There are silt- and mudstones rich in diatom fossils, as well as tuffites abundant in diatoms, while in the southernmost, Kavran sequence of the Viventek suite, there are even diatomites. Basically, diatomite beds are paragenetically connected with tuffs which under- or overlie the layers with biogenic opal. These relations seem to suggest that diatomites accumulated in temporary zones of a volcano-terrigenous deficit, and were quickly sealed off by tephra layers from the bottom waters.

Hence, the reasons for the missing or scant diatom fossils in the Voyampolka volcano-terrigenous strata can be not only their solution under katagenesis [4], but also opal solution at the water-sediment interface and in the upper sediments back at the stage of sedimentogenesis.

Significantly, under katagenic solution, spaces in the rock occupied by diatom shells may stay and be filled with some other substance. Under electron microscope, sections of opoka-like rocks of the Viventek suite showed some rounded and oval cavities that corresponded in size to diatoms'. Some of them were empty, some filled with zeolites and siliceous minerals. These "traces" of diatoms may suggest that initially there were not very many of them.

Acid volcanic glass is more resistant but at that stage of rock alteration it rarely stays undissolved in the sediment. Rich in silt-pelitic glass particles, fine-grain rocks after weak katagenesis may, by appearance and in terms of their SiO_2 concentrations, be very similar to opokas. However, they are radically different in higher contents of TiO_2 and some other petrogenic elements.

Variations in the thickness of beds
The important contribution of synchronous-redeposited tephra follows from the thickness of siliceous mudstones and their associated tuff horizons' ratios.
The intercalation of different rock types has been studied in two major sections of the Voyampolka series (Tochilo and Maynach) with layer-by-layer measurements. All in all, 1830 measurements have been taken, embracing all three natural associations described above. Portions of the Gakkh suite composed of the volcano-terrigenous association of rocks showed narrow variations in the thickness of siliceous mudstones (Fig. 4). The intercalated clayey siltstones are routinely only fractions of a centimetre thick or missing altogether. Then, layers of siliceous mudstones, with the distinct top and base, lie directly on each other.
Sections corresponding to the Gakkh and Viventek suites and composed of the volcano-terrigenous-tephrogenic association show individual or grouped thick beds of siliceous mudstones.
Their thickness may sometimes be 8 to 10 times the average (Fig. 5). Thick beds of siliceous mudstones were found to follow a definite regular pattern, associating with one (underlying) or several (intercalated) tuff layers. In these portions of thesequence, mudstones are persistently alternating with thin (0.5 cm) beds of clayey siltstones.

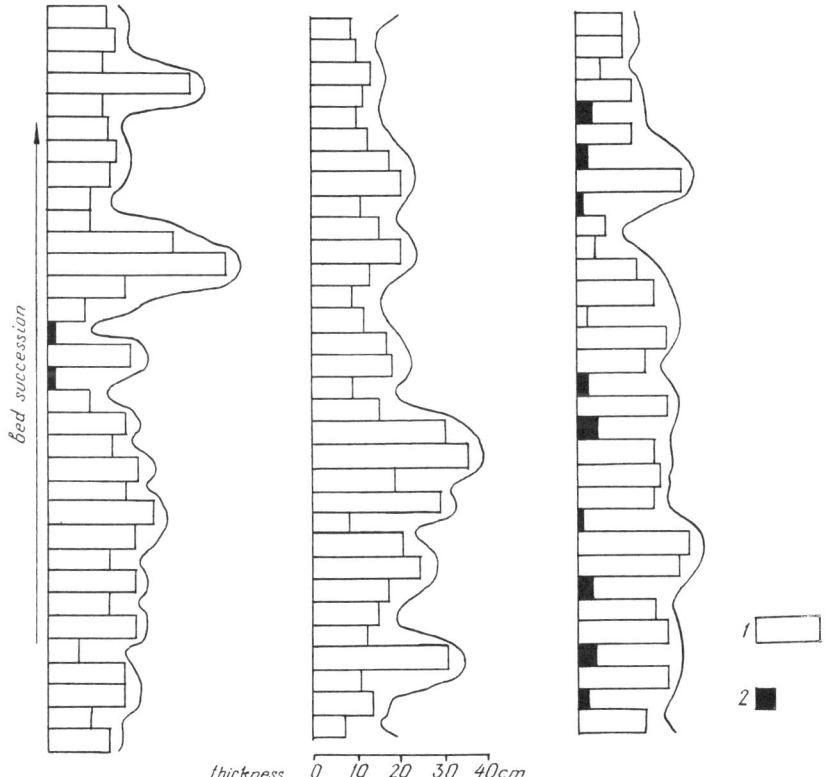

Figure 4. Histograms of thickness of layers of siliceous mudstones and argillaceous siltstones of the Gakkh volcano-terrigenous association. 1 - siliceous mudstones; 2 - argillaceous siltstones and siliceous mudstones, with a major admixture of clay.

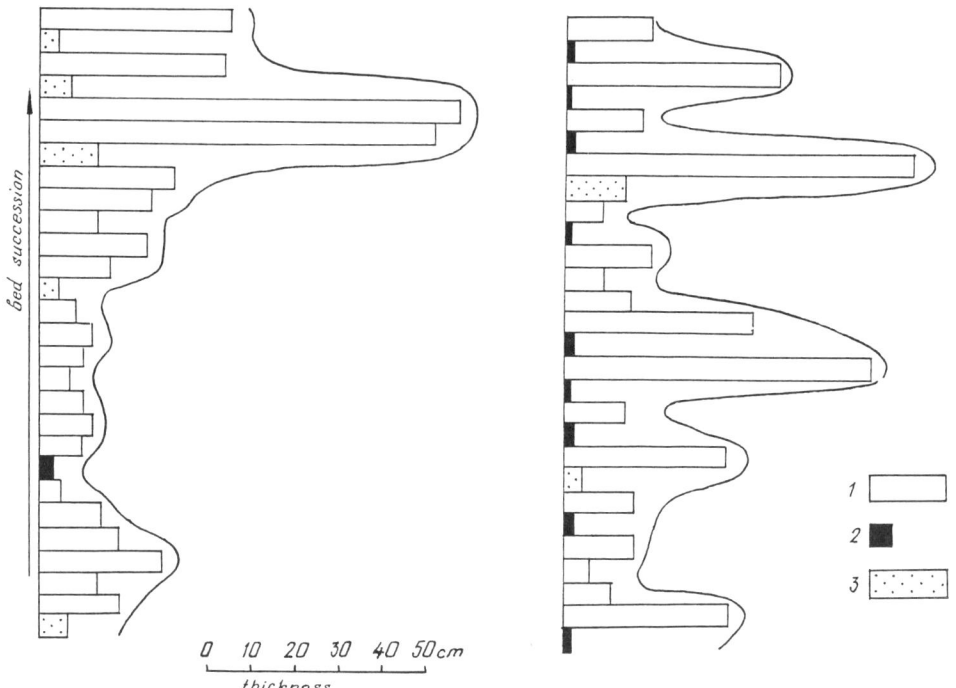

Figure 5. Histograms of thickness of layers of siliceous mudstones, argillaceous siltstones and tuffs of the Gakkh volcano-terrigenous-tephrogenic association. 1 - siliceous mudstones; 2 - argillaceous siltstones; 3 - tuffs.

The above proportions and variations in the thickness of siliceous mudstones and tuffs can be interpreted as resulting from active mobilization of the ash fallen out in the catchment area, and its transportation onto the shelf. There are reasons to believe that the clastics were derived from the volcanic slopes and the catchment area by various streams. On the shelf, the fine-grain material was transported by slow currents in the stratified water column where bottom layers were richer in the suspended fine silt. Siliceous mudstones sometimes show a few millimetres thick lenses of larger silt and sand particles deposited by more high rate jets of bottom waters. A heavier load of clastics and the turbidity of bottom shelf waters might be among the reasons for uneven distribution of mollusc fossils in these strata, and numerous worm trails.

CONCLUSIONS

The basic factor responsible for the flyschoid sequences on the shelf of the West Kamchatka trough, laid down away from eruptive centres, was a discrete input by volcanoes of large masses of loose tephra in the catchment area. Rhythmic bedding of siliceous volcano-terrigenous strata was due to variations in the surface runoff and rates of removal off land of large masses of fine-grain tephra (a climate factor), and its subsequent dispersal by bottom currents over the shelf surface under calm hydrodynamic conditions, favouring distribution of fine-grain material over vast territories (a hydrodynamic factor). With the explosive activity getting more powerful, the amount of volcano-terrigenous sediments (synchronous-redeposited tephra) deposited on the shelf was growing without serious climate oscillations recorded.

Acknowledgements
Thanks are due to Galina Ermakova for the English translation of the present paper.

REFERENCES

1. A.Ye. Shantser. Cenozoic history of Kamchatka: formation and destruction of unstable orogenic rises. In: *Essays on tectonic history of Kamchatka.* pp. 109-164. Nauka, Moscow (1987). (*in Russian*).
2. V.I. Grechin. *Miocene deposits of Western Kamchatka (sedimento- and katagenesis).* Proc. GIN AN SSSR, vol. 282, Nauka, Moscow (1976). (*in Russian*).
3. K.G. Kazakov. Structure and environment of the Voyampolka fan in Western Kamchatka (illustrated by the Maynach sequence). In: *Lithology of Cenozoic Shelf Units.* pp. 111-133. GIN, Moscow (1989). (*in Russian*).
4. N.P. Kuralenko. Basic features of sedimentation in the continent-to-ocean transition zone, illustrated by Miocene strata of Kamchatka. In: *Lithology of Cenozoic Shelf Units.* pp. 134-185. GIN, Moscow (1989). (*in Russian*).
5. A.R. Geptner and N.P. Kuralenko. On the formation of the composition of loose sediments at the feet of active volcanoes, *Litologiya i polez.iskopayemye* **5**, 30-45 (1979). (*in Russian*).
6. A.R. Geptner. Genesis of siliceous rhythmically bedded strata of the shelf of volcanic regions. In: *Geology of oceans and seas, Collection of abstracts.* vol.4, pp. 29-30 (1990). (*in Russian*)
7. G. Bohrmann. Accumulation of biogenic silica and opal dissolution in Upper Quaternary Skagerrak sediments, *Geo-Marine Letters* **6**, 165-172 (1986).
8. D.R. Schink, N.L. Guinasso and K.A. Fanning. Processes affecting the concentration of silica at the sediment-water interface of the Atlantic Ocean, *Jour. Geophy. Res.* **80**, 3013-3031 (1975).

The Origin of Chert in the Monterey Formation of California (USA)

R. J. BEHL[1] and R. E. GARRISON[2]
[1]Marine Science Institute, University of California, Santa Barbara, CA 93106, USA
[2]Earth Sciences Department, University of California, Santa Cruz, CA 95064, USA

Abstract-- Many chert-bearing sequences deposited beneath ancient upwelling systems contain similar structural and stratigraphic features in spite of marked differences in their depositional, diagenetic, and tectonic environments. Examples from the Miocene Monterey Formation of California illustrate that these features relate to the timing and processes of chertification. Field, petrographic, and oxygen isotope relations indicate that pure (>90% silica) cherts formed earlier and by different mechanisms than the well-documented diagenetic sequence of less-pure porcelanites and siliceous shales. Dense, vitreous cherts formed by the diagenetic concentration of silica within the purest biogenic calcareous-diatomaceous sediments that accumulated with only little dilution by terrigenous material in the outermost portions of the Neogene continental borderland.
In the Monterey Formation, opal-CT cherts formed before opal-CT porcelanites, and most quartz cherts formed considerably earlier than the opal-CT to quartz transformation in porcelanites and mudrocks. Opal-CT cherts formed shallowly by pore- and fracture-filling cementation of diatomite at temperatures between 2-33°C. Quartz cherts formed by replacement of opal-CT chert or dolomite, by fracture-filling cementation, or by pore-filling cementation of porcelanite between 36-76°C.
The rate and timing of chertification, with respect to burial and tectonic deformation, controls the development of a number of characteristic chert structures, including breccias, contorted beds, dikes, lineations, and spheroids. Contorted chert beds and breccias formed by volume-conservative or dilatant brittle deformation. Chert dikes and spheroids record early chertification and compaction. Lineations formed by fracturing of chert laminations and silicification of deformed and crenulated diatomite. No unusual states or precursors are required for development of these structures, all of which depend on the extreme brittleness of chert.

Keywords: chert, silica diagenesis, Monterey Formation, siliceous sediments, breccias, oxygen isotopes.

INTRODUCTION

Sedimentary geologists have long debated the origin and significance of chert [1,2] which form an important component of most mountain belts and all ocean basins. Cherts are also significant because: 1. they form geochemically resistant "time capsules" which contain well-preserved chemical and fossil records of the past [3,4]; 2. bedded cherts are a key stratigraphic and lithologic indicator of ancient regions of oceanic upwelling and high primary productivity [5] and/or deposition beneath the calcium carbonate compensation depth [6]; 3. the stratigraphic distribution of chert may reflect secular trends in the global silica cycle [7,8]; 4. cherts form economically important fractured petroleum reservoirs in the Monterey Formation of California [9] and are an important constituent of Precambrian banded iron formations. For all of these reasons it is critical for geologists to understand the mechanisms and environments of chert formation in order to separate diagenetic from depositional influences on their presence in the stratigraphic record.

While it is clear that most Phanerozoic marine cherts are derived from the dissolution of biogenic siliceous tests and reprecipitation of diagenetic silica [10-12], it is still controversial whether the bedding-scale distribution and composition of bedded cherts reflect the primary composition of sediments or the diagenetic concentration of silica [13-16]. Bedded biosiliceous deposits may undergo a diagenetic maturation in which diatomaceous or radiolarian-rich sediments are lithified with complete conservation of their primary chemical compositions. Alternately, extensive mobilization and transportation of silica within sediments may create rocks with new, diagenetically controlled compositions. Probably both mechanisms operate under different conditions to create a wide spectrum of siliceous rock types [14,17]; this study focuses on the extreme compositional and textural end-member – pure, vitreous chert.
In the Monterey Formation of California (USA), pure cherts compose only a few per cent of the total stratigraphic thickness, whereas less-pure porcelanites and siliceous mudstones are abundant. These cherts, however, form important fractured oil reservoirs in the subsurface, and details of their origin are economically significant. Moreover, Monterey cherts display a number of distinctive diagenetic and structural features, such as breccias, dikes, lineations, intraformational folds and chert spheroids, that are not found in other siliceous lithologies, yet are common to pure cherts throughout the world. Because the chertification process transforms highly porous, biosiliceous sediments into exceedingly hard and brittle rocks, there is an important feedback relationship between diagenesis and deformation. Herein, stratigraphic, petrographic, and structural relations are combined with oxygen isotopic data to develop a detailed understanding of the sequence and mechanisms of chertification in the Miocene Monterey Formation. Further details of the oxygen isotope study are presented in Behl [17].

The Monterey Formation
The Miocene Monterey Formation of California is a heterogeneous, chiefly diatomaceous, hemipelagic deposit that accumulated in a number of small, marginal basins of the Neogene California continental borderland between 17 and 5.5 Ma (Early to Late Miocene)[18]. The Monterey sediments are part of the circum-Pacific swath of Neogene diatomaceous sediments [19,20] which reflect major oceanographic changes in deep-water circulation, paleoclimate, and centers of biosiliceous productivity [20-24]. These sediments – originally hemipelagic clay-coccolith-diatom oozes of varied composition – reflect deposition along the continental margin beneath the ancient California Current upwelling system [19-20,22,25-27]. Based on compositional variation, the Monterey Formation has been informally subdivided into stratigraphic members according to a number of schemes; the most commonly cited are the 5 members of Isaacs [25,28-29] and the 3 members of Pisciotto & Garrison [22]. The 4-member stratigraphic division of MacKinnon [30] (Figure 1), a variation on Isaacs' scheme, is the most useful for this study because it is based on the same coastal region of the Santa Maria basin.
During most of the middle to late Miocene, the distal portions of the California borderland (e.g., the Santa Maria basin and outer portions of the Santa Barbara and Pismo basins; Figure 2) accumulated relatively pure hemipelagic calcareous-siliceous sediments, whereas the more proximal basins (e.g., the San Joaquin, Salinas, and Cuyama basins) received more terrigenous input [22,31]. Significant thicknesses of dense, vitreous chert developed only in the purer biogenic sediments of the outboard basins; cherts are rare in the medial basins, and almost nonexistent in the most proximal San Joaquin basin. Relatively high sedimentation rates for biogenous sediments, often exceeding 100 m/m.y. (>400 m/m.y. precompaction sedimentation rates)[29,32], rapidly and deeply buried Monterey sediments down steep geothermal gradients, locally accelerating the rate of silica diagenesis. Unlike most older biosiliceous deposits, all stages of silica diagenesis can still be found in the Monterey Formation, thus allowing study of the entire sequence of chertification.
In addition to their geographic restriction to the outer basins of the Neogene California borderland, pure cherts are limited stratigraphically within the Monterey Formation. Although present throughout most of the Monterey, cherts are best developed in the middle part of the formation [22,33], specifically the Upper Calcareous-Siliceous member of Isaacs [28-29] and MacKinnon [30] (Figure 1). In the Santa Maria basin, these deposits are late Middle to early Late Miocene age (Early Mohnian benthic foraminiferal stage of Kleinpell [34], *D. hustedtii* – *D. lauta* subzones B and C of Barron [35-36]). White [36] showed that depositional onset of

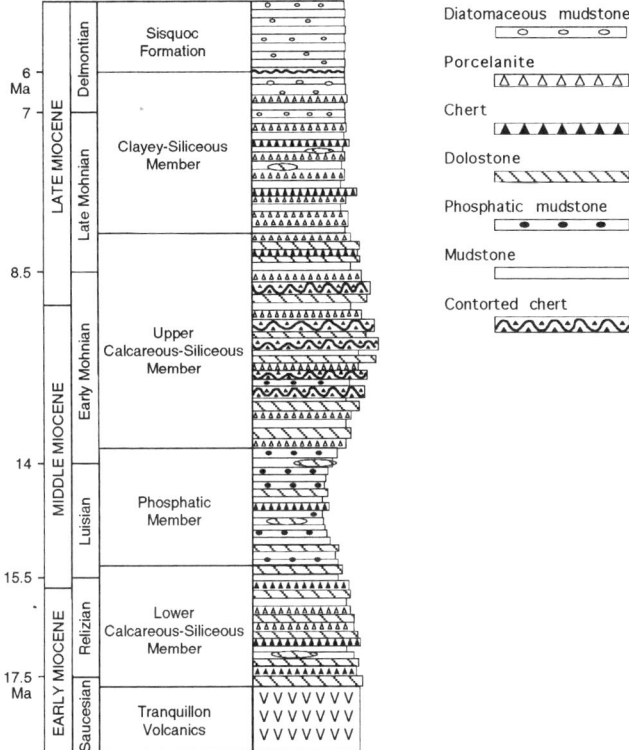

Figure 1. Composite geologic column and stratigraphic subdivision of the Monterey Formation for the coastal Santa Maria-Lompoc area (after MacKinnon [30]).

the cherty facies along the California coast was time transgressive, starting in the north at 14.7 to 15.0 Ma (Bodega and Outer Santa Cruz basins), and following 2 m.y. later in the south (12.7 Ma in the Santa Maria basin). This time period coincides with a gradual eustatic sea level fall from the highest stand of the Neogene [37], and may coincide with increased coastal upwelling and biosiliceous productivity [20,22]. However, lateral changes in composition and depositional environment makes it difficult to generalize Monterey paleoceanography from single sections or regions [38].

Field areas
Due to complex regional tectonism, the Monterey Formation varies laterally in composition and thickness and has been unevenly buried beneath younger sediments. Consequently, different diagenetic stages and features of the Monterey Formation are distributed throughout coastal California in several of the outboard basins of the Neogene, namely the Bodega, Salinas, Pismo, Santa Maria and Santa Barbara basins (Figure 2). This study focuses primarily on the Santa Maria basin because it was likely the most outboard of all Neogene basins now exposed on land and contains the most extensively developed subaerially exposed cherty sequences. Instructive exposures in several diatomite quarries in the Santa Maria basin provided key information about the earliest stages of opal-CT chertification that are unavailable at other locations where the Monterey has been more deeply buried and diagenetically altered. Access to the outcrops mentioned are described in Grivetti [39], Dunham and Blake [40], and Behl [17].

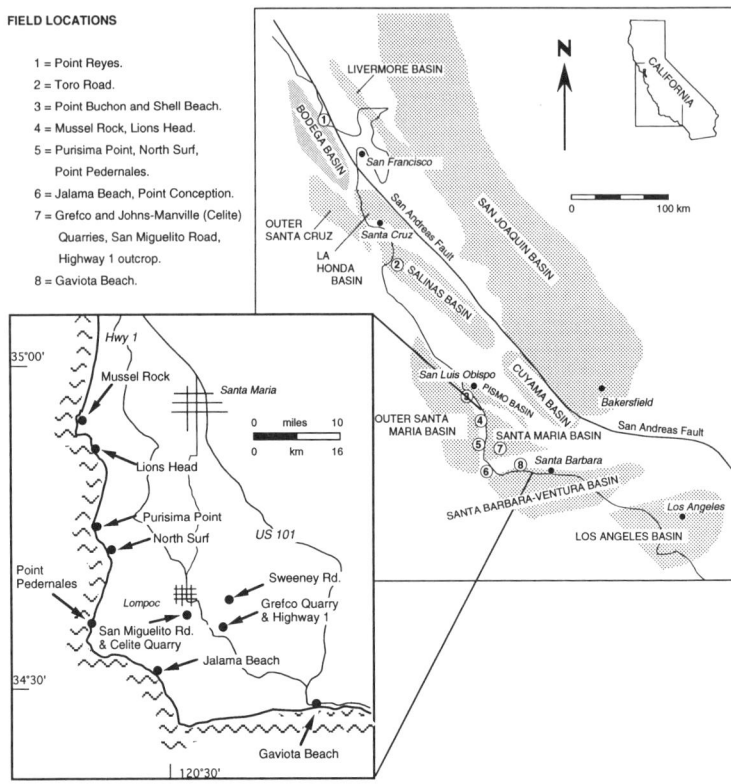

Figure 2. Major Neogene basins of California and field locations. Inset shows details of the Santa Maria area.

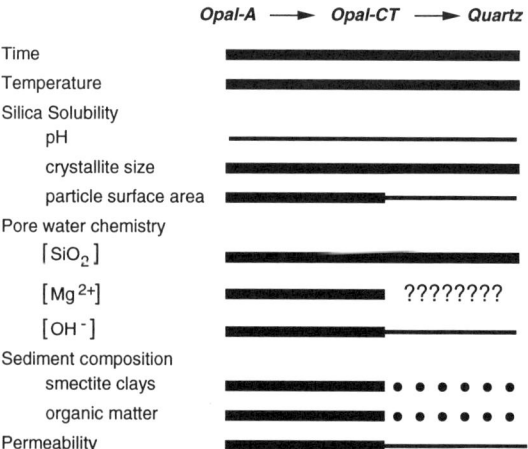

Figure 3. Important variables in silica diagenesis. Bold lines indicate that the effect is very important, fine lines are important, • = possible indirect effect, ? = unknown effect. Modified from Kastner (1985).

SILICA DIAGENESIS

Biosiliceous sediments are chiefly composed of diatoms and radiolarians which settle through the water column from the photic zone [41-42]. These consist of hydrous, X-ray amorphous silica, termed *opal-A* by Jones and Segnit [43]. As opal-A is undersaturated everywhere in the ocean and is unstable over geologic time spans, only a small fraction escapes dissolution in the water column and with time and burial that portion will transform to metastable *opal-CT* (terminology of Jones & Segnit [43]) and ultimately to *diagenetic quartz* by two discrete dissolution/precipitation steps.

This transformation sequence is recorded in the deep sea [44-47], in outcrop [25-26,33,48-52], and has been reproduced in the laboratory [53-57]. Opal-CT precipitates as cryptocrystalline blades and needles, and rarely as fibrous cement. Diagenetic quartz occurs as very fine, equigranular crystallites that are either cryptocrystalline (<1μm), microcrystalline (1-20μm), megacrystalline (>20μm), or fibrous [58]. Fibrous varieties of quartz include normal chalcedony (length-fast), quartzine (length-slow), lutecite (inclined extinction) [58-59], and microflamboyant (subequant domains with undulose or fanning extinction)[60].

The kinetics of the individual phase transformations is chiefly controlled by the increased temperature of burial, but is also influenced by compositional constraints (Figure 3)[55,61-63]. Key among these are: (1) the presence of clay and organic matter inhibits the rate of opal-A to opal-CT transformation [13,26,55,64], and (2) the presence of calcium carbonate increases the rate of opal-CT nucleation [55,62] and may speed quartz formation [13,26,65-66].

The Monterey Formation is an excellent natural observatory of silica diagenesis, as it contains rocks of varied compositions in all stages of silica diagenesis, including opal-A diatomite and diatomaceous mudstone, opal-CT and quartz porcelanite, and opal-CT and quartz chert. Chiefly due to the work of Bramlette [33], Murata and colleagues [48-49,67-68], Hein *et al.* [46], Isaacs [13,25-26], and Pisciotto [50-51], the diagenesis of impure, detritus-bearing diatomaceous sediments (<90% biogenic and authigenic silica), which make up most of the Monterey, is well understood. These are the diatomaceous sediments that transform to porcelanite or siliceous mudstone with burial.

Isaacs [13,26,69] showed that the relative timing of silica phase transformations in different strata depends on the silica:detritus ratio of the original sediments (Figure 4). She argued that

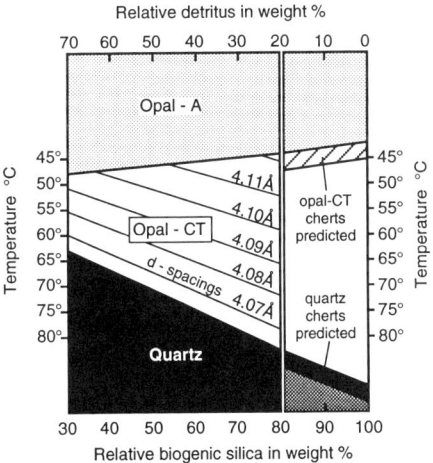

Figure 4. Left side: timing and temperature of silica phase changes with respect to composition in porcelanite and siliceous mudstone after Isaacs [13,25] and Keller and Isaacs [118]. Note that opal-CT d(101) spacings must decrease to <4.07Å prior to conversion to quartz and that detritus-rich sediments initially precipitate opal-CT with smaller d(101) spacing. Right side: linear prediction of timing of chertification as in Isaacs [26].

the changes in the physical properties of siliceous rocks resulted chiefly from compaction and porosity loss during "closed system," *in situ* dissolution and reprecipitation of silica. Although Isaacs considered most pure cherts to have formed by the same *in situ* diagenetic process, she also identified "atypical" quartz cherts that may have formed by the early addition of silica to calcareous-siliceous sediments with high (>8:1) silica:detritus ratios. Other workers have also presented geochemical and petrologic data indicating that some cherts form by the local addition of silica [14-16,48,70]. Our data suggests that *most* pure cherts form by local concentration of silica, a process different from the closed-system diagenesis of porcelanites and siliceous mudstones. Furthermore, a variety of field and oxygen isotopic evidence suggests that cherts form by several different mechanisms over a broader range of burial depths than would be predicted by Isaacs' [13,25-26] general model for siliceous rocks (Figure 4).

METHODS

In this study, we document the stratigraphic setting of cherts, their host lithologies and diagenetic relationships (i.e., cross-cutting or replacement contacts, and differential compaction), and the relative timing of chertification. Petrographic relationships were determined by polarized light, cathodoluminescence, and blue light fluorescence microscopy, scanning electron microscopy (SEM), and transmission electron microscopy (TEM). Mineralogy was determined by petrographic microscope, X-ray diffraction (XRD), and selected area electron diffraction (SAD) with the TEM. Chemical composition of selected samples was determined by energy dispersive X-ray spectrometry (EDS) on the SEM, and by X-ray fluorescence spectrometry (XRF).

In addition to the standard petrologic methods of investigation, we studied a variety of characteristic chert structures (e.g., breccias, contorted beds, lineations, dikes, and spheroids) to better constrain the timing, environment, and mechanisms of chertification and to explore the relation between deformation and diagenesis.

We measured oxygen isotopic compositions of opal-CT and quartz phase cherts at the Berkeley Center for Isotope Geochemistry at Lawrence Berkeley National Laboratory with BrF_5 extraction [71]. Oxygen isotope compositions are reported in the standard δ-notation, where the $\delta^{18}O$ value is the per mil deviation in $^{18}O/^{16}O$ with respect to Vienna Standard Mean Ocean Water (V-SMOW) [72]. Most duplicate analyses deviated from the mean by < 0.3‰, averaging 0.16‰ for quartz and 0.3‰ for opal-CT cherts. Details of our analytical procedures are included in Behl [17] and Behl and Smith (*in preparation*).

LITHOLOGY

Chert vs. porcelanite
The term *"chert"* has been broadly used to include siliceous rocks spanning a wide range of compositions and textures that probably formed by a number of different processes. This study focuses specifically on the purest cherts, which we define texturally in the same way as Bramlette [33], Pisciotto [50-51], and Isaacs [13,26,69] as the fine-grained siliceous rocks that are hard, dense, vitreous (or waxy), and break with a conchoidal fracture. Cherts defined in this way are composed of nearly pure silica – generally between 90-99 weight % SiO_2 – and display stratigraphic and structural styles markedly different from the less-pure siliceous rocks. *Porcelanite* is less hard and dense than chert, has a blocky to splintery fracture, and a matte surface texture, like that of unglazed porcelain. Petrographically, the key difference between chert and porcelanite is clay content and porosity [13,40,69,73]. Cherts can be markedly calcareous (up to 45%) and still be characteristically "cherty" providing the silica:detritus ratio exceeds 8:1 [13]. Chert and porcelanite are distinguished independently of silica phase, i.e., both cherts and porcelanites can be composed of opal-CT or quartz. We do not follow the early DSDP tradition of making a mineralogic distinction between chert (quartz) and porcelanite (opal-CT) because the two silica phases can be virtually indistinguishable in the field. For

Table 1.

Silica phase, lithology, chertification process, and XRF data for major oxides of siliceous rocks from the Monterey Formation, Santa Maria area. Major oxides have a lower reporting limit of 0.01%. Relative standard deviation is <1% of measurements. All Cr_2O_3 and Fe_2O_3 values are less than possible contamination from sample preparation.

Sample	* Si phase	Lithology	^ Chertification Process	Na2O %	MgO %	Al2O3 %	SiO2 %	P2O5 %	K2O %	CaO %	TiO2 %	Cr2O3 %	MnO %	Fe2O3 %	† LOI %	Sum %
CT-BC1	CT	glassy black chert lamination in opal-A diatomite	PFC, E	0.01	0.42	0.14	87.90	0.03	0.04	2.13	0.03	0.03	0.01	0.46	9.00	100.2
CT-BC2	CT	glassy black chert lamination in opal-A diatomite	PFC, E	<0.01	0.11	0.07	90.50	0.02	0.05	0.77	0.03	0.03	0.01	0.44	7.70	99.7
G90-1-Fc	CT	brown chert nodule in opal-A diatomite	PFC	<0.01	0.17	0.92	89.10	0.15	0.08	0.42	0.06	0.03	0.01	0.53	7.85	99.3
G90-1-Fd	A	diatomite host	—	0.02	0.32	2.61	84.7	0.46	0.26	0.36	0.13	0.02	0.01	1.04	9.45	99.4
G91-3b	CT	black chert spheroid in opal-A diatomite	PFC, E	<0.01	0.19	0.49	87.40	0.06	0.08	1.17	0.05	0.03	0.02	0.58	9.65	99.8
G91-3c	CT	black chert nodule surrounding spheroid	PFC	<0.01	0.22	0.51	83.00	0.08	0.08	3.24	0.05	0.02	0.01	0.55	12.20	100.0
G91-3d	A	diatomite host	—	0.08	0.36	1.84	81.60	0.27	0.22	2.40	0.10	0.02	0.01	0.65	12.30	99.8
HWY1-100	Q	black chert, breccia fragment	CTR	<0.01	0.08	0.02	95.60	0.04	0.02	0.28	0.03	0.17	0.02	1.23	2.10	99.6
JMEC-c	CT	brown & black chert bed, irregular diagenetic contact	PFC	0.06	0.24	1.43	88.90	0.05	0.23	0.37	0.10	0.02	0.01	0.40	7.45	99.3
JMEC-d	A	diatomite host	—	0.23	0.40	3.31	85.10	0.07	0.41	0.42	0.15	<0.01	0.02	0.59	9.30	100.0
LH90-06	Q	black chert, dolomite-replacement	DR	<0.01	0.11	0.12	96.40	0.03	0.04	0.31	0.03	0.10	0.02	0.76	1.35	99.3
MUS-1p	CT	dark brown cherty porcelanite nodule	PFC	0.85	0.72	2.73	82.00	0.18	0.54	0.86	0.16	0.03	0.02	1.86	9.55	99.5
MUS-1d	A	clayey diatomite, host	—	2.94	1.07	6.10	69.30	0.36	1.25	1.60	0.33	0.02	0.04	1.65	15.40	100.1
MUS-3c	Q	black chert lamination, dolomite-replacement	DR	<0.01	0.58	0.03	93.90	0.03	0.02	1.04	0.03	0.16	0.02	1.20	2.40	99.4
MUS-3d	CT	siliceous dolomite, host	—	0.02	12.10	0.05	41.80	0.04	0.02	18.30	0.02	<0.01	0.02	0.20	28.00	100.6
MUS90-25	Q	black chert lamination in opal-CT chert	CTR	<0.01	0.13	0.03	96.30	0.03	0.03	0.36	0.02	0.20	0.02	1.43	1.50	100.1
MUS90-33	Q	black chert lamination in opal-CT porcelanite	CTR	<0.01	0.10	0.05	96.40	0.02	0.02	0.27	0.03	0.22	0.02	1.58	1.40	100.1
NS	Q	black chert in opal-CT porcelanite	PFC, L	<0.01	0.06	0.11	96.70	0.03	0.04	0.25	0.03	0.11	0.02	0.85	1.50	99.7
NS-3c	Q	black chert in opal-CT porcelanite	PFC, L	<0.01	0.06	0.06	96.80	0.02	0.03	0.24	0.03	0.21	0.02	1.48	1.20	100.2
NS-3p	CT	porcelanite host	—	1.08	0.65	0.29	79.10	0.02	0.28	0.39	0.03	<0.01	0.01	0.23	18.30	100.4

* Silica phase: A = opal-A, CT = opal-CT, Q = quartz.
^ Chertification process & timing: E = early-formed opal-CT chert, PFC = pore-filling cementation, CTR = "normal" opal-CT replacement, DR = dolomite replacement, L = late quartz chert.
† LOI (loss on ignition at 950°C): includes water, hydroxyl, carbonates (e.g., dolomite & calcite), and organic matter.

Table 2.
Silica phase, lithology, inferred chertification process, and sedimentary components calculated from XRF data of Table 1, Monterey Formation, Santa Maria area.

Sample	* Si phase	Lithology	^ Chertification Process	Detritus %	Silica %	†Sedimentary components Al-silic. %	Detri. qtz %	Apatite %	Dolomite %	Calcite %	Silica/Detritus ratio
CT-BC1	CT	glassy black chert lamination in opal-A diatomite	PFC, E	0.8	94.2	0.6	0.2	0.1	2.0	2.9	111.5
CT-BC2	CT	glassy black chert lamination in opal-A diatomite	PFC, E	0.4	97.9	0.3	0.1	0.0	0.5	1.2	230.2
G90-1-Fc	CT	brown chert nodule in opal-A diatomite	PFC	5.6	93.6	4.2	1.4	0.3	0.3	0.2	16.7
G90-1-Fd	A	diatomite host	–	16.1	83.5	12.1	4.0	1.0	0.2	-0.8	5.2
G91-3b	CT	black chert spheroid in opal-A diatomite	PFC, E	3.0	94.4	2.3	0.8	0.1	0.7	1.7	31.2
G91-3c	CT	black chert nodule surrounding spheroid	PFC	3.2	90.1	2.4	0.8	0.2	0.8	5.7	28.4
G91-3d	A	diatomite host	–	11.5	83.7	8.6	2.9	0.6	0.8	3.5	7.3
HWY1-100	Q	black chert, breccia fragment	CTR	0.1	99.2	0.1	0.0	0.1	0.4	0.2	852.9
JMEC-c	CT	brown & black chert bed, irregular diagenetic contact	PFC	8.7	90.7	6.5	2.2	0.0	0.4	0.3	10.5
JMEC-d	A	diatomite host	–	20.1	79.6	15.0	5.0	-0.1	0.2	0.3	4.0
LH90-06	Q	black chert, dolomite-replacement	DR	0.7	98.6	0.5	0.2	0.1	0.5	0.2	142.8
MUS-1p	CT	dark brown cherty porcelanite nodule	PFC	17.0	80.7	12.8	4.3	0.2	2.1	-0.1	4.7
MUS-1d	A	clayey diatomite, host	–	40.2	56.5	30.2	10.1	0.5	2.1	0.7	1.4
MUS-3c	Q	black chert lamination, dolomite-replacement	DR	0.2	96.7	0.1	0.0	0.1	2.7	0.4	558.3
MUS-3d	CT	siliceous dolomite, host	–	0.3	41.7	0.2	0.1	0.1	55.3	2.6	148.7
MUS90-25	Q	black chert lamination in opal-CT chert	CTR	0.2	98.9	0.1	0.0	0.1	0.6	0.3	572.6
MUS90-33	Q	black chert lamination in opal-CT porcelanite	CTR	0.3	99.0	0.2	0.1	0.0	0.4	0.2	343.7
NS	Q	black chert in opal-CT porcelanite	PFC, L	0.6	98.8	0.5	0.2	0.1	0.2	0.3	156.4
NS-3c	Q	black chert in opal-CT porcelanite	PFC, L	0.3	99.1	0.3	0.1	0.0	0.3	0.3	287.5
NS-3p	CT	porcelanite host	–	2.0	95.6	1.5	0.5	0.0	3.5	-1.1	48.1

* Silica phase: A = opal-A, CT = opa-CT, Q = quartz.
^ Chertification process & timing: E = early-formed opal-CT chert, PFC = pore-filling cementation, CTR = "normal" opal-CT replacement, DR = dolomite replacement, L = late quartz chert.
† Sedimentary components. Calculated from XRF major oxides with the formulae of Isaacs et al. (1983), based on Isaacs (1980). Components normalized to 100% on an organic-matter-free basis.
  Detritus = 5.6 x Al2O3
  Aluminosilicates = 4.2 x Al2O3
  Detrital quartz = Aluminosilicates + 3
  Silica (biogenic & authigenic) = SiO2 - (3.5 x Al2O3)
  Apatite = (P2O5 - (0.032 x Al2O3)) + 0.424
  Dolomite = (MgO - (0.11 x Al2O3)) + 0.219
  Calcite = (CaO - (0.08 x Al2O3) - (0.555 x Apatite) - (0.304 x Dolomite)) + 0.56

example, although most black glassy cherts are composed of quartz phase silica, some are formed of opal-CT. The two rocks can only be tentatively distinguished in the field by subtle differences in density and hardness; they must be examined in the laboratory by petrographic microscope or XRD to make positive mineralogic identifications.

Cherts display a variety of colors in the Monterey, ranging from pale grays and yellow-browns to more common dark browns and black. Light coloration generally reflects surficial weathering and leaching of organic matter and carbonates. Quartz cherts are more likely to be black, chiefly due to the tight, interlocking mosaic of microcrystalline quartz which refracts and absorbs incident light, and to the very low matrix porosity (2-15% [30]) which inhibits weathering of inclusions and the development of surficial roughness that reflects incident light. Thin flakes or fragments of black quartz chert, such as seen in drill cuttings, are mostly transparent yellowish brown.

Composition

We analyzed twenty whole-rock samples of opal-CT chert, quartz chert, diatomite, porcelanite and dolomite from the Santa Maria basin for eleven major oxides by XRF (Table 1). Samples were selected to characterize the composition of representative varieties of chert and to compare the composition of chert with its host or protolith to assess the mechanism of diagenesis. In Table 2, oxide weight % is converted to weight % of sedimentary components (i.e., detritus, silica, apatite, dolomite, and calcite) using the formulae of Isaacs and others [74]. "Silica" is the sum of authigenic and biogenic silica. Loss-on-ignition at 950°C includes water, organic matter, and carbonates; the latter is calculated from the oxide %'s. Sedimentary components are normalized to 100% (not including organic matter).

Opal-CT cherts have 90.1 to 97.9% silica (mean 93.5%), 0.4 to 8.7% detritus (mean 2.7%), 0.3 to 2.0% dolomite (mean 0.8%), and 0.2 to 5.7% calcite (mean 2.0%); apatite never exceeds 0.3%. Quartz cherts have 96.7 to 99.2% silica (mean 98.7%), 0.1 to 0.7% detritus (mean 0.3%), 0.2 to 2.7% dolomite (mean 0.7%), and 0.2 to 0.4% calcite (mean 0.3%). Silica:detritus ratios are 10 to 230 (mean 37) for opal-CT cherts and 143 to 853 (mean 416) for quartz cherts.

PETROGRAPHY

Both opal-CT and quartz cherts are composed of dense, fine-grained matrices of diagenetic microcrystalline silica with thin, primary laminations containing minor amounts of amorphous organic matter, calcareous microfossils, or clay minerals, and early diagenetic laminations of dolomite or authigenic phosphate. Petrographically, calcite and dolomite form up to 20% of some cherts. Opal-CT cherts are composed of a cryptocrystalline groundmass of opal-CT blades or needles too small to be resolved with a petrographic microscope, which are locally arranged into discernible 3-15µm lepispheres (Figure 5ab). In thin-section, opal-CT is generally brown to light yellowish gray in plane-polarized light, and pseudoisotropic (extinct) when viewed with crossed polarizers. The brownish hue may be due to the presence of numerous ultrapores in a similar manner to that of chalcedony [75].

Most quartz cherts are transparent in plane-polarized light except where tinted by disseminated organic matter (Figure 6af). Quartz cherts are chiefly composed of an interlocking mosaic of microcrystalline or microflamboyant quartz (Figure 6) with irregular optical domains. Cryptocrystalline quartz occupies isolated patches or laminations, most commonly where associated with clay minerals. In general, the microtexture of quartz cherts reflects primary compositional variations in its protolith. Quartz crystallite size varies inversely with the concentration of disseminated organic matter, clay minerals, pyrite, and secondary dolomite euhedra (Figure 6b-e), probably reflecting different rates of quartz nucleation and growth in the different chemical microenvironments [76].

Length-fast chalcedony is common as void-filling cement within microfossils and fractures (Figure 6fg). Length-slow chalcedony (quartzine) is rare in the Monterey, but occurs locally as a replacement of barite (cf. Keene [77]) or as discrete bands within length-fast chalcedony

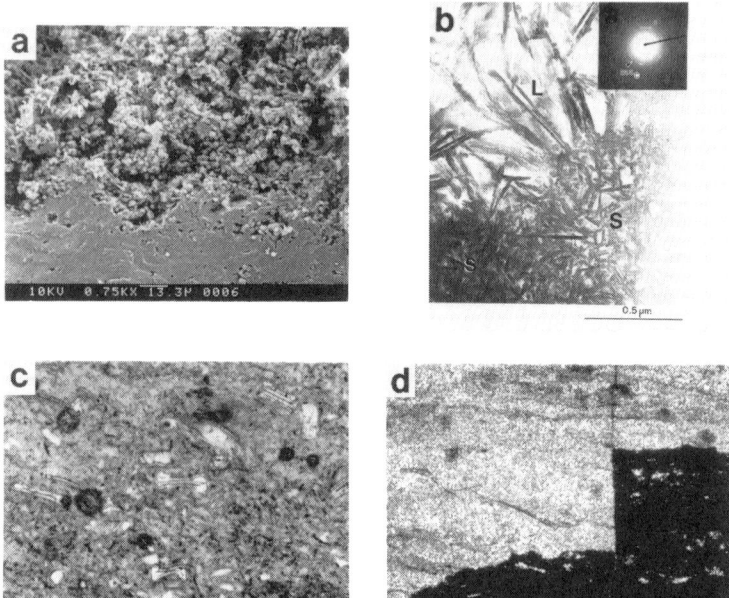

Figure 5. Petrography of opal-CT cherts. A. SEM photomicrograph of coalescing mass of opal-CT lepispheres in diatomite (top) at periphery of opal-CT chert lens. Opal-CT lepispheres cement diatom frustule debris. Continued cementation develops massive opal-CT chert (bottom) with porosity limited to voids less than ~5 μm in diameter. B. TEM cross-sectional image through two opal-CT lepispheres; note larger blades (L) surround densely packed core of smaller blades (S). Selected Area Diffraction ring patterns indicate random orientation of crystallites. C. Photomicrograph of opal-CT chert still preserving robust opal-A diatom frustules. Plane polarized light (PP); field of view (FOV) = 1.5 mm. D. Crossed-nicols (XN) photomicrograph of opal-CT to quartz transformation front in chert. Note that opal-CT is pseudoisotropic (black), quartz is microcrystalline to microflamboyant, and that the transformation front is controlled by the cemented fracture. FOV = 3 mm.

fracture-fillings. With continued diagenesis or metamorphism, both forms of chalcedony tend to neomorphose to equigranular microcrystalline quartz.

FIELD OCCURRENCES

In the Monterey Formation, opal-CT and quartz phase cherts occur as beds, laminations, nodules, lenticular bodies, discordant sheet-veins, breccias, dikes, and spheroids (Figure 7) [25,33,39,40,48,50,69,78]. Simple chert beds typically range from 1 - 50cm in thickness; complex, structurally thickened, chert beds locally exceed 2m. Most chert beds are thinly laminated (0.2 to 5mm thick), and many are interlaminated or interbedded with diatomite, porcelanite, siliceous mudstone, and/or dolomite on scales from millimeters to meters [33,39,50-51].

Chert Structures
Dramatic changes in physical properties (e.g., porosity, density, hardness, and yield strength) accompany the development of chert from diatomite or porcelanite. For example, chertification commonly occurred within diatomaceous sediments still retaining 65% to 80% porosity. Thus, in all situations, chert is markedly harder and stronger than any of the surrounding sediments

Origin of Chert in the Monterey Formation (USA)

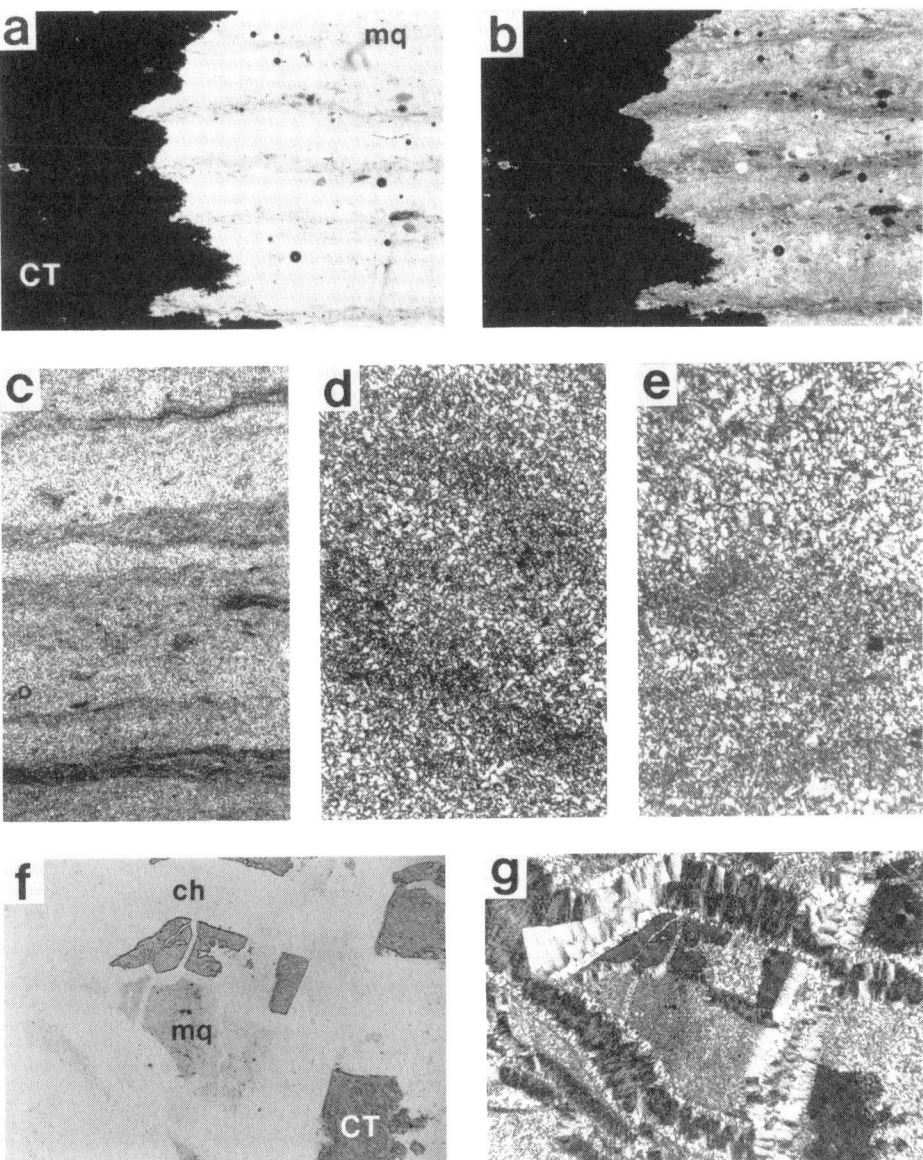

Figure 6. Photomicrographs of quartz cherts. A. Opal-CT to quartz transformation in "late" quartz chert. Opaque, porous opal-CT porcelanite (CT) to left, transparent microcrystalline quartz chert (mq) to right. Note greater advance of chertification front along wispy, organic matter-bearing laminations. PP; field of view (FOV) = 7mm. B. Same as A; XN. C. Size of microflamboyant quartz domains varies inversely with organic content of each thin lamination. Crossed nicols; FOV = 4mm. D,E. Details of variation in microtexture of quartz cherts. Microcrystalline or microflamboyant domains range from 5-30 μm in diameter. XN; FOV = 0.5mm. F. Brecciated and cemented quartz and opal-CT chert. Patches with higher relief are opal-CT chert fragments (CT), lightly stained areas are microcrystalline quartz (mq), and clear areas are chalcedony (ch). PP; FOV = 2mm. G. Same as F, XN.

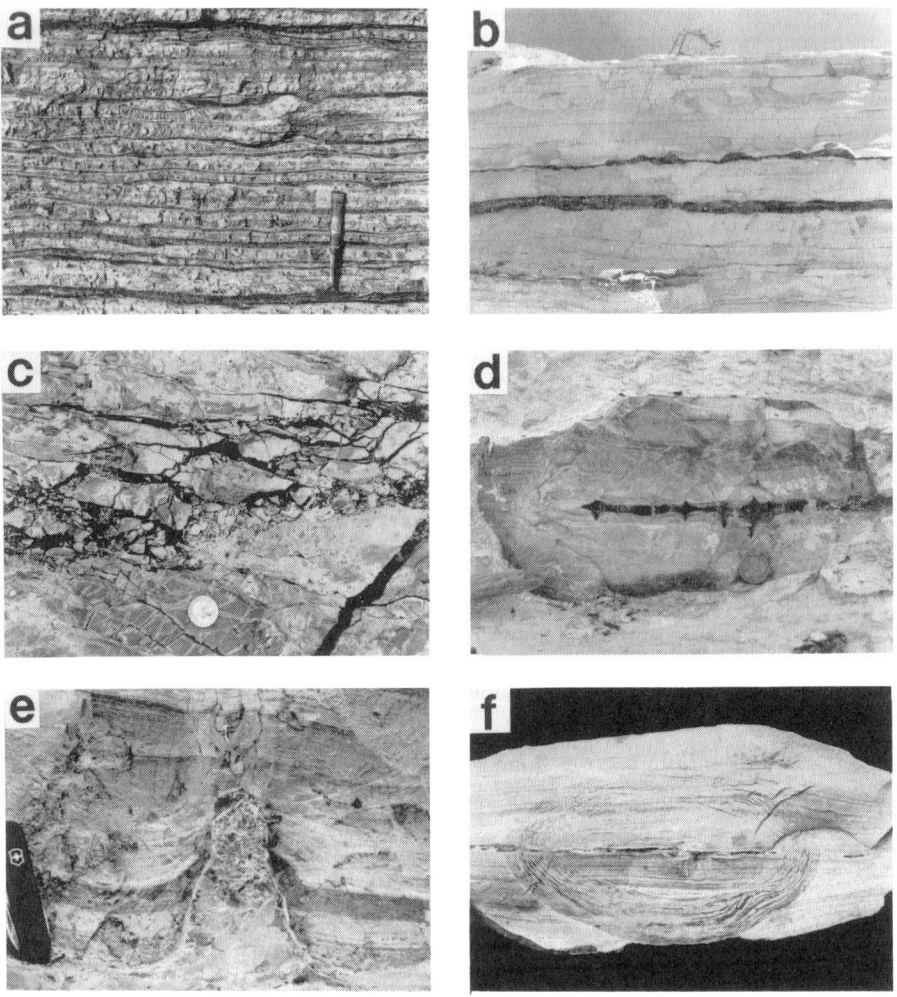

Figure 7. Field occurrences of chert in the Monterey Formation. A. Opal-CT chert beds and lenses interbedded with siliceous (opal-CT) mudstone. Hammer is 30 cm. B. 10 cm-thick opal-CT chert bed and coalesced horizon of nodules in opal-A diatomite. C. Bedding plane view of cross-cutting, tar-cemented, opal-CT chert breccia. Coin is 2.5 cm. D. Black opal-CT chert lamination with "dikes" in later-formed opal-CT chert bed. Over- and underlain by opal-A diatomite; coin is 2 cm. E. Opal-CT chert "dike" in opal-A diatomite. F. Opal-CT chert spheroid in opal-A diatomite.

and under tectonic or burial stress will deform differently than its host rocks. Consequently, the style or degree of deformation is a clue to the diagenetic stage during deformation [79-81] and to the temporal sequence of chertification.

A number of distinctive structures are displayed by cherts in widespread siliceous deposits of contrasting depositional and tectonic settings. These chert structures include breccias, dikes, lineations, contorted beds, and spheroids [cf. 82-84] – all of which occur in the Monterey. The broad distribution of these characteristic structures suggests that they may reflect some of the fundamental features in the process of chertification. Our observations indicate that all of the

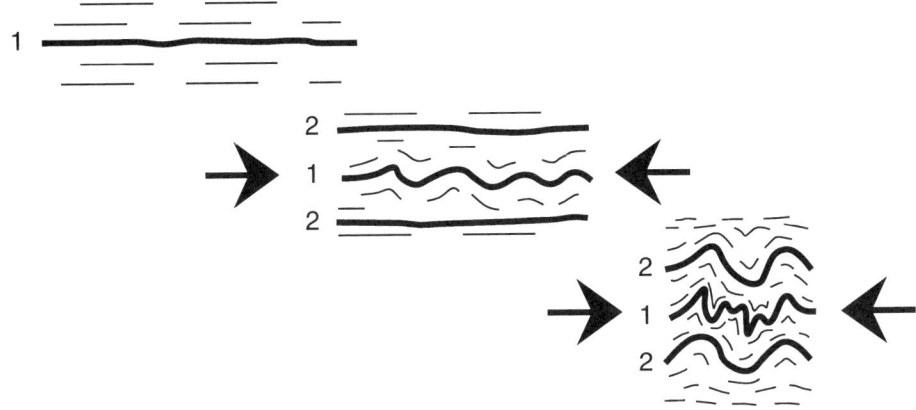

Figure 8. Between similar chert beds in the same structural setting, the bed that has been chertified the longest will display the greater amount of macroscopic buckling and brecciation. Only after a diatomite becomes chert will it experience brittle tectonic deformation. Non-chert beds generally undergo more-subtle, plastic shortening.

structures can be attributed to the contrasting styles of deformation of the chert and host sediment during burial compaction or tectonic strain, and to the temporal relation between deformation and chertification. Unusual properties or precursor states of the chert are not required to explain these structures (cf. [82-86]).

Our interpretations are guided by a few simple rules: 1. chert is always harder and more competent than its precursor sediment or associated sedimentary rocks, therefore it deforms brittlely when strained past its yield point. Consequently, when extended, chert fractures and forms angular boudins, when compressed, it fractures, buckles, and becomes thrust-faulted. In the laminated and thinly bedded Monterey, the spacing of the resulting fractures is generally close (typically 2 mm to 2 cm), giving a high degree of freedom to movement of an entire bed. 2. At any single outcrop, the relative amount of deformation among otherwise similar cherts generally reflects their relative timing of chertification. For example, a fractured and folded chert bed indicates earlier embrittlement (by chertification) than an adjacent undeformed or mildly deformed chert bed (Figure 8). Diatomaceous sediments generally deform plastically under tectonic stresses, but once chertified will only deform brittlely by buckling and brecciation. Therefore, unlike subtly deforming ductile rock, cherts macroscopically display much of their accumulated history of brittle strain. This relationship is useful for determining relative sequences of chertification.

Chert breccias: Opal-CT and quartz chert breccias, which are common in the Monterey Formation and many other siliceous deposits, develop during nearly all stages in the burial history of the sediments [39,79,87]. With potentially long and complex histories of brittle deformation, opal-CT and quartz cherts can develop into intricately fractured and cemented breccias [79], many of which are important as petroleum reservoirs (Figures 7c, 9ab, 13f). Chert breccia fragments are characterized by their sharp, straight to curved edges (Figure 9c); in contrast, brittle deformation that predates chertification is characterized by rougher, irregular fracture surfaces (Figure 9f) typical of porous diatomaceous sediments or porcelanites.

Breccias occur in a variety of forms (Figures 7c, 9). Stratigraphic breccias commonly initiate as individual chert laminations are fractured into numerous rectangular prismatic fragments by tectonic or compactional stresses (Figure 13f-h). As average matrix permeability is exceedingly low for most Monterey lithologies («1 millidarcy) [25,30,88], fracture permeability is critically important as a conduit for dissolved silica. Consequently, brittle deformation and

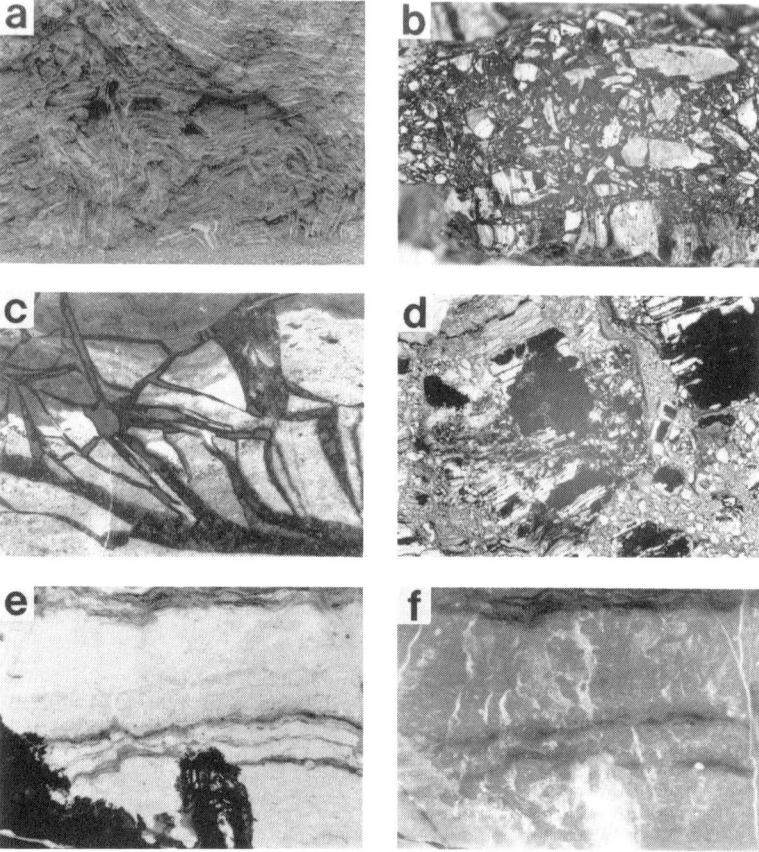

Figure 9. Chert breccias. A. Buckled and brecciated opal-CT chert beds at Sweeney Road. B. Tar-impregnated, cross-cutting opal-CT and quartz phase dilation breccia. Field of view (FOV) = 10cm C. Dolomitic opal-CT chert breccia with chalcedony cement. Note sharp, curved fractures. Fluorescence photomicrograph; FOV = 7mm. D. Slabbed opal-CT dilation breccia cemented with chalcedony and undergoing cross-cutting transformation of opal-CT (white) to quartz (black). FOV = 6cm. E. Photomicrograph (PP) of transparent quartz chert with typical, crinklely organic laminations. Opaque area at bottom left is still opal-CT. FOV = 7mm. F. Fluorescence photomicrograph of E reveals nearly 10% cryptic cementation; rough, irregular fractures suggest the protolith was diatomite when initially fractured and cemented. C, E, and F photographed by J. Stock [89].

brecciation commonly results in repeated cycles of embrittlement, enhanced permeability, fracturing, cementation, and re-embrittlement, which ultimately contributes abundant fracture-filling silica to the total volume of the rock (Figures 6fg, 9cd, 10). This mechanism is an important process in nearly all stages of chertification. In fact, many chert beds that appear undeformed when viewed in outcrop or hand specimen, actually contain 5-40% fracture-filling silica. This "cryptic brecciation" may have occurred at any diagenetic stage, within partially lithified diatomite or quartz chert. In many samples, prior brecciation and cementation is undetectable until examined by special petrographic techniques, such as with blue light fluorescence microscopy (Figure 9ef) [89].

Chert breccias generally record net dilation by the rotation and translation of individual fragments, each rigid enough to support newly opened fractures without collapsing under the lithostatic or tectonic load (Figures 9bcd, 10). If pore fluids contain enough dissolved silica to

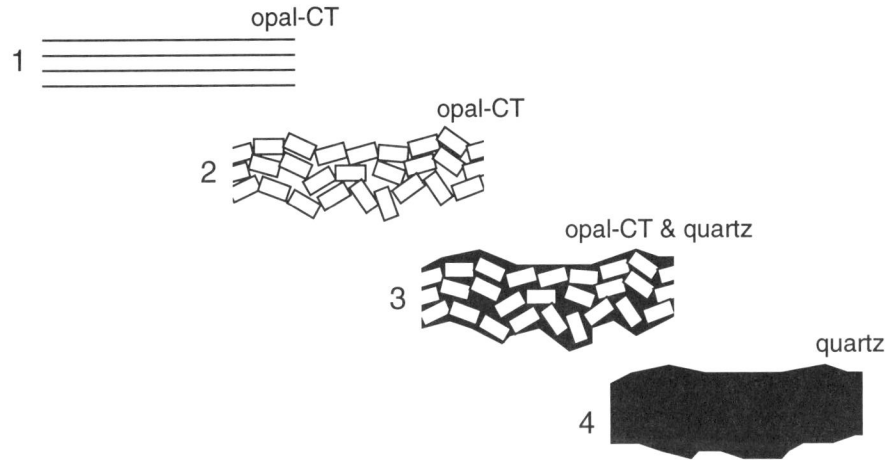

Figure 10. Silica cement is a major component of many chert breccias and beds, even those not obviously fractured. Model for cryptic brecciation and cementation: (1) Formation of thinly laminated opal-CT chert, (2) brecciation, (3) cementation with microcrystalline quartz or chalcedony, and (4) conversion of opal-CT chert fragments to microcrystalline quartz.

be saturated with respect to quartz, chalcedony or opal-CT, then fractures will ultimately be cemented with diagenetic silica. However, if brecciation is followed by an influx of oil, fractures will remain uncemented and chert breccias can provide high-permeability petroleum reservoirs (Figures 7c, 9b).

Chert lineations and dikes: Chert lineations have been previously described in the Cretaceous Mishash and Sayyarim Formations of Israel by Steinitz [83-84], and in the Monterey Formation by Grivetti [39]. In the Monterey, they are small (1-5 mm high), closely spaced ridges that protrude from the bedding surfaces of glassy opal-CT or quartz chert (Figure 11). Most sets of chert lineations have subparallel intersections with bedding surfaces that gently anastomose, rarely deviating more than 10-20° from the principal trend (Figure 11ab). At any one location, lineations maintain a consistent orientation with respect to the chert's bedding surface, but are folded along with bedding (Figure 11d). Compared between different locations, however, the lineation occurs at all possible angles to fold axes [39], indicating that the lineation developed prior to many of the folding events.
Field and petrographic observations suggest that chert lineations form by three different mechanisms (Figure 12); all of which reflect the interconnection of diagenesis with deformation.
Mechanism #1: Many chert lineations form by cementation of closely spaced fractures in thinly bedded chert [39] and preferential silicification of adjacent diatomaceous sediment. Many fracture sets form shallowly, and all greatly increase permeability above that of undisrupted diatomite – 100's millidarcies compared with 0.001 - 1.0 millidarcies [30,90]. In this process, lineation ridges form by cementation of open fractures in the initial chert lamination and pore-filling impregnation of immediately adjacent diatomaceous sediment (Figure 12). Subsequent compaction of unsilicified sediment against projecting lineations warp primary laminations into sub-millimeter-scale folds – a common microstructural feature of cherts.
Mechanism #2 chert lineations are defined by the projecting edges of rotated, angular chert boudins that were deformed in an extensional regime. Subsequent cementation of the extended fragments integrates them into a continuous and lineated chert lamination (Figure 12). In some cases, fragments become separated to form isolated boudins in a plastically deformed matrix.

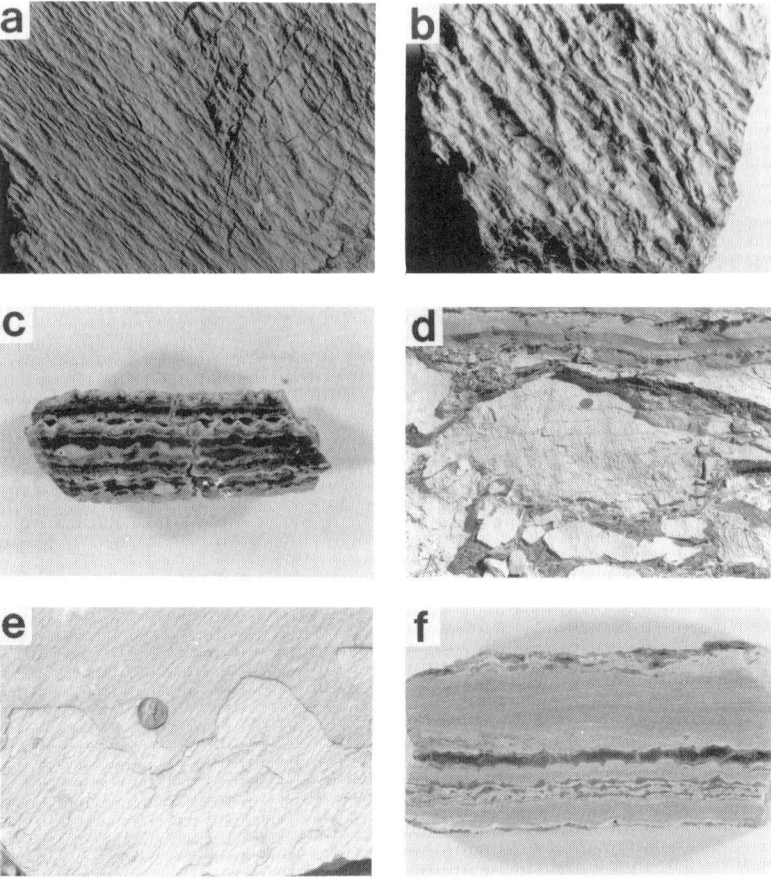

Figure 11. Chert lineations. A, B. Bedding surface view of chert lineations. Field of view (FOV) = 10cm. C. Cross-section of lineated chert. Dolomite is leached from between two chert laminations. FOV = 5cm. D. Folded chert lineations on bedding surface of contorted chert bed. Coin = 2.5cm. E. Crenulation lineation on bedding surface of opal-A diatomite serves as a mold for some chert lineations. Coin = 2cm. F. Polished slab cross-section of lineated chert in dolostone. FOV = 6cm.

In Mechanism #3, chert lineations form by opal-CT impregnation of a preferred lamination in crenulated diatomite. Compressive deformation of diatomaceous sediment prior to or during chertification is demonstrated by a pervasive, crenulation lineation (<0.5 mm amplitude, 1-2 mm wavelength) on most bedding surfaces in the Grefco and Johns-Manville diatomite quarries (Figure 11e) that corresponds exactly with the smallest-scale chert lineations. Microstructural analysis of diatomite suggest that the crenulations result from significant horizontal contraction (approximately 20% at the Grefco Quarry [17]). This degree of shortening is otherwise inconspicuous as there is only a minimal change in the physical properties of the diatomite (e.g., it still retains 65-75% porosity). Silicification of a preferred bed or lamination within the crenulated diatomite will be molded by the template of the pre-existing lineations (Figure 12).

Chert dikes are closely related structures to chert lineations, that have been previously described in the Cretaceous of Israel [83-84]. Like lineations, chert dikes are linear ridges or bulges that

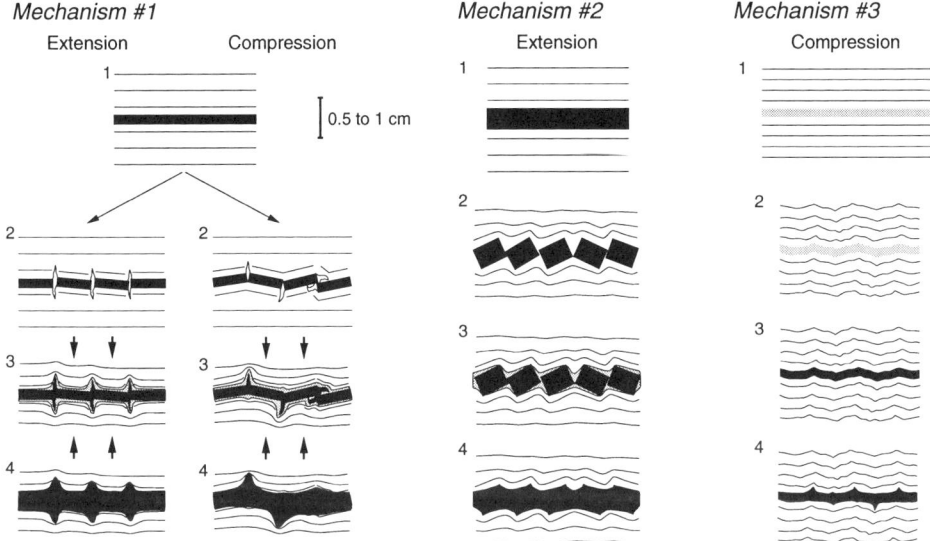

Figure 12. Mechanisms for development of chert lineations. Mechanisms #1 and #2 reflect brittle fracturing and cementation of early formed opal-CT cherts. Mechanism #3 reflects silicification of a preferred layer (light gray) molded by crenulated diatomite. Fine horizontal lines represent primary lamination, solid black represents opal-CT chert, and gray stippling indicates incipient silicification of diatomite. See text for detailed discussion.

project from the upper or lower surfaces of chert beds, but can be significantly larger and give the appearance of penetration through overlying or underlying strata (Figure 7de). In the Monterey Formation, the larger structures are relatively rare, but some extend outward 5-20 cm from the bedding surfaces. Steinitz [83] attributed lineations and dikes to vertical, plastic flow of the chert's precursor material during expansion of an unspecified nature. We interpret these features as early-formed siliceous veins or vertically oriented nodules that develop by cementation of fractures and silicification of adjacent sediment along early faults or joints in a similar fashion to lineation mechanism #1. Axial veins are present in all opal-CT dikes and are still detectable in many of the recrystallized quartz phase structures. Some dikes occur along large-scale joints or faults that cross-cut several meters of strata, but most form along 3 - 10 cm-high, *en echelon* sets of intrastratal microfaults which are common extensional features of the Monterey [33,39,91-92] and many other fine-grained marginal deposits [93-95].

As many cherts form very shallowly (<100 m burial [17]), features that appear to be displacive are more simply explained by differential compaction of diatomite around ridges or prominences of virtually incompressible chert (Figure 7e). Diatomaceous oozes are the most porous of all marine sediments, ranging from 55% to 92% porosity [90,96-99]. Consequently, compaction of an initial thickness of sediment from ~87% to 30% porosity (typical for opal-A to opal-CT phase rocks [100]) will produce up a five-fold shortening [14].

Chert spheroids: Chert spheroids are relatively rare, nodule-like structures present in opal-CT or quartz phase cherts; they display a distinctive concentric banding that cross-cuts primary lamination (Figure 7f) [17,78,101]. This banding consists of two-parts: (1) opal-CT or quartz phase *lithologic chert bands* which contain minor amounts of primary heterogeneities (detritus, calcite, organic debris), as do most cherts. These bands alternate with (2) pure opal-CT, microcrystalline quartz, or chalcedony *spheroidal bands* which are the void-filling cement of curved, spheroidal fractures. Chert spheroids form by the brittle fracture of diatomite around the convex surface of a strain-resistant chert nodule during compaction or tectonic shortening.

Field, petrographic, and oxygen isotope data suggest that chert spheroids are shallowly formed opal-CT nodules that formed in less than 1 m.y. and with less than 100 meters of burial [17,101]. Chert spheroids, or spheroidal fractures, are locally associated with faults, fractures, and chert dikes.

Contorted (excessively folded) chert beds: Many chert beds in the Monterey Formation are intensely contorted, yet occur between unfolded layers of diatomite, porcelanite, mudstone, or dolostone (Figures 7ef, 13) [33,39-40,79,102-103]. In the Santa Maria area, contorted cherts are characterized by upright, symmetric to asymmetric folds, generally parallel to large-scale tectonic structures but with no structurally consistent sense of vergence [39]. Folding is accomplished by buckling and brecciation of the chert beds into small, tabular fragments which are free to independently rotate and slide (Figures 13f-h, 19ab)[39-40,79]. Continued diagenesis (fracture- and pore-filling cementation and opal-CT to quartz transformation) obliterates most textural evidence of brittle deformation, and consequently, many folded cherts appear to have been ductilely folded (Figure 13gh). This style of intraformational folding is restricted to chert and not displayed by other lithologies in the Monterey Formation.

Considering that the chert folds are brecciated, rarely recumbent, and almost never truncated by adjacent beds [39], they could not be synsedimentary slump-folds [104-105] which were later silicified (cf. Redwine [106]). Similarly, the evidence for brittle deformation in the Monterey cherts is difficult to reconcile with plastic volume expansion during silicification, as suggested for similar (but larger scale) contorted cherts in the Cretaceous of Israel [84]. Furthermore, the amount of strain recorded by macroscopic folding of contorted opal-CT cherts is nearly equal with the shortening shown by the microstructural deformation of adjacent opal-A diatomite, indicating that folding could not be due to soft-sediment deformation incurred preferentially by the chert's precursor, or to diagenetic dilation during chertification [17]. Instead, contorted chert beds in the Monterey record the different deformational styles of brittle, strain-resistent chert and ductile diatomaceous sediments when tectonically shortened [39,79]. Consequently, the degree of folding displayed by adjacent chert beds is an indicator of their relative timing of chertification.

CHERTIFICATION

At least two end-member processes of silica diagenesis operated in the Monterey Formation: (1) *in situ* dissolution and reprecipitation of silica, resulting in porosity loss by compaction (siliceous mudstones, porcelanites, and some cherts); this is a compositionally conservative process. And (2), silicification by pore- and fracture-filling cementation or carbonate-replacement, in which silica is transported in an open system. Our data suggest that the process of *chertification* requires the concentration of diagenetic silica to approach the characteristic "cherty" properties of extreme hardness (6-7), brittleness, waxy to vitreous luster, and conchoidal fracture.

Individual Monterey cherts formed by different processes, including: transformation of opal-A to opal-CT, transformation of opal-CT to quartz, pore-filling cementation of diatomite or porcelanite with opal-CT or quartz, the replacement of dolomite by quartz, and cementation of dilation breccias. These processes occured over a wide range of burial depths which are constrained by field relationships and oxygen isotopes.

Carbonate workers have shown that silicification (silica impregnation and replacement) occurs preferentially at sites of increased permeability, foraminifer content, and organic content [66,107-108]. Similarly, cherts develop in biosiliceous sediments where permeability is highest, such as at fractures, in disrupted beds, and among the coarsest microfossils [73,101].

Opal-CT Cherts
In the Monterey Formation, the earliest opal-CT cherts form laminations and thin beds (0.3-2 cm thick) and nodules in the purest calcareous-diatomaceous sediments or along avenues of enhanced permeability. These cherts are typically black and glassy (unless weathered), and are

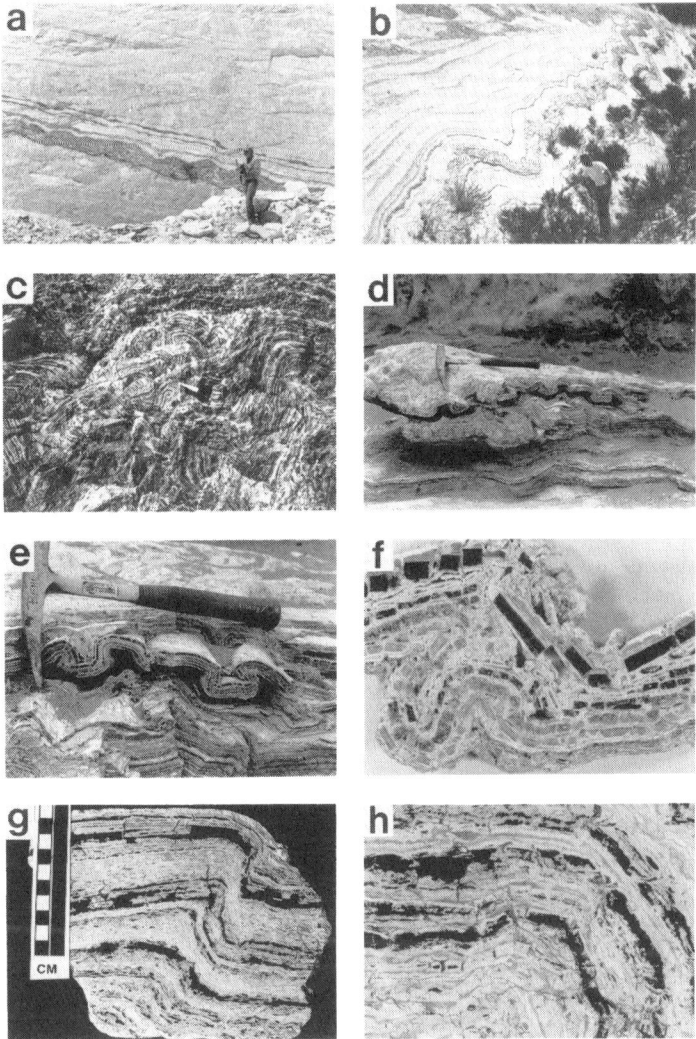

Figure 13. Contorted cherts. A,B. Folded opal-CT cherts over- and underlain by unfolded opal-A diatomite and diatomaceous mudstone. C. Intensely folded, structurally thickened, bed of contorted opal-CT cherts. Key is 7cm. D. Package of folded opal-CT and quartz phase chert overlying gently buckled opal-CT cherts and porcelanites. Hammer is 30cm. E. Detail of D. F. Chert is folded by brecciation into a myriad of fragments, each free to translate or rotate to accommodate the strain. In this sample the fragments are largely uncemented and still retain a high degree of fracture porosity. Transformation from opal-CT (light gray/white) to quartz (black) occurred in individual fragments, from the center outwards; at this point the quartz chert fragments still retain a thin rim of light gray opal-CT. FOV = 7cm. G. That opal-CT cherts folded by brecciation generally becomes obscured with continued diagenesis. Here, post-deformational transformation of opal-CT (white) to quartz (black) followed preferred horizons, cutting across individual fragments and obliterating most macroscopic evidence of brittle deformation. This chert would appear to be ductilely deformed after quartzification has gone to completion. H. Detail of G. FOV = 8cm.

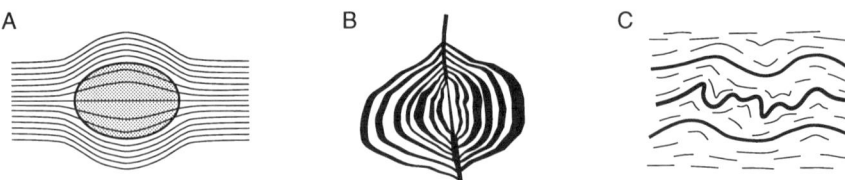

Figure 14. Opal-CT cherts that formed by the early concentration of fracture- or pore-filling silica cement can be recognized by a number of field criteria, including: (A) differential compaction on a bedding- or lamination-scale, (B) spheroidal fractures, and (C) greater tectonic deformation than neighboring, later-formed, chert beds.

easily mistaken for quartz cherts. Early chertification is indicated in the field by differential compaction features or by a greater degree of brittle deformation than other, adjacent chert beds (Figure 14; note difference between central and overlying chert beds in Figure 13a). Later-formed "intermediate" opal-CT cherts also develop before opal-A converts to opal-CT in most diatomaceous strata. However, these cherts form thicker (5-20 cm), generally less intensely deformed beds in noncalcareous diatomaceous strata. Intermediate cherts locally form extensive beds or coalesced lenses that encase the earliest formed and brecciated opal-CT cherts (Figure 7df). The last-formed opal-CT cherts (termed "late" opal-CT cherts) develop as the first part of the opal-A to opal-CT transformation in less-pure siliceous sediments (cf. Figure 4). Late opal-CT cherts generally form continuous beds and laminations instead of discontinuous lenses, nodules, or complex breccias. Clay-rich diatomaceous sediments transform to porcelanite or siliceous mudstone, never to chert [26,33,40,48,69]. Field relations show that all opal-CT cherts (early, intermediate, and late varieties) form before less-pure diatomaceous sediments undergo transformation to opal-CT porcelanite or siliceous mudstone.

Most opal-CT cherts form by the local addition of silica in nodules, lenses, beds, laminations, or at fractures (Figure 5ac). Addition of silica is demonstrated by the compositional differences between opal-CT chert nodules and adjacent diatomite [14-16,48,70 and this study]. Porosity relations and chemical compositions indicate that 2 - 4 times as much silica had to be added to that originally present in a unit volume of diatomaceous sediment to form opal-CT chert [17]. Unfortunately, only nodules can be directly compared along bedding with their protoliths, and their chertification may not be representative of other evenly bedded cherts. Yet all bedded opal-CT cherts that we measured similarly exceeded the concentration of silica in adjacent diatomite (cf. Tables 1 and 2). Moreover, many opal-CT chert beds or laminations display textural evidence of nonuniform silica cementation such as having thicker laminations than adjacent diatomite, lateral changes in lamination thickness, and silica cemented tears and fractures (i.e., lineations and dikes) (Figure 7). It is therefore evident that most pure opal-CT cherts form with the concentration of silica by pore-filling cementation.

In addition to pore-filling cementation, the development and cementation of fractures play a key role in chertification (see Chert Structures section). We find that brittle structural deformation is an integral part of the chertification process in the Monterey Formation, not an unrelated structural topic (Figures 9c-f, 10, 12, 13f-h). Although generally obscured by later diagenesis, void-filling cementation of torn diatomite, fractured chert, or interstratal cracks (due to independent jostling of adjacent, rigid chert laminations) contributes a significant volume of silica to most chert beds (typically 5% to 40% by volume; Figures 9ef, 6ef). Field relations show that relative thickening (dilation) of contorted and brecciated chert beds is a consistent feature of all but the last-formed cherts. Thus the physical properties of chert (extreme hardness and brittleness) play a critical role in the further development of more chert. Consequently, tectonic or slope-related deformation of any siliceous sequence that still contains labile opal-A or opal-CT can result in an increased relative amount of chert by cyclic dilation brecciation and recementation.

Quartz Cherts
Similar to the timing relationship between opal-CT cherts and porcelanites, nearly all quartz cherts formed before the opal-CT to quartz conversion in associated porcelanite. This is demonstrated in many outcrops in the Santa Maria basin where black glassy quartz cherts are interbedded or interlaminated with opal-CT porcelanite; this diagenetic sequence is opposite that suggested for quartz cherts by the data of Isaacs [13,25-26] (cf. Figure 4) in which the purest siliceous rocks are the last to convert from opal-CT to quartz.
Field and petrographic criteria allow distinction of four kinds of quartz chert. Most of these cherts inherit the primary and diagenetic structures of their precursors; for instance, quartz cherts will still display the deformed laminations of a slumped or microfaulted opal-A diatomite bed and, similarly, will preserve differential compaction features that developed around a pre-existing lens of opal-CT chert.

"Normal" opal-CT-replacement quartz chert: Most quartz cherts in the Monterey Formation formed by the *in situ* transformation of opal-CT cherts. That is, the concentration of silica required to form a dense chert took place with the formation of the opal-CT precursor, consequently, little or no additional silica is *required* at this later stage to form quartz cherts. Nonetheless, compositional data (Table 2) show that the average quartz chert has significantly higher silica:detrital ratios than the average opal-CT chert (416 and 37, respectively), suggesting that additional silica was emplaced at the phase transformation or during earlier, but now cryptic, brecciation and cementation.
Initial conversion of opal-CT to quartz follows preferred chert laminations, even where folded or fractured (Figures 13gh, 15b). Microscopically, the transformation generally starts near the center of each opal-CT chert fragment and expands outwards towards the periphery as a continuous reaction surface (Figures 13f-h, 15a-e, 16ef). The transformation to opal-CT apparently slows near the fragment's edge or at cemented fractures, consequently this stage of diagenesis, where opal-CT chert fragments have mostly converted to quartz yet still retain a thin rim of opal-CT, is commonly displayed in hand sample and thin-section (Figure 15, 22ef). Less frequently, opal-CT chert is replaced by microcrystalline quartz growing from a large number of disseminated nucleation sites (Figure 15de). These patches rapidly coalesce into continuous microcrystalline quartz chert and consequently this transient stage is rarely preserved.

Shale-associated quartz cherts: Shale-associated quartz cherts are a subset of "normal" quartz cherts which also formed chiefly by the replacement of pre-existing opal-CT cherts. However, they also probably gain additional silica from interlaminated clay- and organic matter-rich mudrocks which still contain higher-solubility opal-CT after the cherts have converted to quartz. As cryptocrystalline opal-CT (in the mudstone) is three times as soluble as microcrystalline quartz in chert (20-30 ppm and 6-10 ppm at 25°C, respectively [61]), diffusive transport from mud-rich to chert layers would allow growth of quartz crystallites at the expence of opal-CT. This mechanism may be important for diagenetic enhancement of primary bedding in the Monterey Formation [14,79]. Enhanced rates of opal-CT dissolution in the mudrocks may also result from pressure solution, locally intensified by compaction between irregular protruding surfaces of overlying and underlying quartz chert (e.g., chert lineations or nodular beds). Evidence of burial-load or tectonically induced pressure solution, such as bedding-parallel stylolites and high-angle clay seams, are common in porcelanite and siliceous mudstone [50,70,79,109], but rare or absent in cherts. Differential compaction, silicification of pressure shadows in mud-rich laminations, and incorporation of clay- and organic-rich seams into chert layers are additional indicators of pressure solution and reprecipitation of silica [79,81].

Dolomite-replacement quartz cherts: Quartz cherts in the Monterey Formation also formed by replacement of dolomite [17,39,49-50]. In outcrop, quartz cherts replace thick dolostone beds and nodules in irregular cross-cutting patterns (Figure 16a) similar to carbonate-replacement flints in the the Cretaceous chalk of Europe or the Pacific basin [73]. Quartz cherts also replace heterogeneous siliceous dolomites or dolomitic cherts, composed of thinly interlaminated

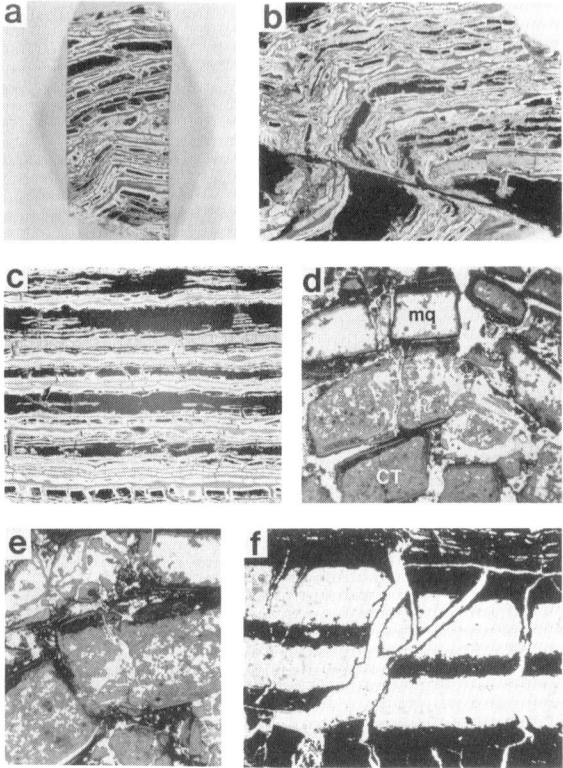

Figure 15. "Normal" opal-CT-replacement quartz cherts. A-C are slabs. D-F are photomicrographs. A. The centers of opal-CT (white) chert fragments are first replaced by microcrystalline quartz (black), leaving the rims as opal-CT. 5cm high. B. As individual fragments are completely replaced the transformation front extends along preferred laminations, incorporating the breccia fragments. FOV = 5cm. C. Progressive opal-CT to quartz transformation produces continuous and thicker layers of black quartz chert. FOV = 6cm. D, E. Replacement of folded and brecciated opal-CT chert with microcrystalline quartz. Individual fragments are in nearly all stages of the transformation. Gray fragments are opal-CT (CT), fragments near center of both photomicrographs display incipient, patchy replacement of opal-CT with quartz, and transparent (white) fragments are entirely converted to microcrystalline quartz (mq) except for their outermost rims. PP; FOV = 8mm for D, 5mm for E. F. Opal-CT chert laminations (dark) largely replaced by microcrystalline quartz (light), except for at rims in contact with chalcedony cemented fractures (light). XN; FOV = 12mm.

dolomite and opal-CT, with replacement occuring first in the more siliceous layers, then cutting discordantly across the purer dolomite (Figure 16bc).

Late quartz chert: Dense, vitreous, black quartz cherts form by pore-filling cementation and replacement of relatively pure opal-CT porcelanites in many locations in the Santa Maria area [39] (Figure 6ab). Late quartz cherts also require a significant local addition of silica. We consider these to be late-forming quartz cherts because of the lack of differential compaction features across diagenetic fronts between porous porcelanite and dense chert. "Late" probably means at the cessation of continued burial and, in some cases, during uplift. The porcelanite protolith is an unusual type of siliceous rock, because it is composed of virtually pure opal-CT (Table 2, sample NS-3p), yet, judging by its hardness and density, it retains relatively high

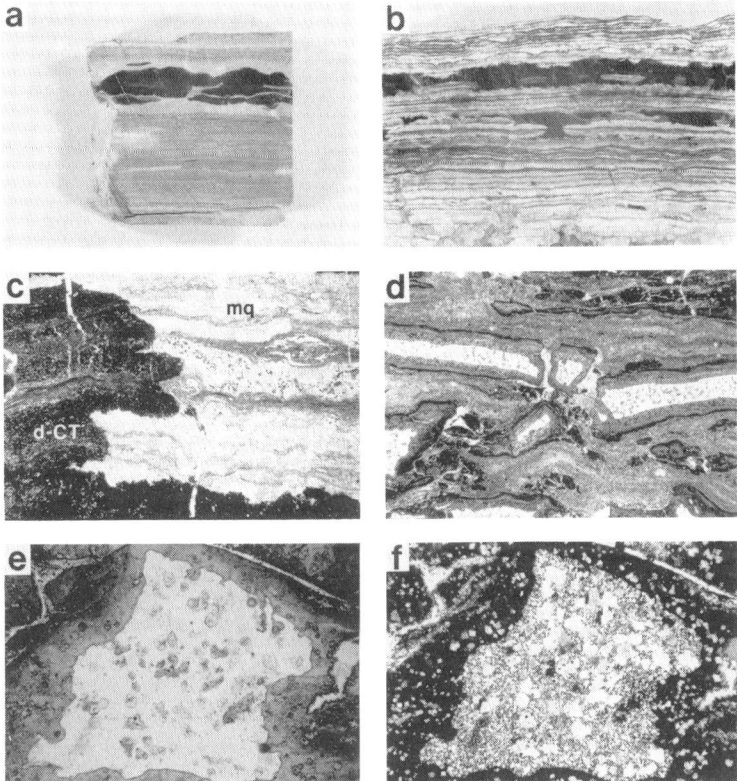

Figure 16. Dolomite-replacement quartz cherts. A. Irregular replacement of laminated dolostone bed with quartz chert (black). Width of specimen is 4 cm. B. Laminar and cross-cutting quartz replacement of thinly laminated dolomite (gray) and dolomitic opal-CT chert (light gray to white). Width is 8 cm. C. XN Photomicrograph of B showing replacement of dolomite and opal-CT (d-CT) by microcrystalline quartz chert (mq). FOV = 11mm. D,E. Interlaminated chert and dolomite, in which the chert has deformed brittlely, while the dolomite flowed plastically. Central parts of chert fragments have converted to microcrystalline quartz (clear). Dolomite euhedra in both opal-CT and quartz parts of the chert are much larger than crystallites in the dolomite groundmass. PP; FOV = 9mm in D, 2mm in E. F. Same as E, but with XN.

porosity. Unlike other matte textured, soft porcelanites, it contains only 2% detritus. It is unclear why these sediments did not form dense opal-CT chert during earlier diagenesis.

OXYGEN ISOTOPIC COMPOSITION OF CHERTS

We analyzed 47 carefully selected samples for their oxygen isotopic compositions in order to verify and quantify the chertification sequence inferred from the described stratigraphic, structural, and petrographic relations. Documentation of the stratigraphic and diagenetic setting of each sample allows interpretation of the broad range of $\delta^{18}O$ values in terms of process and timing of chertification. O-isotope data is presented in tabular form in Behl [17].

General isotopic results

Opal-CT cherts (n=5) range from $\delta^{18}O$ = 34.6 to 38.2‰, recording higher $\delta^{18}O$-values than previously reported for opal-CT chert from the Monterey (Figure 17). These high values directly reflect our focus on those opal-CT samples for which there is field evidence for early chertification (i.e., chert spheroids and vitreous, fractured chert laminations in calcareous-diatomaceous sediments). Because of the limited number of opal-CT analyses, our data is combined with that of Pisciotto [50-51], who also sampled from the Santa Maria area; his opal-CT chert measurements (n=3) ranged from $\delta^{18}O$ = 30.2 to 33.7‰, giving an integrated range of 30.2 to 38.2‰ (n=8).

Quartz cherts (n=42) vary from $\delta^{18}O$ = 24.2 to 31.2‰, which is in good agreement with the complete range reported in previous studies. The four types of quartz chert have overlapping, but distinct, isotopic ranges (Figure 17). $\delta^{18}O$ of shale-associated cherts ranges from 28.2-31.2‰, dolomite-replacement quartz cherts from 24.5-30.1‰, normal opal-CT-replacement quartz cherts from 24.2-30.7‰, and late quartz cherts from 24.8-29.8‰. A single "early formed" quartz chert clast from an intraformational conglomerate at Gaviota Beach has $\delta^{18}O$ = 30.3‰, precisely in agreement with measurements from the same outcrop by Haimson [110].

The broad ranges of the dolomite-replacement, normal, and late quartz cherts are chiefly due to a few unrepresentative outlier values (Figure 17). Three extremely low values probably reflect their stratigraphic position in the phosphatic member of the Monterey at the Lions Head section. Early phosphogenesis and dolomitization in this member, in which planktonic carbonate was remobilized and incorporated into authigenic carbonate fluorapatite and dolomite [111-112], could have driven shallow pore waters towards more negative values [113]. If so, the reduced $\delta^{18}O$ values of phosphatic-member cherts would reflect equilibration with ^{18}O-depleted pore water instead of higher temperatures-of-formation. These "outlier" values are disregarded as unrepresentative in our interpretation of chertification temperatures.

Conversely, two brecciated samples of opal-CT-replacement chert and one late chert have anomalously high $\delta^{18}O$ values. We interpret these compositions as due to precipitation of fracture- and pore-filling quartz at reduced temperatures during tectonic uplift. One other opal-CT-replacement chert had anomalously high $\delta^{18}O$ values although it did not appear petrographically or stratigraphically unusual; this sample is also considered anomalous for the temperature calculations.

Temperatures of chert formation

Assumptions: The $\delta^{18}O$ of chert depends on temperature-dependent isotopic fractionation during precipitation and the composition of the pore water with which it equilibrated. Unlike the calcite-water system, there are many different equations that relate temperature to oxygen isotope fractionation for the silica-water system; these give up to 13°C different temperatures for the same input values. In this study, temperatures-of-formation are calculated with an intermediate expression [114] that yields geologically reasonable values.

To estimate the isotopic composition of Monterey pore waters we modified the pore water evolution model of Winters & Knauth [115] to reflect abrupt decreases in porosity at the bulk silica phase changes at 44°C and 82°C (see [17] for further discussion). This pore-water model maintains an effectively infinite water/rock ratio for opal CT chert formation in highly porous diatomaceous sediments (< 44°C) and an intermediate ratio during quartz chert formation in chiefly opal-CT phase rocks (at 44° to 82°C). Boundary conditions of $\delta^{18}O_{seawater}$ = -0.7‰ SMOW [116] and bottom-water temperature = 7°C (B.P. Flower, 1992, personal commun.; [117]) set $\delta^{18}O_{pore\ water}$ = -0.7‰ during formation of opal-CT cherts and +1‰ during equilibration of quartz cherts.

Temperature of chertification: The isotopic composition of opal-CT cherts indicates that chertification occurred between 2-33°C (Figure 18). The earliest-formed opal-CT cherts (i.e., chert spheroids and deformed, glassy black chert laminations) precipitated at <6°C, essentially

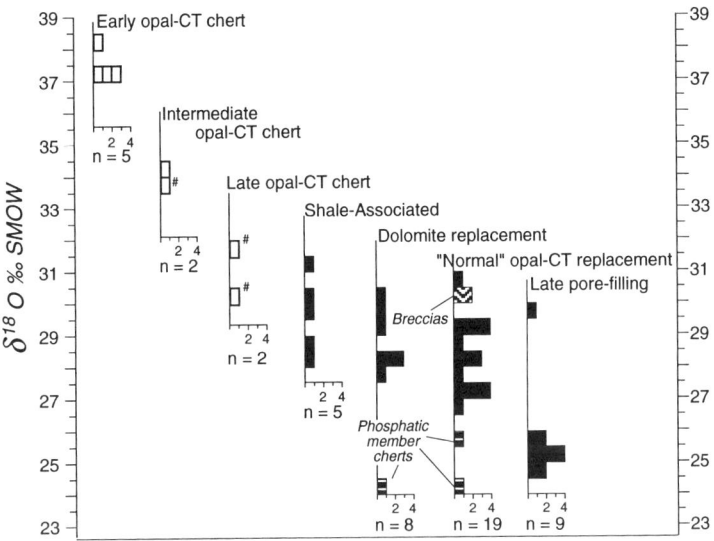

Figure 17. Oxygen isotopic compositions of Monterey Formation opal-CT (white) and quartz (black) cherts plotted in 0.5‰ intervals by field classification. # = oxygen isotope data point from Pisciotto [50-51].

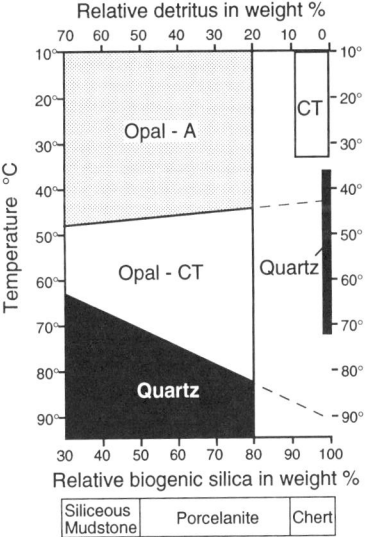

Figure 18. Diagram showing the timing and temperature of silica phase changes, modified from Keller & Isaacs [118] to include data on cherts from this study. Composition and temperatures-of-formation of opal-CT and quartz cherts are respectively represented by white and black rectangles at right. Compare with Figure 4.

the same temperature as bottom water during deposition of the upper Monterey sediments. This low-temperature (shallow burial) origin is supported by porosity, stratigraphic, and physical properties analysis of chert spheroids [17,101]. Late opal-CT cherts precipitated between 25-33°C, before the opal-A to opal-CT transformation in porcelanites and mudrocks at about 44°C. Quartz cherts formed between 36-76°C. The temperatures-of-formation of shale-associated quartz cherts ranged from 36-52°C, dolomite-replacement quartz cherts ranged from 42-53°C, "normal" opal-CT-replacement quartz cherts precipitated at 45-61°C, and 8 of 9 late quartz cherts formed between 65-72°C.

This data shows that all of the quartz cherts in this study formed at significantly lower temperatures (and consequently shallower burial) than predicted by the models of Isaacs [13,26] and Keller & Isaacs [118] (Figure 18). The oxygen isotope derived sequence is verified by the stratigraphic distribution of quartz cherts interbedded with opal-CT mudrocks and porcelanites at Mussel Rock, Purisima Point, and Point Pedernales.

SYNTHESIS

By combination of field, petrographic and oxygen isotopic analyses, this study shows that pure cherts (>90% silica) in the Monterey Formation follow a number of diagenetic pathways that are different than that of porcelanites and siliceous mudrocks. This data allows definition of a complete sequence of chertification within a quantitative temperature framework. Monterey cherts formed only in the purest biogenous (diatomaceous or calcareous-diatomaceous) sediments that accumulated undiluted by terrigenous components beneath high productivity, upwelling systems. Detritus-free chert protoliths are only found in the outermost basins, or portions of basins, of the Neogene California continental borderland. In the Santa Maria basin, the best developed cherts are stratigraphically limited to the lowstand between 11Ma and 9 Ma.[36], suggesting that the strength of coastal upwelling increased with the latitudinal thermal gradient [119] or that diatomaceous sediments (with low grain density) were more efficiently winnowed into the deep, outer basins by intensified bottom currents [38].

Most opal-CT cherts formed by the local concentration of silica as pore-filling or fracture-filling cement within porous opal-A diatomite, as opposed to the *in situ* silica phase transformation responsible for the diagenesis of porcelanite and siliceous mudstone. Opal-CT cherts formed rapidly, with only shallow burial, at temperatures between 2-33°C. The first-formed, early opal-CT cherts developed in calcareous-siliceous sediments and at sites of enhanced permeability, such as at fractures or in beds containing larger microfossils (e.g., foraminifers, sponge spicules, or radiolarians). Later-formed opal-CT cherts are located in the purest diatomaceous strata and in slightly less-pure sediments adjacent to high-permeability, brecciated early cherts. Only after the formation of opal-CT cherts did opal-CT porcelanites form from less-pure diatomaceous sediments. Because fracture permeability is of key importance for the concentration of silica in low-permeability diatomite and porcelanite, brittle deformation in response to tectonic and synsedimentary stress is a critical factor in chertification. The relative timing of opal-CT chertification and tectonic deformation controls the development of a variety of characteristic chert structures, including breccias, contorted beds, dikes, lineations, and spheroids; all of which form by a combination of brittle deformation and silica cementation.

Quartz cherts also form over a range of depths and temperatures by a variety of processes, several of which also require local addition of silica, namely dolomite-replacement and late pore-filling quartz cementation. Almost all quartz cherts form before the transformation of opal-CT to quartz in associated porcelanite and siliceous mudrocks, opposite to the sequence suggested by the data of Isaacs [13,26]. An overlapping sequence of different kinds of quartz chert (shale-associated, dolomite-replacement, normal opal-CT-replacement, and late pore-filling) form between 36-76°C.

This study shows that bedded cherts record a wide range of diagenetic settings and burial depths, not simply the geochemistry of the primary depositional environment. Our findings indicate that the composition and physical properties of chert is principally the result of postdepositional processes, i.e., chemical diagenesis and structural deformation. Even though the extreme concentration of silica is a diagenetic process, chertification occured almost exclusively in preferred layers controlled by the composition (calcareous-diatomaceous or pure diatomaceous sediments) and/or permeability of the primary sediments. Furthermore, the relationship between the site and sequence of chertification and the timing of deformation has ramifications for petroleum geologists; whereas early-formed and brecciated cherts form important high-permeability reservoirs in the Monterey Formation that are key to the economic exploitation of heavy oils, later chertification only reduces the effectiveness and potential of a fractured reservoir.

Acknowledgements
This work is part of doctoral research at the Earth Sciences Department of the University of California, Santa Cruz by RJB under the guidance of REG. Acknowledgement is made to the Donors of The Petroleum Research Fund, administered by the American Chemical Society, for the support of this research (Grant 18439-AC2 to R.E. Garrison). Additional support was generously provided to RJB by an Achievement Rewards for College Scientists Scholarship, a Dissertation Year Fellowship from the Institute of Marine Sciences (UC Santa Cruz), and by a grant from Mobil Oil Exploration and Production. Gideon Steinitz's detailed work on the cherts of Israel stimulated much of this investigation of chert structures as windows to the diagenetic process. Jonathan Krupp and Eugenio Gonzáles provided invaluable support with electron microscopy, photography, and sample preparation. Oxygen isotope analyses were performed at the Berkeley Center for Isotope Geochemistry in collaboration with Brian M. Smith. We thank David Jenkins for his kind grant of access to the Grefco Palo Colorado diatomite quarry. An earlier version of this manuscript benefited from reviews by C.M. Isaacs and M.L. Delaney.

REFERENCES

1. R. Hesse. Origin of chert, I. Diagenesis of biogenic siliceous sediments, *Geosci. Can.* **15**, 171-192 (1988).
2. R. Hesse. Silica diagenesis: origin of inorganic and replacement cherts, *Earth-Sci. Rev.* **26**, 253-284 (1989).
3. L.P. Knauth and S. Epstein. Hydrogen and oxygen isotope ratios in nodular and bedded cherts, *Geochim. Cosmochim. Acta* **40**, 1095-1108 (1976).
4. R.W. Murray, M.R. Buchholtz ten Brink, D.C. Gerlach, G.P. Russ, III, and D.L. Jones. Rare earth, major, and trace elements in chert from the Franciscan Complex and Monterey Group, California: Assessing REE sources to fine-grained marine sediments, *Geochim. Cosmochim. Acta* **55**, 1875-1895 (1991).
5. J.R. Hein and J.T. Parrish. Distribution of siliceous deposits in space and time. In: *Siliceous Sedimentary Rock-Hosted Ores and Petroleum*. J.R. Hein (Ed.). pp.10-57, Van Nostrand Co., New York (1987).
6. H.R Grunau. Radiolarian cherts and associated rocks in space and time, *Eclog. Geol. Helv.* **58**, 157-208 (1965).
7. M. Steinberg. Biosiliceous sedimentation, radiolarite periods and silica budget fluctuations, *Ocean. Acta* **SP**, 149-154 (1981).
8. B. McGowran. Silica burp in the Eocene ocean, *Geology* **17**, 857-860 (1989).
9. W.E. Crain, W.E. Mero and D. Patterson. Geology of the Point Arguello field. In: *Cenozoic Basin Development of Coastal California: Rubey Volume VI*. R.V.Ingersoll and W.E. Ernst (Eds). pp.407-426, Prentice-Hall, Englewood Cliffs (1987).
10. S.W. Wise, B.F. Buie and F.M. Weaver. Chemically precipitated sedimentary cristobalite and the origin of chert, *Eclog. Geol. Helv.* **65**, 157-163 (1972).
11. S.W. Wise and F.M.Weaver. Chertification of oceanic sediments, Int. Ass. Sediment., vol.1, pp.301-326. Blackwell Scientific, Oxford (1974).
12. R.E. Garrison. Radiolarian cherts, pelagic limestones, and igneous rocks in eugeosynclinal assemblages. Int. Ass. Sediment., vol.1, pp.367-399. Blackwell Scientific, Oxford (1974).
13. C.M. Isaacs. Influence of rock composition on kinetics of silica phase changes in the Monterey Formation, Santa Barbara area, California, *Geology* **10**, 304-308 (1982).
14. R. Tada. Compaction and cementation in siliceous rocks and their possible effect on bedding enhancement. In: *Cycle and Event in Stratigraphy*. G. Einsele, W. Ticken and A. Seilacher (Eds). pp.480-491, Springer-Verlag, Berlin (1991).
15. R.W. Murray, M.R. Buchholtz ten Brink, D.C. Gerlach, G.P. Russ, III and D.L. Jones. Rare earth, major, and trace elements in chert from the Franciscan Complex and Monterey Group, California: Assessing the influence of chemical fractionation during diagenesis, *Geochim. Cosmochim. Acta* **56**, 2657-2671 (1992).

16. R.W. Murray, D.L. Jones and M.R. Buchholtz ten Brink. Diagenetic formation of bedded chert: evidence from chemistry of the chert-shale couplet, *Geology* **20**, 271-274 (1992).
17. R.J. Behl. *Chertification in the Monterey Formation of California and Deep-Sea Sediments of the West Pacific.* Unpublished Ph.D. thesis, University of California, Santa Cruz, California (1992).
18. J.A. Barron. Paleoceanographic and tectonic controls on the deposition of the Monterey Formation and related siliceous rocks in California, *Paleocean. Paleoclim. Paleoecol.* **53**, 27-45 (1986).
19. J.C. Ingle, Jr. Cenozoic depositional history of the northern Continental Borderland of southern California and the origin of associated Miocene diatomites. In: *Guide to the Monterey Formation in the California coastal area, Ventura to San Luis Obispo.* C.M. Isaacs (Ed.).vol.52, pp.1-8. Pac. Sec. Amer. Assoc. Petrol. Geol., Los Angeles (1981).
20. J.C. Ingle, Jr. Origin of Neogene diatomites around the north Pacific rim. In: *The Monterey Formation and Related Siliceous Rocks of California.* R.E. Garrison, R.G. Douglas, K.E. Pisciotto, C.M. Isaacs and J.C. Ingle, Jr. (Eds). vol.15, pp.159-179. Pac. Sec. SEPM, Los Angeles (1981).
21. J.P. Kennett. *Marine Geology.* Prentice-Hall, Englewood Cliffs (1980).
22. K.A. Pisciotto and R.E. Garrison. Lithofacies and depositional environments of the Monterey Formation, California. In: *The Monterey Formation and Related Siliceous Rocks of California.* R.E. Garrison, R.G. Douglas, K.E. Pisciotto, C.M. Isaacs and J.C. Ingle, Jr. (Eds). vol.15, pp.97-122. Pac. Sec. SEPM, Los Angeles (1981).
23. G. Keller and J.A. Barron. Paleoceanographic implications of deep-sea hiatuses, *Bull. Geol. Soc. Am.* **94**, 590-613 (1983).
24. F. Woodruff and S.M. Savin. Miocene deepwater oceanography, *Paleoceanography.* **4**, 87-140 (1989).
25. C.M. Isaacs. *Diagenesis in the Monterey Formation Examined Laterally Along the Coast Near Santa Barbara, California.* Unpublished Ph.D. thesis, Stanford University, Stanford (1980).
26. C.M. Isaacs. Outline of diagenesis in the Monterey Formation examined laterally along the Santa Barbara Coast, California. In: *Guide to the Monterey Formation in the California coastal area, Ventura to San Luis Obispo.* C.M. Isaacs (Ed.).vol.52, pp.25-38. Pac. Sec. Amer. Assoc. Petrol. Geol., Los Angeles (1981).
27. C.M. Isaacs, K.A. Pisciotto and R.E. Garrison. Facies and diagenesis of the Miocene Monterey Formation, California: a summary. In: *Siliceous Deposits in the Pacific Region.* A. Iijima, J.R. Hein and R. Siever (Eds). Developments in Sedimentology, vol.36, pp.247-282. Elsevier, Amsterdam (1983).
28. C.M. Isaacs. Lithostratigraphy of the Monterey Formation, Goleta to Point Conception, Santa Barbara Coast, California. In: *Guide to the Monterey Formation in the California coastal area, Ventura to San Luis Obispo.* C.M. Isaacs (Ed.). vol.52, pp.9-23. Pac. Sec. Amer. Assoc. Petrol. Geol., Los Angeles (1981)
29. C.M. Isaacs. Compositional variation and sequence in the Miocene Monterey Formation, Santa Barbara coastal area, California. In: *Fine-Grained Sediments: Deep Water Processes and Environments.* D.K. Larue and R.J. Steel (Eds). pp.117-132. Pac. Sec. Amer. Assoc. Petrol. Geol., Los Angeles (1983).
30. T.C. MacKinnon. Petroleum geology of the Monterey Formation in the Santa Maria. In: *Oil in the California Monterey Formation. Field Trip Guidebook.* T.C. MacKinnon (Ed.). vol.T311, pp.11-27. American Geophysical Union, Washington, D.C. (1989).
31. R.E. Garrison. Pelagic and hemipelagic sedimentary rocks as source and reservoir rock. In: *Deep Marine Sedimentation: Depositional Models and Case Histories in Hydrocarbon Exploration and Development.* G.C. Brown, D.S. Gorsline and W.J. Schweller (Eds). vol. 66, pp.123-149, Pac. Sec. SEPM, Los Angeles (1990).
32. R.E. Garrison and J.C. Ingle, Jr. Sedimentology, depositional environments, and biostratigraphic studies. In: *The Geochemical and Paleoenvironmental History of the Monterey Formation: Sediments and Hydrocarbons. V. 2, Data Volume.* pp.21-98. Global Geochemistry Corp., Canoga Park, California (1985).
33. M.N. Bramlette. The Monterey Formation of California and the origin of its siliceous rocks, *U.S. Geol. Surv. Prof. Paper*, **212**, 1-57 (1946).
34. R.M. Kleinpell. *Miocene Stratigraphy of California.* Amer. Assoc. Petrol. Geol., Tulsa (1938).
35. J.A. Barron. Updated diatom biostratigraphy for the Monterey Formation of California. In: *Siliceous microfossils and microplankton studies of the Monterey Formation and modern analogs.* R.E. Casey and J.A. Barron (Eds). vol.45, pp.105-119. Pac. Sec. SEPM, Los Angeles (1986).
36. L.D. White. *Chronostratigraphic and Paleoceanographic Aspects of Selected Chert Intervals in the Miocene Monterey Formation, California.* Unpublished Ph.D. thesis, University of California, Santa Cruz, California (1989).
37. B.U. Haq, J. Hardenbol and P.R. Vail. Chronology of fluctuating sea levels since the Triassic (chart version 31A), *Science* **235**, 1156-1166 (1987).
38. J.S. Hornafius. Facies analysis of the Monterey Formation in the northern Santa Barbara channel, *Amer. Assoc. Petrol. Geol. Bull.* **75**, 894-909 (1991).
39. M.C. Grivetti. *Aspects of stratigraphy, diagenesis, and deformation in the Monterey Formation near Santa Maria - Lompoc, California.* Unpublished M.A. thesis, University of California, Santa Barbara, California (1982).

40. J.B. Dunham and G.H. Blake. Guide to coastal outcrops of the Monterey Formation of western Santa Barbara county, California. In: *Guide to coastal outcrops of the Monterey Formation of western Santa Barbara county, California.* J. B. Dunham (Ed.). vol.53, pp.1-36. Pac. Sec. SEPM, Los Angeles (1987).
41. S.E. Calvert. Composition and origin of North Atlantic deep sea cherts, *Contri. Min. Petr.* **33**, 273-288 (1971).
42. S.E. Calvert. (1974) Deposition and diagenesis of silica in marine sediments. Int. Ass. Sediment., vol.1, pp.273-299. Blackwell Scientific, Oxford (1974).
43. J.B. Jones and E.R. Segnit. The nature of opal: I. Nomenclature and constituent phases, *J. Geol. Soc. Australia* **18**, 56-68 (1971).
44. G.R. Heath and R. Moberly, Jr. Cherts from the western Pacific, Leg 7, Deep Sea Drilling Project. In: *Init. Repts. DSDP.* vol.7, pp.991-1007. U.S. Govt. Printing Office, Washington, D.C. (1971).
45. J.B. Keene. Cherts and porcellanites from the North Pacific. In: *Initial Reports DSDP.* vol.32, pp.429-507. U.S. Govt Printing Office, Washington, D.C. (1975).
46. J.R. Hein, D.W. Scholl, J.A. Barron, M.G. Jones and J. Miller. Diagenesis of late Cenozoic diatomaceous deposits and formation of the bottom simulating reflector in the southern Bering Sea, *Sedimentology* **25**, 155-181 (1978).
47. V. Riech and U. von Rad. Silica diagenesis in the Atlantic Ocean: diagenetic potential and transformations. In: *Deep Drilling Results in the Atlantic Ocean: Continents, Margins, and Paleoenvironments.* M. Talwani, W. Hay and W.B.F. Ryan (Eds). Maurice Ewing Ser., vol.3, pp.315-340. Am. Geophys. Union, Washington (1979).
48. K.J. Murata and R.R. Larson. Diagenesis of Miocene siliceous shales, Temblor Range, California, *J. Res. U.S. Geol. Surv.* **3**, 553-566 (1975).
49. K.J. Murata, I. Friedman and J.D. Gleason. Oxygen isotope relations between diagenetic silica minerals in Monterey Shale, Temblor Range, California, *Am. J. Sci.* **277**, 259-272 (1977).
50. K.A. Pisciotto. *Basinal sedimentary facies and diagenetic aspects of the Monterey Shale, California.* Unpublished Ph.D. thesis, University of California, Santa Cruz, California. (1978).
51. K.A. Pisciotto. Diagenetic trends in the siliceous facies of the Monterey Shale in the Santa Maria region, California, *Sedimentology* **28**, 547-571 (1981).
52. R. Tada and A. Iijima. Petrology and diagenetic changes of Neogene siliceous rocks in northern Japan, *J. Sed. Pet.* **53**, 911-930 (1983).
53. R.M. Carr and W.S. Fyfe. Some observations on the crystallization of amorphous silica, *Amer. Mineral.* **43**, 908-916 (1958).
54. W.G. Ernst and S.E. Calvert. An experimental study of the recrystallization of porcelanite and its bearing on the origin of some bedded cherts, *Amer. J. Sci.* **267**-A, 114-133 (1969).
55. M. Kastner, J.B. Keene and J.M. Gieskes. Diagenesis of siliceous oozes: I. Chemical controls on the rate of opal-A diagenesis -- an experimental study, *Geochim. Cosmochim. Acta* **40**, 1041-1059 (1977).
56. S. Mizutani. Progressive ordering of cristobalitic silica in the early stage of diagenesis, *Contr. Min. Petr.* **61**, 129-140 (1977).
57. M. Kastner and J.M. Gieskes. Opal-A to opal-CT transformation: A kinetic study. In: *Siliceous Deposits in the Pacific Region.* A. Iijima, J.R. Hein and R. Siever (Eds). Developments in Sedimentology, vol.36, pp.211-227 Elsevier, Amsterdam (1983).
58. R.L. Folk and J.S. Pittman. Length-slow chalcedony: a testament for vanished evaporites, *J. Sed. Pet.* **41**, 1045-1058 (1971).
59. O.W. Flörke, H. Graetsch, B. Martin and K. Roller. Nomenclature of micro-crystalline and non-crystalline silica minerals, based on structure and microstructure, *Neues Jahr. Mineral.* **163**, 19-42 (1991).
60. K.L. Milliken. The silicified evaporite syndrome – two aspects of silicification history of former evaporite nodules from southern Kentucky and northern Tennessee, *J. Sed. Pet.* **49**, 245-256 (1979).
61. L.A. Williams, G.A. Parks and D.A. Crerar. Silica Diagenesis, I. Solubility controls, *J. Sed. Pet.* **55**, 301-311 (1985).
62. L.A. Williams and D.A. Crerar. Silica Diagenesis, II. General Mechanisms, *J. Sed. Pet.* **55**, 312-321 (1985).
63. M. Kastner. Carbonate and silica diagenesis. In: *The Geochemical and Paleoenvironmental History of the Monterey Formation: Sediments and Hydrocarbons. V. 2, Data Volume.* pp.178-207. Global Geochemistry Corp., Canoga Park, California (1985).
64. N.W. Hinman. Chemical factors influencing the rates and sequences of silica phase transitions: Effects of organic constituents, *Geochim. Cosmochim. Acta* **54**, 1563-1574 (1990).
65. R. Greenwood. Cristobalite: its relationship to chert formation in selected samples from the Deep Sea Drilling Project, *J. Sed. Pet.* **43**, 700-708 (1973).
66. Y. Lancelot. Chert and silica diagenesis in sediments from the central Pacific. In: *Initial Reports DSDP; Leg 17*. E.L. Winterer, J.I. Ewing, *et al.* (Eds). vol. 17, pp. 377-405. U.S. Govt Printing Office, Washington (1973).

67. K.J. Murata and J.K. Nakata. Cristobalitic stage in the diagenesis of diatomaceous shale. *Science* **184**, 567-568 (1974).
68. K.J. Murata and R.G. Randall. Silica mineralogy and structure of the Monterey Shale, Temblor Range, California, *J. Res. U.S. Geol. Surv.* **3**, 567-572 (1975).
69. C.M. Isaacs. Field characterization of rocks in the Monterey Formation along the coast near Santa Barbara, California. In: *Guide to the Monterey Formation in the California coastal area, Ventura to San Luis Obispo.* C.M. Isaacs (Ed.).vol.52, pp.39-53. Pac. Sec. Amer. Assoc. Petrol. Geol., Los Angeles (1981).
70. H.K. Brueckner and W.S. Snyder. Chemical and Sr-isotopic variations during diagenesis of Miocene siliceous sediments of the Monterey Formation, California, *J. Sed. Pet.* **55**, 553-568 (1985).
71. R.N. Clayton and T.K. Mayeda. The use of bromine pentafluoride in the extraction of oxygen from oxides and silicates for isotopic analysis, *Geochim. Cosmochim. Acta* **27**, 43-52 (1963).
72. J.R. O'Neil. Appendix: Terminology and standards. In: *Stable Isotopes in High Temperature Geological Processes.* J.W. Valley, H.P. Taylor, Jr., and J.R. O'Neil (Eds). *Min. Soc. Amer., Reviews in Mineralogy,* **16**, 561-570 (1986).
73. R.J. Behl and B.M. Smith. Silicification of deep sea sediments and the oxygen isotope composition of diagenetic siliceous rocks from the western Pacific, Pigafetta and East Mariana Basins, ODP Leg 129. *Proc. ODP, Scientific Results.* vol.129, pp.81-117. Ocean Drilling Program, College Station, TX (1992).
74. C.M. Isaacs, M.A. Keller, V.A. Gennai, K.C. Stewart and J.E. Taggart, Jr. Preliminary evaluation of Miocene lithostratigraphy in the Point Conception COST well OCS-CAL 78-164 No. 1, off southern California. In: *Petroleum generation and occurrence in the Miocene Monterey Formation, California.* C.M. Isaacs and R.E. Garrison (Eds). vol.33, pp.99-110. Pac. Sec. SEPM, Los Angeles (1983).
75. R.L. Folk and C.E. Weaver. A study of the texture and composition of chert, *Amer. J. Sci.* **250**, 498-510 (1952).
76. R. Hesse. Selective and reversible carbonate-silica replacements in Lower Cretaceous carbonate-bearing turbidites of the Eastern Alps, *Sedimentology* **34**, 1055-1077 (1987).
77. J.B. Keene. Chalcedonic quartz and occurrence of quartzine (length-slow chalcedony) in pelagic sediments, *Sedimentology* **30**, 449-454 (1983).
78. N.L. Taliaferro. Contraction phenomena in cherts, *Bull. Geol. Soc. Am.* **45**, 189-232 (1934).
79. W.S. Snyder, H.K. Brueckner and R.A. Schweickert. Deformational styles in the Monterey Formation and other siliceous sedimentary rocks. In: *Petroleum Generation and Occurrence in the Miocene Monterey Formation, California.* C.M. Isaacs and R.E. Garrison (Eds). vol.33, pp.151-170. Pac. Sec. SEPM, Los Angeles (1983).
80. H.K. Brueckner, W.S. Snyder and M. Boudreau. Diagenetic controls on the structural evolution of siliceous sediments in the Golconda allochthon, Nevada, U.S.A, *J. Struct. Geol.* **9**, 403-417 (1987).
81. W.S. Snyder. Structure of the Monterey Formation: stratigraphic, diagenetic, and tectonic influences on style and timing. In: *Cenozoic Basin Development of Coastal California: Rubey Volume VI.* R.V. Ingersoll and W.E. Ernst (Eds). pp.321-347, Prentice-Hall, Englewood Cliffs (1987).
82. Y. Kolodny. Petrology of siliceous rocks in the Mishash Formation (Negev, Israel), *J. Sed. Pet.* **39**, 166-175 (1969).
83. G. Steinitz. Chert "dike" structures in Senonian chert beds, southern Negev, Israel, *J. Sed. Pet.* **40**, 1241-1254 (1970).
84. G. Steinitz. Enigmatic chert structures in the Senonian cherts of Israel, *Bull. Geol. Surv. Israel* **75**, 1-46 (1981).
85. J.J. Fagan. Carboniferous cherts, turbidites, and volcanic rocks in northern Independence Range, Nevada, *Geol. Soc. Amer. Bull.* **73**, 595-612 (1962).
86. J.H. Fink and Z. Reches. Diagenetic density inversions and the deformation of shallow marine chert beds in Israel, *Sedimentology* **30**, 261-271 (1983).
87. L.J. Regan and A.W. Hughes. Fractured reservoirs of Santa Maria district, California, *Amer. Assoc. Petrol. Geol. Bull.* **33**, 32-51 (1949).
88. W.C. Belfield, J. Helwig, P. La Pointe and W.K. Dahleen. South Elwood oil field, Santa Barbara Channel, California, a Monterey Formation Fractured Reservoir. In: *Petroleum Generation and Occurrence in the Miocene Monterey Formation, California.* C.M. Isaacs and R.E. Garrison (Eds). vol.33, pp.213-221. Pac. Sec. SEPM, Los Angeles (1983).
89. J. Stock. A Coupled Diagenetic and Micro-Structural Study of Cherts from the Miocene Monterey Formation, California. Unpublished senior thesis, University of California, Santa Cruz, California (1992).
90. J.C. Marsters and H.A. Christian. Hydraulic conductivity of diatomaceous sediment from the Peru continental margin obtained during ODP Leg 112. *Proc. ODP, Scientific Results.* vol.112, pp.633-638. Ocean Drilling Program, College Station, TX (1990).
91. A. Seilacher. Fault-graded beds interpreted as seismites, *Sedimentology* **13**, 155-159 (1969).
92. R.E. Garrison and P.C. Ramirez. Conglomerates and breccias in the Monterey Formation and related units as reflections of basin margin history: In: *Conglomerates in Basin Analysis: A Symposium Dedicated to*

A.O. Woodford. I.P. Colburn, P.L. Abbott and J. Minch (Eds). vol.62, pp.189-206. Pac. Sec. SEPM, Los Angeles (1989).
93. N. Lundberg and J.C. Moore. Macroscopic structural features in Deep Sea Drilling Project cores from forearc regions. In: *Structural Fabrics in Deep Sea Drilling Project Cores from Forearcs.* J.C. Moore (Ed.). vol.166, pp.13-44. Geol. Soc. Amer. Mem., Boulder (1986).
94. N. Lindsley-Griffen, A. Kemp and J.F. Swartz. Vein structures of the Peru margin. *Proc. ODP, Scientific Results.* vol.112, pp. 3-16. Ocean Drilling Program, College Station, TX (1990).
95. P.R. Hill and J.C. Marsters. Controls on physical properties of Peru continental margin sediments and their relation to deformation styles. *Proc. ODP, Scientific Results.* vol.112, pp. 623-632. Ocean Drilling Program, College Station, TX (1990).
96. E.L. Hamilton. Variations of density and porosity with depth in deep sea sediments, *J. Sed. Pet.* **46**, 280-300 (1976).
97. P.F. Barker and J.P. Kennett, *et al. Proc. ODP, Init. Repts.* vol.113. Ocean Drilling Program, College Station, TX (1988).
98. W.R. Bryant and F.R. Rack. Consolidation characteristics of Wedell Sea sediments: results of ODP Leg 113. *Proc. ODP, Scientific Results.* vol.113, pp.211-223. Ocean Drilling Program, College Station, TX (1990).
99. J.S. Compton. Porosity reduction and burial history of siliceous rocks from the Monterey and Sisquoc Formations, Point Pedernales area, California, *Bull. Geol. Soc. Am.* **103**, 625-636 (1991).
100. C.M. Isaacs. Porosity reduction during diagenesis of the Monterey Formation, Santa Barbara coastal area, California. In: *The Monterey Formation and Related Siliceous Rocks of California.* R.E. Garrison, R.G. Douglas, K.E. Pisciotto, C.M. Isaacs and J.C. Ingle, Jr. (Eds). vol.15, pp.257-272. Pac. Sec. SEPM, Los Angeles (1981).
101. R.J. Behl. Chert spheroids of the Monterey Formation; early diagenetic structures of biosiliceous sediments, *Sedimentology* (1993, in press)
102. W.P. Woodring and M.N. Bramlette. Geology and Paleontology of the Santa Maria District, California. *U.S. Geol. Surv. Prof. Paper* **222**, 1-185 (1950).
103. K.A. Pisciotto. Notes on Monterey rocks near Santa Maria, California. In: *Guide to the Monterey Formation in the California coastal area, Ventura to San Luis Obispo.* C.M. Isaacs (Ed.).vol.52, pp.73-81. Pac. Sec. Amer. Assoc. Petrol. Geol., Los Angeles (1981).
104. J. Helwig. Slump folds and early structures, Northeastern Newfoundland Appalachians, *J. Geol.* **78**, 172-187 (1970).
105. C.G. Elliott and P.F. Williams. Sediment slump structures: a review of diagnostic criteria and application to an example from Newfoundland, *J. Structural Geol.* **10**, 171-182 (1988).
106. L.E. Redwine. Hypothesis combining dilation, natural hydraulic fracturing, and dolomitization to explain petroleum reservoirs in Monterey Shale, Santa Maria area, California. In: *The Monterey Formation and Related Siliceous Rocks of California.* R.E. Garrison, R.G. Douglas, K.E. Pisciotto, C.M. Isaacs and J.C. Ingle, Jr. (Eds). vol.15, pp.221-248. Pac. Sec. SEPM, Los Angeles (1981).
107. A.H.F. Robertson. The origin and diagenesis of cherts from Cyprus, *Sedimentology* **24**, 11-30 (1977).
108. G. Gao and L.S. Land. Nodular chert from the Arbuckle Group, Slick Hills, SW Oklahoma: a combined field, petrographic and isotopic study, *Sedimentology* **38**, 857-870 (1991).
109. L.A. Williams. Lithology of the Monterey Formation (Miocene) in the San Joaquin Valley of California. In: *Monterey Formation and Associated Coarse Clastic Rocks, Central San Joaquin Basin, California.* L.A. Williams and S.A. Graham (Eds). vol. 25, pp.17-36. Pac. Sec. SEPM, Los Angeles (1982).
110. M. Haimson. *Oxygen Isotope Studies of Silica in the Monterey Formation, California..* Unpublished M.S. thesis, Arizona State University, Tempe, Arizona (1982).
111. M. Kastner, K. Mertz, D. Hollander and R.E. Garrison. The association of dolomite-phosphorite-chert: causes and possible diagenetic sequences. In: *Dolomites of the Monterey Formation and other Organic-Rich Units.* R.E. Garrison, M. Kastner and D.H. Zenger (Eds). vol.41, pp.75-86. Pac. Sec. SEPM, Los Angeles (1984).
112. R.E. Garrison, M. Kastner and Y. Kolodny. Phosphorites and phosphatic rocks in the Monterey Formation and related Miocene units, coastal California. In: *Cenozoic Basin Development of Coastal California: Rubey Volume VI.* R.V.Ingersoll and W.E. Ernst (Eds). pp.348-381, Prentice-Hall, Englewood Cliffs (1987).
113. D.P. Schrag, D.J. DePaolo and F.M. Richter. Oxygen isotope exchange in a 2-layer model of ocean crust, *Earth Planet. Sci. Let.* **111**, 305-317 (1992).
114. A. Mathews and R.D. Beckinsale. Oxygen isotope equilibration systematics between quartz and water, *Amer. Mineral.* **64**, 232-240 (1979).
115. B.L. Winters and L.P. Knauth. Stable isotope geochemistry of carbonate fracture fills in the Monterey Formation, California, *J. Sed. Pet.* **62**, 208-219 (1992).

116. D.A. Feary, P.J. Davies, C.J. Pigram and P.A. Symonds. Climatic evolution and control on carbonate deposition in northeast Australia, *Global Planet. Change* **3** and *Palaeo. Palaeo. Palaeo.* **89**, 341-361 (1991).
117. J.C. Ingle, Jr. Paleobathymetric, paleoceanographic, and biostratigraphic studies. In: *The Geochemical and Paleoenvironmental History of the Monterey Formation: Sediments and Hydrocarbons. V. 2, Data Volume.* pp.88-161. Global Geochemistry Corp., Canoga Park, California (1985).
118. M.A. Keller and C.M. Isaacs. An evaluation of temperature scales for silica diagenesis in diatomaceous sequences including a new approach based on the Miocene Monterey Formation, California, *Geo-Marine Let.* **5**, 31-35 (1985).
119. N.J. Shackleton and J.P. Kennett. Paleotemperature history of the Cenozoic and the initiation of Antarctic glaciation: oxygen and carbon isotope analyses in DSDP Sites 277, 279 and 281. In: *Initial Reports DSDP.* vol.29, pp.743-755. U.S. Govt Printing Office, Washington, D.C. (1975).

Sedimentation and diagenesis in paleo-upwelling zones of epeiric sea and basinal settings: A comparison of the Cretaceous Mishash Formation of Israel and the Miocene Monterey Formation of California

Y. KOLODNY[1] and R.E. GARRISON[2]
[1]*The Institute of Earth Sciences, The Hebrew University, Jerusalem 919045, Israel* and [2]*Earth Sciences Department, University of California, Santa Cruz, CA 95064, U.S.A.*

Abstract-- The Upper Cretaceous Mishash Formation of Israel and the Miocene Monterey Formation of California are products of deposition in high fertility coastal upwelling regions and contain similar lithologic assemblages characterized by abundant biosiliceous, phosphatic and organic-rich rocks. Despite this similarity, the two units contrast markedly in depositional environments, sedimentation rates, thicknesses, burial depths and geothermal gradients. These differences can be attributed to different tectonic settings, and they resulted in divergent diagenetic histories. Whereas thick sections of hemipelagic and pelagic biogenic Monterey sediments were deposited in deeply subsided basins along a fractured transform plate margin, much thinner and more slowly deposited Mishash deposits accumulated on a gently deformed shallow-water shelf located on a passive margin. Phosphogenesis was widespread in both units, but sedimentary phosphorites are abundant only in the Mishash because the shallow-water setting of this unit was more conducive to reworking and concentration of phosphatic particles. Moreover, whereas the oxygen isotopic values of the phosphate ions in Mishash apatites are remarkably uniform and mainly record the depositional or early diagenetic signal, the values from the Monterey are much more variable and appear to reflect later diagenetic temperatures of microbial-mediated recrystallization during burial. Similarly, Mishash authigenic carbonates are uniformly depleted in ^{13}C and formed in the zone of sulfate reduction, while comparable Monterey carbonates have more variable ^{13}C values and formed both during early diagenetic sulfate reduction as well as later during burial in the zone of methanogenesis. Oxygen isotopes in the diagenetic silica phases (opal-CT and microcrystalline quartz) have similar ranges of values in both units (22 to 35 ‰), but geological and geochemical constraints dictate different interpretations; in the case of the Monterey, the low values apparently reflect elevated temperatures of burial diagenesis, whereas in the Mishash low values (along with boron concentration analyses) indicate recrystallization in fresh water. Large petroleum deposits occur in the Monterey but not in the Mishash because the former unit was more deeply buried in basins with generally higher geothermal gradients.

Keywords: upwelling, Upper Cretaceous, Miocene, Israel, California, diagenesis, light stable isotopes

INTRODUCTION

The association of biosiliceous rocks, black shales, and phosphorites occurs throughout the Phanerozoic stratigraphic record. Economically, it is significant because it houses important phosphorite deposits as well as petroleum source beds. Scientifically, it is important as a signal of high-fertility conditions associated with paleo-upwelling zones. Primary sediments deposited in such zones are chemically labile because they contain large amounts of organic matter along with unstable opal-A. Consequently such sediments may experience substantial diagenesis which complicates interpretation of the lithologic signal of upwelling. Additionally, temporal and spatial variations in the intensity of upwelling, and hence of the fertility of surface waters, cause variations in sedimentation rates, in oxygen levels of bottom waters, and in primary sediment composition and subsequent diagenesis, and in the overall character of sedimentary successions. Lastly, differences in tectonic settings for this association may result in variability in subsidence and sedimentation rates as well as basin configuration and the relation of basins to important parameters such as low oxygen water masses and sea level changes.

In this article we explore the above noted kinds of variations through a comparative analysis of two units associated with paleo-upwelling systems in distinctly different settings. The Upper Cretaceous (Campanian) Mishash Formation of Israel formed in a shallow, warm epeiric sea influenced by coastal upwelling along the slowly subsiding passive southern margin of the Tethys during a period of gradual tectonic flexure. The Miocene Monterey Formation of California, in contrast, formed in generally rapidly subsiding deep basins which were influenced on the one hand by the transform-margin tectonics of the San Andreas system, and on the other by upwelling of cold, nutrient-rich waters linked to an eastern boundary current, the California Current.

MISHASH FORMATION

Tectonic and paleoceanographic settings
The Mishash Formation, as well as its stratigraphic equivalents in southern Israel, the Sayyarim Fm. and in Jordan, the Amman Formation, are all part of a Late Cretaceous sedimentary sequence deposited in the southeastern reaches of the Tethys, that flooded the northern, passive, margins of the ancient Arabo-Nubian continent. The province stretches across the region surrounding the present day Mediterranean Sea: from Turkey in the north through Syria, Jordan, Israel, Egypt, Tunisia, Morocco to Togo. Depositional environments of this sequence were controlled by an interplay between sea-level changes of the Tethys, the tectonic setting of the area and paleoceanography of that part of the sea. In particular, throughout Late Cretaceous time the dominant feature of the region is the transition from continental deposition in the southeast to continental slope deposits in the northwest [1,2]. The cherty formations of the Campanian, the Mishash and the Amman dominate the landscapes of the northern Sinai, the Negev, and large parts of Jordan (Fig. 1 [3]).

A stable tectonic regime prevailed in the region since the Jurassic till the Late Turonian, resulting in an extensive belt of shallow water carbonates, forming the Judea Group [4]. During Late Turonian to early Late Coniacian time a change in the tectonic regime occurred: the incipient formation of the "Syrian Arc" began [2,5]. This is a series of folds which extends from Syria through most of Israel to Egypt. The structures grew while sedimentation was in progress, resulting in the formation of basins and sills and to the appearance of different sedimentological regimes on the newly formed highs as opposed to the lows. Facies changes between synclines and anticlines mark the Senonian sediments in the region (Fig. 2). When the rate of sedimentation exceeded sea level rise and subsidence on structural highs, shallowing resulted sometimes in omission surfaces and even in exposure and erosion[6-9]. The location of the region in the southeastern part of the latitudinal Tethys was a second factor that determined its depositional environment. The paleolatitude of the region was 8 -15° N; winds were blowing from the NEE with a continent to their left, caused Ekman transport to the NW of surface water, and upwelling of deeper, nutrient rich water which flooded the morphologically undulating broad shelf [9]. The imprint of a high fertility environment persisted through a large part of the Campanian into the Maastrichtian when the Ghareb Formation was deposited in the basins. The lower Ghareb Formation consists of bituminous chalks and marls, their organic content reaching in places 30%. On elevated (anticlinal) structures of the northern Negev the bituminous chalk grades into phosphorite [2].

Based mainly on the distribution of epi- and meso-pelagic planktic and epi- and endo-benthic foraminifera as well as of diatoms radiolarians, coccolithophorids, megafossils and *Beggiatoa*-like sulfur-reducing bacterial mats, Reiss [9] proposed a comprehensive depositional model for the Mishash and associated units (Fig. 3). This involves alternating periods of retarded upwelling and low photic zone productivity during sealevel highstands, yielding carbonate-rich sediments, and times of strong upwelling and high photic zone fertility which produced dysaerobic to anaerobic bottom waters and biosiliceous, phosphatic and organic-rich sediments during intervals of sealevel lowstands. Peloidal phosphorites of economic significance in the Negev region are interpreted as products of winnowing and downslope transport.

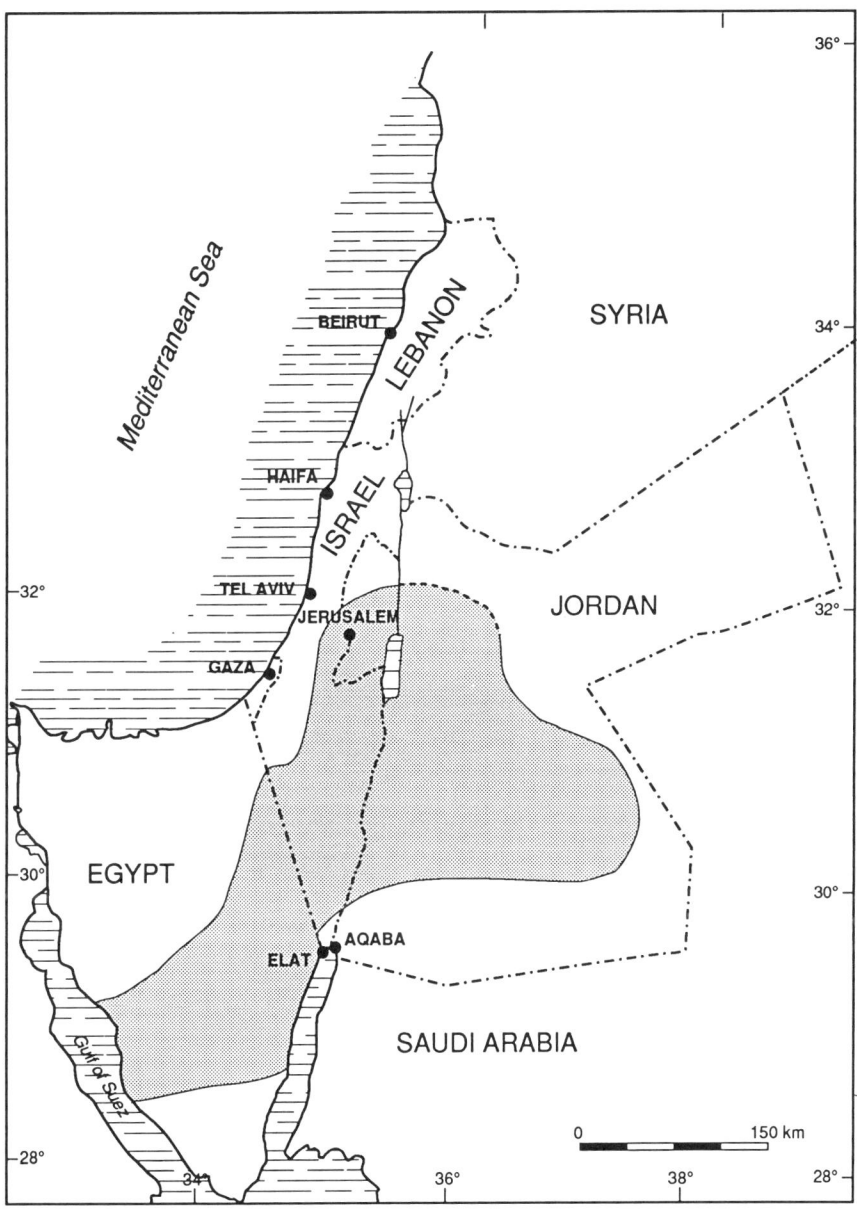

Figure 1. Distribution map of the cherty-phosphatic facies of the Mishash Formation and correlative units. Modified after Flexer [7].

Stratigraphic and facies divisions
In southern Israel (Northern Negev, Judean Desert) the Mishash Fm. overlies the chalks of the Menuha Fm. of Santonian to Early Campanian age (Figs. 2 and 3). The Menuha Fm. - Mishash Fm. sequence has its equivalent in the throughout chalky En Zetim Fm. towards the

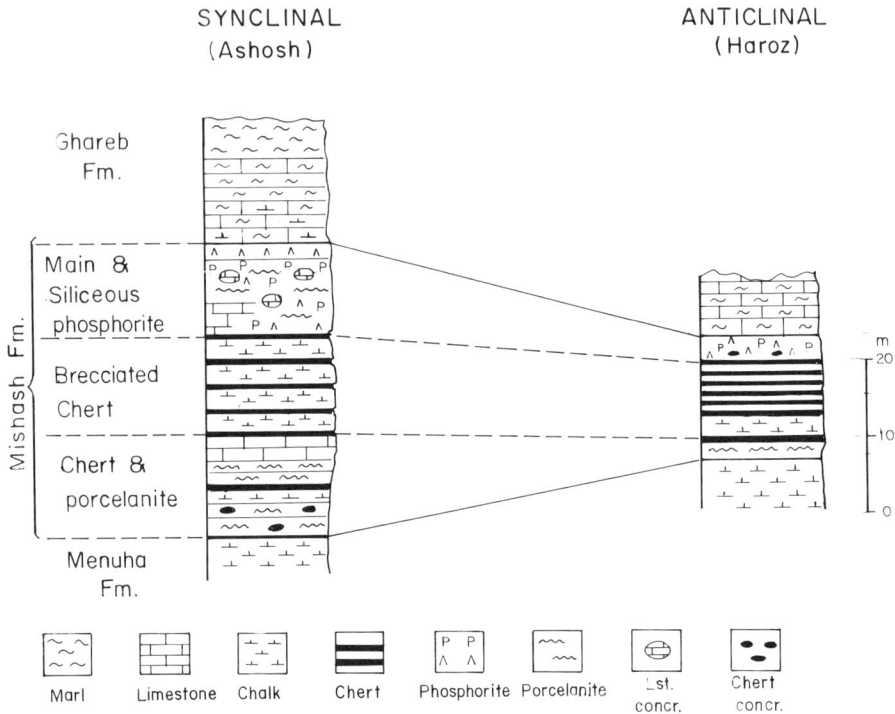

Figure 2. Schematic columnar sections in the Ashosh and Haroz facies of the Mishash Formation. After Bartov et al. [10].

open shelf (northern Israel), and in the Sayyarim Fm., an alternation of chalks, cherts and sandstones towards the continent (the southern Negev [10]). The cherty - chalky - phosphoritic facies extends over a belt of 50 to 350km wide, trending SW to NE from the central Sinai through southern Israel into Jordan (Fig. 1).

In general the lower part of the Mishash Fm. contains abundant siliceous rocks whereas the upper part is more phosphate-rich. However, both siliceous and phosphatic rocks occur throughout the unit, and several lithostratigraphic divisions of the Mishash Fm. have been proposed [11-13] (Figs. 2 and 3).

The undulating structural-morphological relief in the Late Coniacian is reflected in both thickness differences of the Menuha Fm. and facies changes in the Mishash Fm. [11,13-15] The Menuha Formation was deposited mainly in the basins, wedging out towards the structural highs. This filling did not flatten the relief to a level surface. The character of the Mishash Formation is still clearly dependent on its paleostructural location. On anticlinal areas a thin, mainly chert sequence occurs, termed by Bentor and Vroman [14] the Haroz Facies. In synclines a cherty - chalky, much thicker sequence is found, which has been termed the Ashosh Facies. It has been pointed out by Kolodny [11], Steinitz [16], and Soudry and coworkers [13] that during transition from paleostructural high to low, not all parts of the section are affected in the same way. Thus, the lower, cherty members of the Mishash Formation. are dominant on anticlines, whereas the phosphatic unit is thicker in synclines. It is also clear from field evidence, that the structural elements that govern the facies distribution are not identical to present-day structures.

The structural evolution of the northern Negev can be tentatively summarized by the following steps [8,13]: 1. Creation of the Post Turonian relief; 2. Deposition of the Menuha Fm. preferentially in synclines, which partly flattened the relief; 3. Deposition of the porcelanitic and cherty part of the Mishash Formation; 4. Rejuvenation of the structure, and deposition of

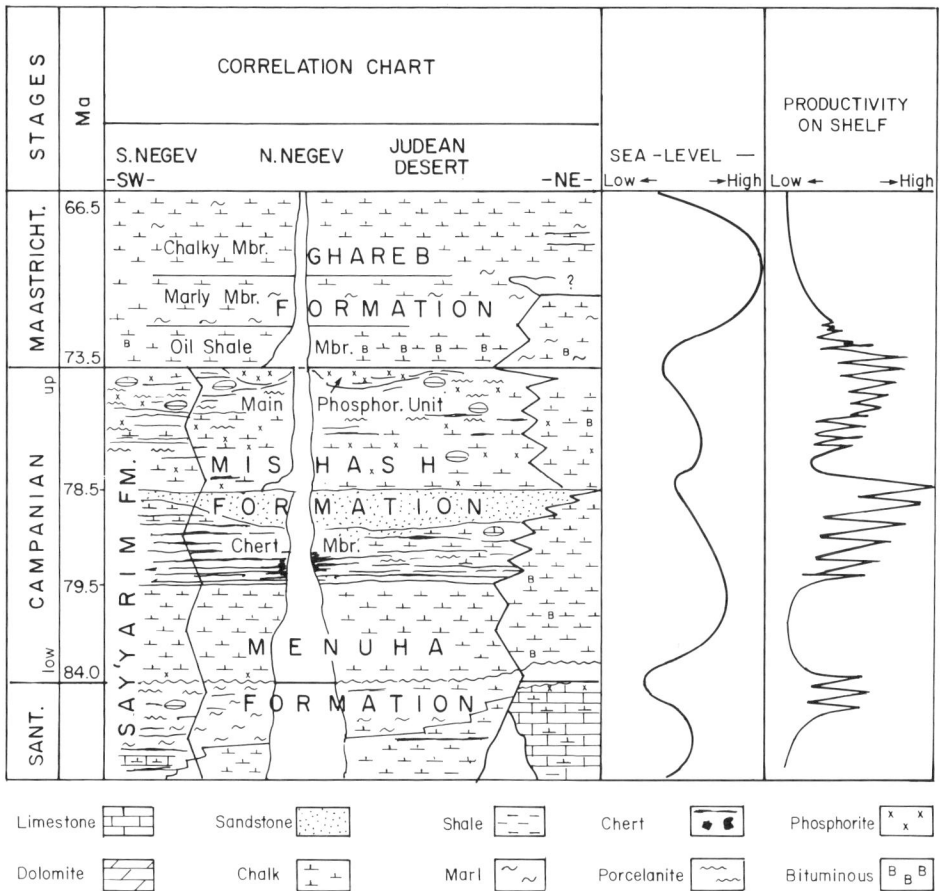

Figure 3. Correlation chart of Senonian formations in Israel, after Reiss [9]. Note that whereas thickness is schematically linear, the time marks are not. Thus whereas the thickness of the chert and phosphatic members is about equal, it took five times longer to form the upper phosphate member as it took to deposit the chert member (approx. 5 Ma vs. approx. 1 Ma).

the phosphatic part of the section mainly in synclines while non-deposition or subaerial alteration occurred on the highs.

The Siliceous Facies. Two mineralogical - petrological types of siliceous rocks are abundant in the Mishash Formtion: porcelanites and quartzose cherts. The distinction between the two terms is not the same in Israel and in California. In the Mishash context the definition of porcelanite follows the DSDP tradition. It describes as porcelanites siliceous, non-detrital sedimentary rocks the main mineralogical component of which is opal-CT. In the Monterey on the other hand porcelanites are defined by their field and hand specimen appearance of unglazed porcelain regardless of whether the main diagenetic silica phase is opal-CT or quartz. Porcelanite occurs mainly in two stratigraphic sites in the Mishash Formation [17]: porcelanitic layers are abundant, usually "sandwiching" homogeneous brown chert beds at the base of the sequence, immediately overlying the Menuha chalk; porcelanite is again found in the uppermost part of the Formation, where it is usually phosphatic, often associated with numerous apatitic fish scales. Whereas under the microscope the main constituent of porcelanites is isotropic, making birefringent components of these rocks, such as quartz infilled foraminifera test more conspicuous, the rocks show the well known XRD pattern of opal-CT with the 4.1Å and 4.3Å peak doublet [18] . As mentioned above, porcelanite occurs in the Mishash Formation both at the base and top of the formation. The two occurrences are separated by up to several tens of meters of section. No systematic change in opal-CT concentration with stratigraphy has been observed.

Cherts. The rocks which dominate the landscapes of the Mishash Formation, are the abundant chert layers. Texturally, two main types of chert have been distinguished [15,19]: a. homogeneous cherts and b. "heterogeneous cherts".

Most of the cherty layers both at the base of the Mishash Formation, adjacent to porcelanites and at the top of the sequence, in association with phosphatic rock, are of homogeneous, brown, gray, white or black cherts. These consist of microcrystalline quartz, with veins and voids (mainly fossil infillings) being filled by fibrous chalcedony and megaquartz. In contrast to those the cherts of the Chert Member (Brecciated Chert Member), are mostly heterogeneous, in the sense that one can always distinguish in them two textural components: "fragments" floating in a "matrix". Based on geometrical criteria, a distinction can be made between textures which are the result of irregular replacement of limestone by quartz ("spotty cherts") and those in which chert laminae of a first generation silica precipitation were torn apart and brecciated at some stage when the "matrix" was either siliceous and mechanically incompetent or yet remained calcareous. In many cases the "matrix" of heterogeneous cherts shows preferred orientation of quartz crystallites [20] sometimes referred to as "mass polarization" [21]. The first silicification stage was followed by a second silicification, which resulted in the entire rock being composed of microcrystalline quartz. The two silicification events were not only separated in time as evidenced by textural analysis; in most instances the environments in which these two events occurred were rather different: Kolodny and coworkers [22] have shown that whereas the first silicification step, the one which produced the chert "fragments" occurred usually in a normal marine environment, the second silicification step often recorded an intrusion of fresh water into the depositional environment. The first stage was apparently early diagenetic; the solutions which equilibrated with the microcrystalline quartz cannot be isotopically distinguished from $\delta^{18}O$ and temperatures of normal seawater ($\delta^{18}O$ of "fragments" is +31 to +33‰$_{SMOW}$); $\delta^{18}O$ of opal-CT of porcelanites (31.5 to 33.3‰) and of homogeneous cherts (29 to 32 ‰) are concordant with these values. The second silicification event occurred in an ^{18}O depleted environment: $\delta^{18}O$ of the chert matrix ranges between 21 and 33‰ (Fig. 4). The interpretation of the $\delta^{18}O$ results as reflecting salinity changes in the environment of diagenesis is supported by analyses of the distribution of boron between fragments and matrix [22].

Spheroids. Cherty, spherical, onion-like structures were first described by Taliaferro [24] from the Monterey and Franciscan Formations of California, and termed by him spheroids. Their shape varies between spherical and disc-like, their size between a few centimeters and half a meter. Their concentric, onion like appearance is caused by alternating broad (2-5 cm), brown bands of microcrystalline quartz, and transparent narrow bands of fibrous chalcedony. In the Mishash Formation spheroids are confined to a rather narrow stratigraphic range - at the base of the formation, in the Brown Chert and Porcelanite Mbr. [19]. Taliaferro [24] ascribed these structures to an entire family of "contraction phenomena" in cherts. Recently Behl [25] presented a new interpretation of their origin: he suggested that spheroids form as a result of the sharp contrast in mechanical strength between the hard dense chert and the weaker, surrounding porcelanite. Accommodation of differential deformation of the chert nodule and host, combined with the continued growth of the nodule result in episodic failure and subsequent precipitation of void filling silica into open spaces.

Most of the authors who studied the Mishash Formation considered the problem of the origin of its siliceous rocks. It is an expression of the current consensus that the most likely mechanism for concentrating silica for the deposition of porcelanites and cherts is biogenic concentration. The support for such mechanism is in fact not direct: it is the high productivity assemblage of nutrient dependent products (Si-P-C, cherts-phosphorites-organic rich rocks), coupled with the difficulty in finding an alternative explanation that suggests that silica secreting organisms must have been the first step in concentrating silica from sea water probably as opal-A. This initial sediment must have undergone at least two steps of dissolution - reprecipitation, from opal-A to opal-CT and then to quartz, before diagenesis has been completed [26]. It is hence not surprising that finding of siliceous organisms in the Mishash Formation was rare. Significantly both the marine diatoms and radiolarians which have been identified, were found in porcelanites [27,28] and chalks [29], both rock types that have a higher probability of not having undergone both dissolution-precipitation steps.

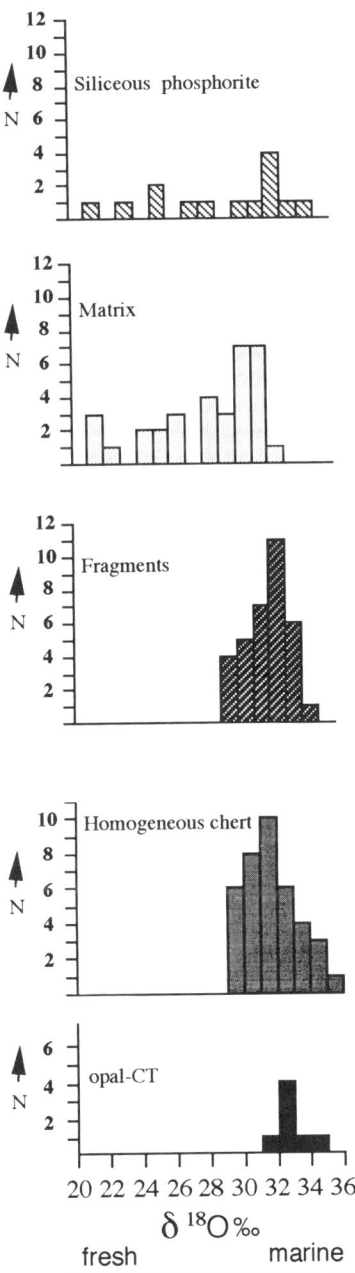

Figure 4. Histograms of $\delta^{18}O$ values in the various silica phases and textural components of cherts in the Mishash Formation. Data from Vengosh et al. [30].

The abundance of silica throughout the Mishash Formation, and the mobility of silica in diagenetic conditions resulted in quartz being an ubiquitous replacement phase. It is most obvious as replacing carbonate fossil shells, but also forms replacement stringers in limestones, frequently resulting in the "spotty" texture and in "linear structures" [15]. There is textural evidence that quartz is a replacement phase in many phosphorites in which apatitic ovules or bones float in quartzose (or opal-CT) matrix. It seems that in many of these cases quartz has early diagenetically replaced a pre-existing carbonate matrix. Only after all the carbonate matrix has been silicified does the replacement process affect the apatitic components as well [19]. Recrystallization and mobilization of quartz proceeded much later in the rock history, evidenced both by void infillings, cementation and veinlets.

Isotopic criteria distinguish between the different silica generations [30]. Whereas microcrystalline quartzose matrices in silicified phosphorites have "normal marine" $\delta^{18}O$ values (31 to 33‰), coarse quartz matrix in siliceous phosphorites has an isotopic signature similar to the late diagenetic chert breccia matrix or to calcitic spar infillings in fossils (20 to 30‰).

Authigenic carbonates: The carbonate components of the Mishash Formation are probably the least studied of this sequence. In numerous field descriptions they are described as "chalks". A closer examination reveals however that although some of them are chalky indeed (mainly in the lower part of the Mishash Formation), in most cases these are soft carbonates, in contrast to carbonates in the underlying Menuha Formation; these soft Mishash carbonates are barren or nearly devoid of nannofossils but commonly rich in epibenthic foraminifera [13]. Most of the carbonates occur as strata between siliceous rocks and phosphorite; most limestones are biomicrites or microsparites. Many authigenic carbonates have a complex diagenetic history: thus Soudry [31] showed that in many phosphorites apatitic grains are replaced by micritic calcites. On the other hand in many instances [23] it can be shown that early diagenetic dolomitization was followed by de-dolomitization, interrupted by silicification.

A structure characteristic of the upper parts of the Mishash and Sayyarim formations are the large carbonate concretions [32]. There is ample evidence for an early diagenetic origin of these: compaction - contortion of over- and underlying beds, excellent preservation of soft bodied fauna, burrowing on the top of the concretions. These concretions are also the sharpest example of another isotopic signature typical of the entire Mishash Formation: the considerable depletion of all its diagenetic carbonates in ^{13}C (Fig. 5). $\delta^{13}C$ values of the carbonate concretions varies between -10 and -15‰ (vs. PDB). Non concretionary authigenic carbonates of the Mishash Formation are also ^{13}C-depleted; values as light as -11‰ are common [23]. The phenomenon has been ascribed to a transfer of ^{12}C from the reduced organic matter carbon reservoir to the carbonate reservoir during diagenetic oxidation of organic matter by sulfate reduction (e.g. [33]). To cause sulfate reduction in the sediment the sea floor must be at least dysaerobic. The relationship between anoxia and high organic carbon concentrations have recently been questioned [34]. It seems very likely that the occurrence of high productivity (required for silica and phosphate deposition), lack of oxygen during diagenesis (required for low $\delta^{13}C$ values) and high organic carbon in the lower Ghareb Formation overlying the Mishash, is more than coincidental.

Phosphatic facies: Phosphorites are the economic justification for many of the studies of the Mishash Formation, as much as hydrocarbons were the basis for much of the research on the Monterey Formation.

The phosphatic, upper part of the Mishash Formation section, is mainly developed in the Ashosh facies. Along several of the synclinal axes of the northern and central Negev, these phosphorites are economically mined: the Zefa' field, Oron, and Zin valley. The deposit at Makhtesh HaQatan has recently been exhausted, whereas the one at Arad was never exploited because the town of Arad is built on the deposit.

In the different fields the phosphatic sections are represented by different alternations of phosphorites, phosphatic chalks and limestone concretions, and siliceous rocks. Nathan *et al.* [12] distinguished five discrete phosphorite beds, potentially fit for economic mining, best developed in the Zin valley section. They numbered beds 0 to III as the "interchert beds" whereas bed IV is the "main phosphorite"; these they attempted to correlate, applying field criteria to sections across the northern and central Negev. The mined phosphorites in the Negev have P_2O_5 concentrations varying between 25% and 30%.

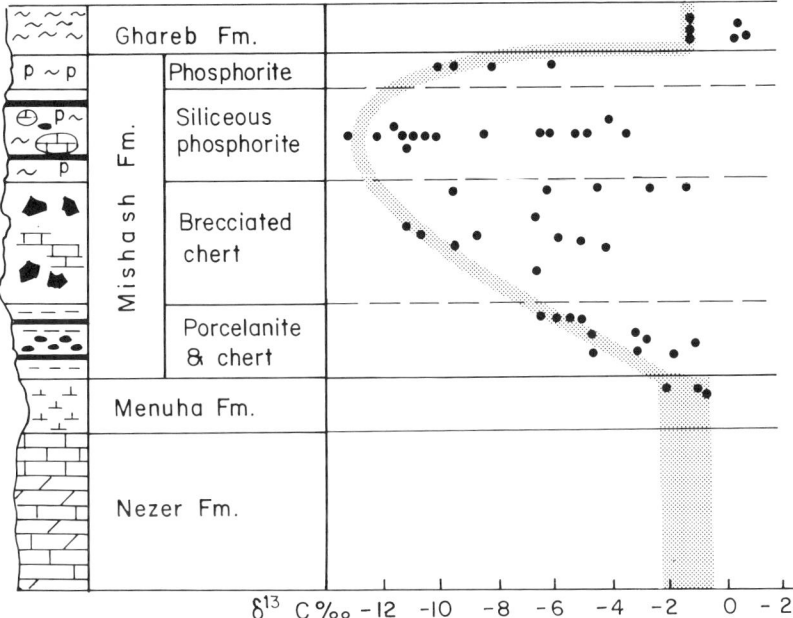

Figure 5. Schematic distribution of $\delta^{13}C$ values in carbonates in the Mishash Formation and the over- and underlying formations. After Kolodny [23].

Soudry and Champetier [35] distinguish six main lithofacies in the Mishash Fm., including the two phosphatic facies of a fine sandy phosphorite and a coarse phosphorite.
The principal mineral in the phosphorites is carbonate fluorapatite (CFA or francolite). Two major textural types were distinguished according to the nature of the apatitic grains in the rock: a. peloidal phosphorites; pelloids (sometimes termed ovules or ovulites) range in size between 80 and 500µm (mostly 150 - 400µm). These are rounded grains, cloudy due to a enrichment in organic matter, isotropic under crossed nicols, very commonly containing foraminiferal tests as "cores". Soudry [36] and Soudry and Champetier [35] showed petrographic and SEM evidence for strong microbial participation in the formation of these peloids. b. Bone phosphorites in which fish bone fragments as well as fragments of other marine vertebrates make up the bulk of the grains. Soudry [36] also showed textural evidence for phosphomicritization of bones by algae. The result of this process is the transformation of birefringent bones into isotropic pelloids. The strong influence of bottom dwelling microbial communities upon the formation of both peloid grains and phosphomicritic matrix in the highest grade phosphorite has now been demonstrated very convincingly.
The geochemistry of the phosphatic rocks of the Mishash Formation has been extensively studied. Nathan et al. [12] investigated the major and minor elements of these rocks Avital et al. [37] studied the U distribution within apatitic grains and showed that optically isotropic pelloids and isotropic bone fragments are about 1.6 times enriched in U as compared with the birefringent bones. Apparently this enrichment of U in the isotropic grains must have something to do with the above mentioned phosphomicritization process.
Shemesh et al. [38] and Shemesh and Kolodny [39] analyzed the isotopic composition of oxygen in the phosphate, as well as that of oxygen and carbon in the carbonate of francolite. These studies can be summarized in several points:

   a. There is no significant difference between $\delta^{18}O$ of bone fragments and ovulites.

   b. The isotopic composition of the main economic phosphorite beds is on the average 19.5‰. The range of Israeli phosphorites of the Mishash Formation is 16.8 to 20.3‰ (Fig. 6). It is interesting to note that when cherts, carbonates and phosphorites of the Mishash Formation were sampled in the Ef'e - Oron area, the narrow distribution of $\delta^{18}O$ in phosphates was contrasted by a much wider distribution of cherts and carbonates, thus supporting the

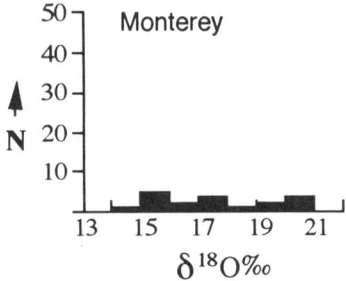

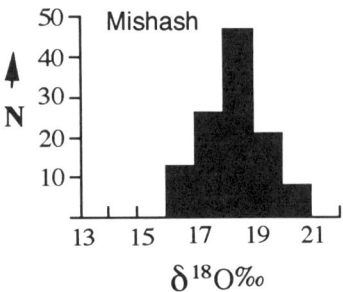

Figure 6. Comparison of distribution of $\delta^{18}O$ values in phosphate oxygen in the Monterey and Mishash Formations. Data from Kastner et al. [95], Vengosh et al. [30)], Shemesh and Kolodny [39].

initial assumption of Kolodny *et al.* [40] that $\delta^{18}O$ in phosphate is more resistant to post depositional changes. The isotopic composition of well-identified fish remains as well as of fossil marine reptiles from the Mishash Formation is in the same, above mentioned range [41].

c. Several phosphorite samples from other parts of the Cretaceous - Tethyan phosphogenetic province: Turkey, Jordan, Egypt, Tunisia and Morocco yielded values in the same range as the phosphorites of southern Israel. Applying the phosphate - water isotopic paleotemperature equation of Longinelli and Nuti [42] and assuming isotopic composition of Senonian water as -1‰, one arrives at temperatures of equilibration of 22 to 29°C between the apatite and environmental water [39].

d. Systematic sampling of phosphorites for isotopic analyses was conducted in Mishash sections in the Galilee, at Arad, the Zin valley, and at Taba south of Elat. In all of these the top of the section, coinciding with the economic phosphorite deposit is marked by an 1.5‰ increase in $\delta^{18}O$. This "heavy" peak has been interpreted as marking deposition of apatite from water of either lower temperature or higher salinity (both expressed as higher σ_T). Thus the "heavy" peak has been interpreted as marking the onset of more intense upwelling along the Tethyan coast. Such an interpretation has been supported by REE and $^{144}Nd/^{143}Nd$ analyses of Grandjean *et al.* [43] on fish bones from West African phosphorites.

MONTEREY FORMATION

Tectonic and paleoceanographic settings
Noted principally for its widespread biosiliceous deposits, the Monterey Formation is one expression of similar deposits that occur around much of the Pacific rim [44,45]. Monterey sediments accumulated mainly in relatively small transtensional or "pull-apart" basins (Fig. 7)

that developed along the San Andreas transform fault system in middle to late Cenozoic time [46]. Atwater [46] and Blake *et al.* [47] demonstrated that these basins developed at different times, consequently the exact succession of lithofacies varies somewhat from basin to basin (see also Graham [48]). Most of these basins show a deepening upward succession of facies, beginning with Late Oligocene non-marine to shallow marine clastic sediments that give way upward to the pelagic and hemipelagic rocks of the Monterey and associated units. Late Oligocene to Early Miocene subsidence was rapid, with rates of 100 to over 300 m/m.y. [49, 50]. Hemipelagic and pelagic Monterey deposition began rather abruptly between 18 and 17 Ma in many basins, possibly due to a combination of rapid subsidence and eustatic sea level rise which retarded the influx of terrigenous sediment [51].

The character of Miocene lithic fill was determined by a combination of tectonic, geographic, and paleoceanographic factors. Rapid lateral changes in thickness and facies suggest the Monterey basins were small, had very narrow shelves and deep central basinal portions (up to 1500 meters deep according to Ingle [49]), and possessed pronounced topography much like the steep-sided basins of the present California borderland [52]. Monterey and coeval facies thus include basin floor, slope, submerged banktop, and shelfal lithofacies [52-54], the latter largely removed by erosion that accompanied subsequent uplift of the basin fill in late Cenozoic time. Distal basins, located far offshore and shielded from terrigenous sediment by intervening basins and submarine banks, accumulated relatively pure pelagic deposits. Proximal basins, in contrast, received more detritus-rich hemipelagic sediments along with, during times of sea level lowstand and/or tectonism, influxes of silty to sandy deposits. These differences, along with the varied subsidence histories of different basins, account in large part for the large variations in lithology and thickness of the Monterey Formation between basins and even within different parts of the same basin; the Monterey, for example, is a few hundred meters thick in parts of some basins and over 2000 meters in others. An additional factor was coastal upwelling associated with the California Current. As discussed below, the intensity of upwelling varied during the Miocene leading to variations in productivity, sediment composition and oxygen levels in the Monterey basins.

Stratigraphic and facies divisions
The Monterey Formation in many basins is subdivided into several facies or informal members based on variations in the amounts of biosiliceous components (chiefly diatoms), biocalcareous components (foraminifers and coccolithophorids), authigenic phosphates, organic matter and terrigenous detritus. The most general subdivsions comprise a lower calcareous-siliceous facies, a middle phosphatic facies, and an upper siliceous facies [Fig. 8; 52,55], although more detailed subdivisions have been proposed for individual basins (e.g. by Canfield [56] for the Santa Maria Basin and by Isaacs [57] for the Santa Barbara Basin). Pisciotto and Garrison [52] attributed the change from generally calcareous sediments (comprising the calcareous-siliceous and phosphatic facies) in the lower Monterey to highly siliceous deposits in the upper part of the unit to intensification of California Current flow, increased coastal upwelling and enhanced biosilceous productivity accompanying the expansion of Antarctic glaciation and attendant global cooling during the Middle Miocene. However, as noted by Barron [58], Isaacs [59] and Isaacs and Lagoe [60], the onset of biosilceous (diatomaceous) sedimentation in the eastern Pacific and along the California margin actually preceded the mid-Miocene glacial expansion in several places. Moreover, the boundaries between the Monterey facies are diachronous between many basins, and the calcareous and phosphatic facies do not occur in some basins. This casts doubts on a simple model in which major paleoceanographic changes left a precise imprint on these continental margin sediments. Instead, local factors such as differential preservation of siliceous and calcareous components, along with varying conditions of phosphatization (e.g. in oxic vs. suboxic basins), may have been superimposed on the major paleoceanographic changes.

It is, however, a fact that the upper part of the Monterey is everywhere highly siliceous (Fig. 8), and this doubtless reflects the vigorous coastal upwelling and very high biosiliceous productivity which followed intensification of Antarctic glaciation in the Middle Miocene. The earlier onset of diatomaceous sedimentation in some localities along the eastern margin of the Pacific, noted above, may have different explanations. A number of workers [58,61,62], for example, have suggested that the beginning of Miocene biosiliceous sedimentation in the northeast Pacific might have resulted from the increased production of North Atlantic Deep

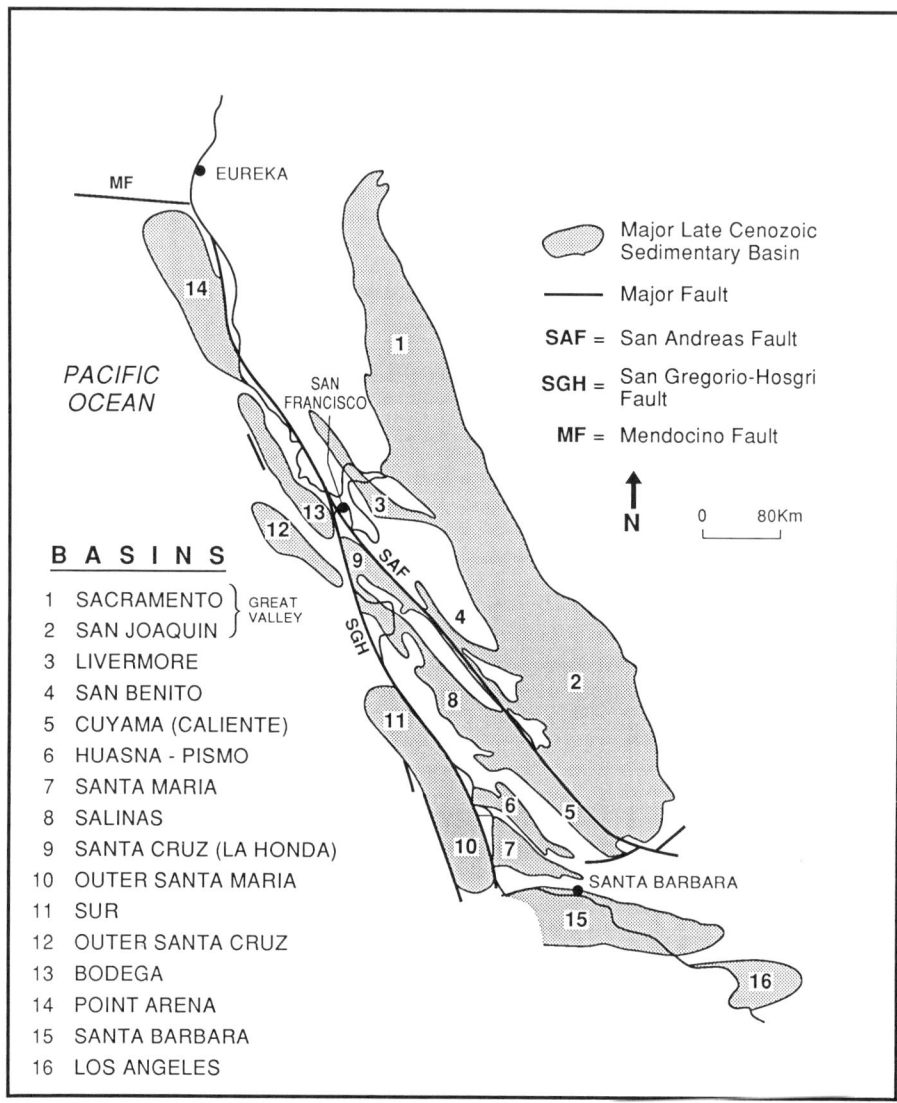

Figure 7. Major Late Cenozoic basins of California. The more distal of these basins (to the west and southwest) contain Neogene successions, including the Monterey Formation, deposited beneath upwelling zones. Modified from Graham [48].

Water in the Early Miocene, an event which led to dissolution of biosiliceous sediments in the Atlantic and funneling of silica- and nutrient-enriched water from the South Atlantic into the Indian and Pacific Oceans. In this scenario, the intensified upwelling in the Pacific following the mid-Miocene Antarctic glacial expansion might be viewed as a spike on the overall trend of Neogene high fertility conditions in the Pacific brought about by this transfer of water masses. This sequence of events seems borne out by the studies of Isaacs [51] in the Santa Barbara

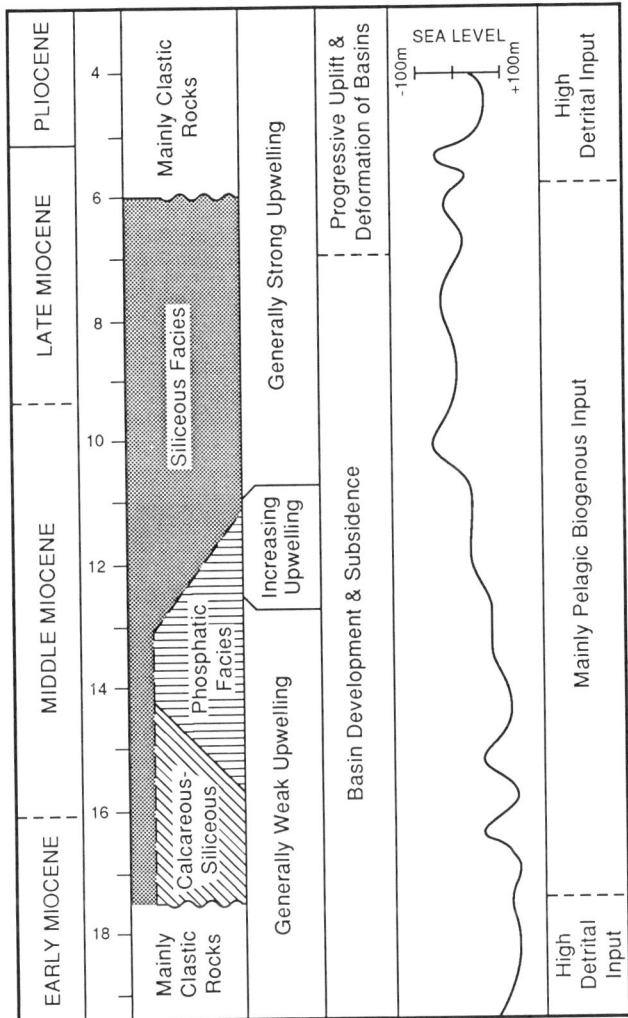

Figure 8. Generalized vertical succession of facies in the Monterey Formation in relation to tectonic and paleoceanographic-eustatic events. This diagram applies to distal basins which received mainly pelagic and hemipelagic sediments. Modified from MacKinnon [77].

Basin showing marked increases in biosiliceous sediment accumulation rates starting at about 11 Ma, coinciding with a marked strengthening of the California Current and attendant upwelling according to the faunal and floral studies of Barron and Keller [62].

Monterey siliceous facies: Sedimentology. Relatively few sedimentological studies have been carried out on Monterey facies, including the siliceous facies. Bramlette [63], Pisciotto [55] and Hieshima [72] described widespread varve-like laminations that probably record seasonal sedimentation events in low oxygen environments that precluded large burrowing animals. Also prominent in parts of the siliceous facies are small-scale slump and slide structures as well as thin beds (3-10 cm thick) of structureless siliceous sediment with sharp basal contacts and gradational upper contact [55,72]. The latter beds appear to be diatom oozes and muds that were redepositied fine-grained turbidites. The abundance of these structures suggest that considerable Monterey deposition occurred in slope or base-of-slope environments. In the Santa Barbara-Ventura Basin, Hornafius [73] presented evidence for preferential winnowing of diatom frustules from bank-tops and resedimentation in adjacent basins where they became the precursor sediments for chert-rich Monterey sequences.

As noted above, low oxygen (anaerobic) bottom conditions were widespread during Monterey deposition and indicate the presence of a well developed oxygen minimum zone, but massive, burrowed parts of the Monterey suggest local environments ranging from dysaerobic to aerobic which supported a robust infauna [59,74,75]. Alternations between burrowed and laminated intervals indicate cyclic fluctuations in oxygenation of bottom waters [76]. This and other kinds of sedimentary cycles are common in the Monterey Formation [52,55,63,77], but relatively little detailed study has been conducted on these features. A promising new approach is the application of sequence stratigraphic concepts to the Monterey; this attempts to identify sequences, parasequences, and systems tracts utilizing variations in lithology, microfossils, organic matter, and gamma radiation [78,79].

Diagenesis. Monterey siliceous facies comprise opal-A diatomites, opal-CT and quartz porcelanites and mudrocks, and quartz cherts and porcelanites. Bramlette [63] was the first to document that Monterey porcelanites and cherts are burial diagenetic alterations of diatom-rich sediments. Subsequent studies by Murata and his co-workers [64-66], Pisciotto [55,67], Isaacs [57,59,70], Haimson [68], Keller and Isaacs [69], and Behl [25] have clarified many aspects of the processes which convert soft, porous diatomaceous ooze to dense, hard and brittle porcelanite and chert. The phase changes from opal-A to opal-CT to quartz are solution-reprecipitation reactions that are time-temperature dependent; other important variables include permeability, pore water chemistry, and sediment composition [26]. Impure diatomaceous sediments with generally less than 80% biogenic or authigenic silica (the rest being carbonate or detrital components) become converted to opal-CT or quartz porcelanites or mudrocks; estimated temperatures of the phase transitions for sediments of this composition are 45 to 50°C for the opal-A to opal-CT transformation and 60 to 85°C for the opal-CT to quartz transformation [55,57,67,69,70]. Work by the previously cited authors has shown that these phase transformations occur through *in situ* dissolution and reprecipitation of silica resulting in "chemical compaction" and porosity loss. Monterey cherts are opal-CT or quartz-bearing, glassy rocks with generally greater than 90% authigenic silica. In a recent study, Behl [25; see also the chapter by Behl and Garrison, this volume] showed that the purest Monterey cherts form by addition of silica (as opposed to the "closed system" of porcelanite and siliceous mudrock diagenesis). Utilizing oxygen isotope thermometry, his work demonstrated that opal-CT cherts in the Monterey formed early (0-350 m burial depth), rapidly (in less than 4 m.y.), and at relatively low temperatures (2-33°C) by pore- and fracture-filling cementation of diatomite. Quartz cherts, in contrast, formed by a variety of mechanisms within a probable temperature range of 36-76°C, a probable range of burial depths of 385-1210 m, and within 4 to 12 m.y. after deposition of the host sediment. Monterey cherts contain a number of distinctive structures that are similar or identical to those described by Steinitz [15,71] in the Mishash Formation. These include intrastratal folds, "dikes", lineations, and spheroids, all shown by Behl [25] to record contrasts in the physical response of the cherts and their enclosing host rocks to burial and tectonic deformation.

Carbonates in the Monterey: Carbonates are present in the Montcrey as siliceous limestones and marls (mainly in the lower calcareous-siliceous facies) and as authigenic dolomite layers and lenses. The siliceous limestones and marls contain common if poorly preserved foraminifers and coccoliths, and silica is present mainly as authigenic opal-CT or quartz, suggesting the original sediment was a foram-coccolith-diatom ooze or mud. Pisciotto and Garrison [52] linked this lithology to comparatively low fertility, low productivity conditions during the early stages of Monterey deposition prior to the intensification of upwelling along the California margin. As noted previously, however, the calcareous-siliceous facies is not present in some basins and only poorly developed in others, hence other factors may have been involved. Moreover, limited petrographic studies suggest that some of the calcite in these rocks has an authigenic as well as biogenic origin. We offer the suggestion that the lower calcareous-siliceous facies may record environments which combined moderate fertility and productivity (compared to very intense upwelling later in the Miocene) with local low oxygen conditions which favored biogenic calcite preservation, authigenic calcite formation, and dissolution of biogenic silica.

Dolomite is the most widespread authigenic carbonate in the Monterey, occurring in all three facies as isolated concretions, concretionary layers and lenses, decimeter-thick beds, brecciated zones and as disseminated crystals in other lithologies. Most conspicuous are concretionary and lenticular beds of dolomite that are restricted to specific stratigraphic horizons and commonly show a rhythmicity that can be correlated in some sections with the Milankovitch precession and obliquity cycles [80]. This latter observation led Garrison [81] to suggest such dolomitic layers may mark flooding surfaces at parasequence boundaries formed during sealevel rises.

Monterey dolomites are calcium-rich (49-56 mol% $CaCO_3$) and have iron contents as high as 16 mol% $FeCO_3$. They have a very wide range of carbon isotope values ($\delta^{13}C_{PDB}$ = - 26 to + 21 ‰) which was interpreted by Pisciotto [82,83] as the result of dolomitization in different zones of organic matter decay. Light-carbon dolomites may thus record derivation from light-carbon dioxide in shallow zones of microbial oxidation and anaerobic sulfate reduction, whereas heavy-carbon dolomites may form during later burial diagenesis in the zone of methanogenesis.

In contrast to their carbon isotopes, Monterey dolomites display a narrow range of oxygen isotope values ($\delta^{18}O_{SMOW}$ = 23 to 38 ‰), but no consensus exists about the interpretation of these values. Some workers, utilizing experimental fractionation expressions of dolomite-water and assuming the water $\delta^{18}O_{SMOW}$ = 0.0 ‰, have attempted to calculate temperatures of formation and burial depths for Monterey dolomites, with estimates of the former in the range of about 10°C to 80°C yielding extrapolated burial depths of a few to over 550 meters. [55, 83, 84]. Several workers, however, have emphasized problems with this approach, including uncertainties about the dolomite-water expression, continued dolomite growth during progressive burial, and re-equilibration with pore waters during dolomite recrystallization at depth [85,86]. In addition, Friedman and Murata [87] found parallel trends in the oxygen isotopes of Monterey dolomites in the San Joaquin Basin and contemporaneous benthic foraminifera from the North Pacific; they interpreted this parallelism to mean that the dolomite formed just beneath the seafloor from in pore waters that were isotopically similar to the bottom waters and preserved its original isotopic composition during subsequent burial.

Phosphatic Rocks: Phosphatic rocks in the Monterey Formation and correlative units are of several types. The most abundant, termed phosphatic marlstones by Garrison *et al.* [88,89] and pristine phosphates by Föllmi and Garrison [90] and Föllmi *et al.* [91], consist of light-colored peloids and nodules of generally friable carbonate fluorapatite (CFA or francolite) that lie within a host rock of organic-rich (up to 30% Total Organic Carbon) shales or porcelanites that are usually laminated. This is the chief lithology of the phosphatic facies and is most abundant within the middle part of the Monterey. Most phosphate nodules and peloids of this type appear to be *in situ* ; their petrology and field occurrences indicate they formed by CFA cementation and replacement of the host sediment. At a number of localities, however, identical nodules can be observed in reworked conglomeritic beds suggesting that they formed during early diagenesis at or just below the sea floor where they could be uncovered and reworked. The organic-rich nature and laminated character of the host rock suggest these phosphates formed in low-oxygen environments, a supposition supported by the presence of sulfur-oxidizing bacterial mats in the host rocks of some of these phosphates [92].

A second kind of Monterey phosphate comprises nodules and thin (10-40 cm) conglomeratic hardground beds of dense CFA which is generally a dark color. These phosphates commonly form or occur within condensed intervals in all parts of the Monterey. Their petrology (e.g. accretionary growth structures) and sedimentary structures (e.g. graded bedding) suggest they are products of complex cycles of phosphatization, exhumation and reworking on the sea floor, reburial and renewed phosphatization [90,91]. Some phosphate beds of this kind appear to mark the boundaries of depositional sequences or parasequences and may have formed in distal basins during changes in sea level [81].

Pelloidal phosphates similar to those in the Mishash Formation constitute a third type of Monterey phosphate. These occur mainly within proximal basin shelfal units (e.g. the Santa Margarita Formation of the Cuyama Basin) that are correlative with the Monterey, but thin turbidite beds composed of phosphate pelloids that were derived from shelfal areas occur within deep-water basinal facies of the Monterey in several areas [88,89]. The shelfal pelloidal phosphates occur in decimeter- to meter-thick layers that are interbedded with finer-grained sediments such as silty shales and siliceous mudstones; the phosphate beds commonly have erosional basal contacts, ripup clasts of the underlying lithology, and graded bedding suggesting they represent high energy deposition, which in some cases may have involved event depostional processes such as shelfal turbidity currents [88,89,93]. Phosphate beds of this type are present in other Neogene shelfal areas within the upwelling region of the eastern Pacific, including the present Peru margin [94], and they appear to record the effects of relatively rapid changes in sea level on broad, shallow shelves in high fertility regions.

Microbial structures are present in all three phosphate varieties, mostly commonly as phosphatized bacterial cells or cyanobacterial coatings, but what role, if any, these organisms played in the phosphatization is unclear. Reimers *et al.* [92], noting that modern bacterial mats show phosphorus enrichment compared to planktonic organic matter, proposed that authigenic CFA formation and genesis of pristine phosphates might be favored in sediments containing abundant phosphorus-bearing mats of this type. It is possible that similar relationships may exist for the other types of phosphates in the Monterey, but this matter has not been investigated in any detail.

Microbial-rich phosphatic marlstones and shales in the Monterey are typically organic-rich (TOC values up to 34% according to Isaacs [59]) and probably constitute major petroleum source rocks. Compared to the Mishash Formation, little geochemical work has been carried out on Monterey phosphates. Kastner *et al.* [95] summarized stable isotopic studies on Monterey francolites for oxygen isotopes of the phosphate ion and for the oxygen and carbon isotopes of the carbonate ion. $\delta^{18}O$ in phosphate ions range between 14.9 and 20.8 ‰, and $\delta^{18}O$ in carbonate ions between 19.5 and 32.2 ‰ (SMOW). These wide ranges contrast markedly with the narrow ranges of Mishash francolites (Fig. 6), and this, along with the strong correlation between the two $\delta^{18}O$ values, suggests that, in the Monterey, both ions re-equilibrated with pore water oxygen in moderate to deep burial diagenetic environments and that the isotopic re-equilibration of the oxygen in the phosphate ion was enzymatically catalyzed. In contrast to the Monterey francolite results, $\delta^{18}O$ values for phosphate in fossil fish bones from the Monterey and related units are mostly higher and have a more restricted range (19.9-20.8 ‰), suggesting these bones equilibrated with Miocene water masses at temperatures between 11 and 25° C and were not subsequently recrystallized during burial diagenesis. $\delta^{13}C$ values of Monterey francolites are negative (mostly between -1 and -8 ‰ (PDB)), values consistent with anaerobic decay in an organic-rich, water-dominated system.

SUMMARY OF DEPOSITIONAL AND DIAGENETIC HISTORIES

Although the Mishash and Monterey formations have similar assemblages of lithofacies, they display pronounced differences in many other characteristics, as summarized in the following paragraphs and in Table 1.

Mishash Formation
Deposition of the Mishash Formation occurred in a sequence of intra-shelf shallow basins where episodic differential movements and sedimentary infilling resulted in the formation of a low undulating relief on the seafloor, with paleoslopes ranging from 0.1 to 1° [16]. This extremely low relief, when coupled with periodic sea-level oscillations [2] must have resulted in a change from basinal marine sedimentation to high energy reworking conditions to subaerial exposure and interaction with fresh water (Fig. 9) [23]. Indeed Steinitz [96] showed that many of the cherts in the Mishash Formation replaced pre-existing evaporites - anhydrite and/or gypsum; "ghosts" of evaporitic phases are rare but widely distributed throughout the Mishash cherts. When comparing this information to the depletion in ^{18}O in heterogeneous chert matrix, Kolodny *et al.* [21] suggested that the Mishash environment was schizohaline, a model in good agreement with the one suggested for chert formation by Knauth [97].

The variable schizohaline nature of the Mishash environment is most strongly revealed by the record preserved in the diagenetic rather than the depositional phase of the assemblages, as convincingly demonstrated by the multi-phase isotopic analyses performed on these rocks by

Vengosh et al. [30]. These authors identified three isotopically concordant assemblages (see Fig. 4):

(a) a depositional assemblage, formed in normal sea water ($\delta^{18}O_w = -1‰$) has been recorded by benthic foraminifer (*Nodosaria*) skeletons ($\delta^{18}O_{SMOW} = +27.5$ to $29.5 ‰$) and apatitic biodetritus ($\delta^{18}O = +18$ to $19.5 ‰$). The latter phase may also actually reflect early

Table 1.
Comparison of depositional and diagenetic parameters for the Mishash and Monterey Formations.

	Mishash Fm.	Monterey Fm.
Tectonic setting	Passive margin	Transform margin
Basin size	small (nx10^1 by nx10^1 km)	small (nx10^1 by nx10^2 km)
Basin subsidence	slow(1 - 10m/my)	rapid (>10^2 m/my)
Durations	6 m.y. (73.5-79.5 Ma)	11-12 m.y. (6-18 Ma)
Thicknesses	0.1- 10x10^{2}m	0.1-3 x 10^3 m
Sedimentation (decompacted)		
rates	slow (1-40 cm/10^{3}yrs)	high (>400 cm/10^{3}yrs)
Water depths	10^0-10^2 m	10^2-10^3 m
Burial depths	10^2 m	10^2-10^3 m
Geothermal gradients	uniform (20-30°C/km)	variable (20-70°C/km)
Carbonate diagenesis ($\delta^{13}C$)	- $\delta^{13}C$, sulfate reduction	+ and - $\delta^{13}C$, sulfate reduction & methanogenesis
$\delta^{18}O_{phos.}$	peloids = fish debris	peloids << fish debris
Temperatures of phosphatization (from $\delta^{18}O_{phos.}$)	30°C	16-68°C

diagenetic conditions which were probably not too different from the depositional ones (see below and [98]).

(b) an early diagenetic assemblage. Conditions recorded by these phases were usually close to normal marine, but at times were evaporitic. Thus calcite in micritic cements (+26 to 27.5 ‰), opal-CT (+31.5 to 33.3 ‰), quartz in homogeneous cherts and in most fragments in chert breccias (+31 to 32 ‰) reflect conditions indistinguishable from normal marine. Some of the more ^{18}O-enriched chert fragments (+33 ‰) probably reflect evaporitic conditions which are also recorded by relicts of evaporite minerals.

(c) a later diagenetic assemblage which apparently records a rather short-lived fresh water phase in the history of Mishash diagenesis. Calcitic spar infillings of fossils (+20 to 26.6 ‰), silica in the matrices of some chert breccias (+21 to 33 ‰) and coarse quartz in silicified phosphorites (+20 to 30 ‰) are evidence for such an episode.

Changes in fertility were a major factor which determined the depositional features of the Mishash Formation. Siliceous rocks of the Lower Mishash and Sayyarim Formations alternate with biomicrites rich in planktic epipelagic foraminifera and coccoliths. This alternation has been explained by Reiss [9] in a way analogous to that suggested by Ingle [49] for the Monterey, namely by changes in photic zone fertility. The abundance of benthic foraminifera which monitor food supply is, according to Berger and Herguera [99] the best indication of high productivity in the overlying water. Thus the silica to form cherts and porcelanites was deposited in an environment of intense upwelling and fertility. If the water column was shallow enough to prevent dissolution of a large part of this silica on its way to sediment, porcelanites and cherts are the resultant product. In such a sea the expansion of the oxygen minimum zone high into the water column eliminates the mesopelagic and epipelagic foraminifera.

In the Phosphoritic Member of the Mishash Formation, fertility was still higher than normal though not as high as during deposition of the siliceous rocks. The abundance of vertebrate fossil remains in the phosphatic beds is witness to the well developed food chain in the environment which formed these sediments.

Monterey Formation

In contrast to the Mishash, most Monterey deposition occurred along a fractured continental margin in deeply and rapidly subsided basins [49, 50] where the effects of sea level changes were much less pronounced and energy levels tended to remain low over long time spans. Some evaporites are present in proximal non-marine to shoreline facies which are correlative to the Monterey, but they are exceedingly rare [52]; all evidence points to deposition in marine waters of normal salinity and temperature for most of the Monterey. The Monterey contains, however, a more diverse lithologic assemblage as well as more pronounced lateral facies and thickness variations compared to the Mishash. This diversity appears due mainly to two factors: 1) Local paleogeographic variations; for example the contrast between proximal basins (where biosiliceous rocks are typically clastic-rich siliceous mudstones and where coarser clastics such as sandstones are interbedded with the biosiliceous rocks) and distal basins (where pelagic cherts are common). Contrasting environments in steep-sided basins (e.g. outer shelf, slope, basin floor or banktop) may also account for some of the lateral variations in thickness and facies. 2) Fluctuations in paleoceanographic conditions, for example changes from periods of strong upwelling and high fertility that yielded biosiliceous facies as to times of less intense upwelling and lower productivity when pelagic sediments were more carbonate-rich.

Also in contrast to the Mishash, in the Monterey there was apparently much less of a linkage between the diagenetic and depositional environments. Whereas the isotopes of Mishash carbonates and cherts record diagenesis in or near environments that varied from normal marine to schizohaline, comparatively rapid accumulation of Monterey sediments more quickly isolated them from seafloor conditions which in any case were more uniform than those of the Mishash. Burial diagenesis thus affected Monterey sediments to a much larger degree because of their greater and more rapid burial and also because of the higher geothermal gradients in some Monterey basins. The most obvious manifestations of this are the thick sections of opal-CT and quartz porcelanites and cherts in the Monterey. But it is also reflected 1) in the isotopic compositions of some authigenic dolomites that record dolomitization within rapidly buried sediments in the zone of methanogenesis, and 2) in the $\delta^{18}O$ values for both the phosphate and carbonate ions in Monterey francolites suggesting isotopic re-equilibration at moderate to deep burial depths.

Both the Monterey and Mishash formations contain widespread phosphatic rocks, but economic peloidal phosphorites, common in the Mishash of the Negev, are scarce in the Monterey. The apparent reason is that extensive shelfal environments, the sites of winnowing and reworking which concentrated peloidal phosphate grains in the Mishash, were very rare in the Monterey basins.

CONCLUSIONS

Though their tectonic, depositional and diagenetic settings differed substantially, the Mishash and Monterey formations have remarkably similar lithologic assemblages of biosiliceous, carbonate and phosphatic rocks. This similarity attests to the importance of paleoceanographic conditions, particularly upwelling, in determining pelagic lithofacies in continental margin environments. Thus a number of features which can be directly related to the high productivity are common to the Mishash and the Monterey. For example, we see similarity in those phenomena which are related to the very presence of biosiliceous sediments: chert breccias,

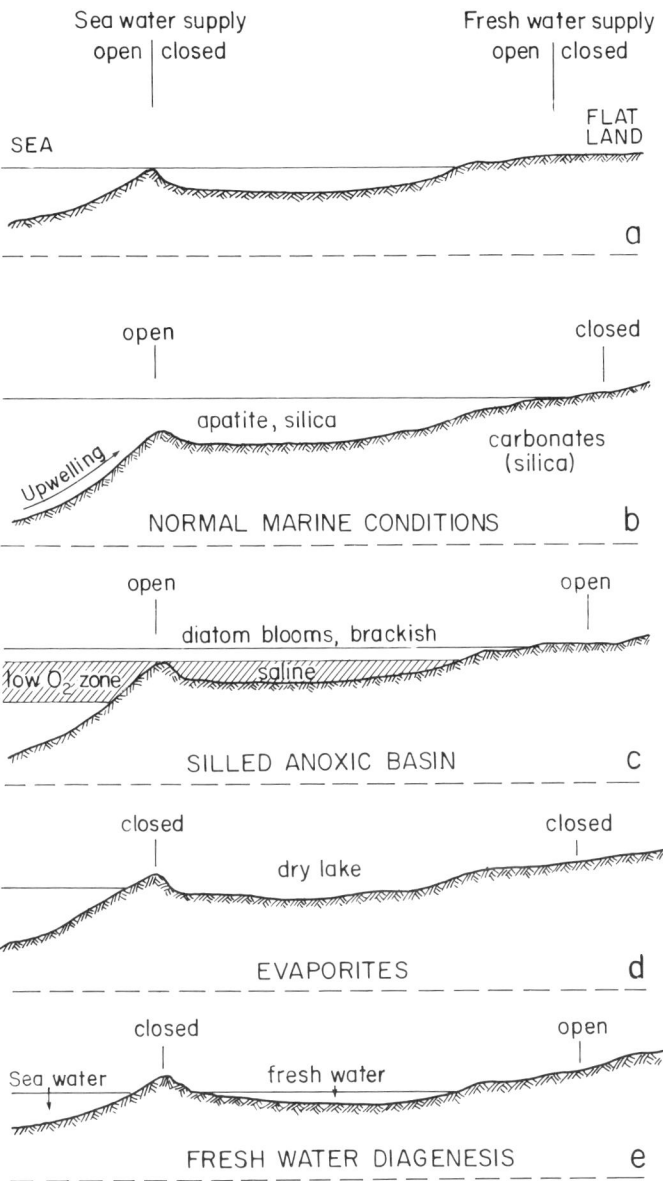

Figure 9. Scheme of a sequence of sedimentary environments which would result from alternations in open/close mode of two hypothetical "valves" to a Mishash basin: the "open sea valve" controlling connection to the ocean and supplying normal sea water and the "fresh water valve" which controls runoff from the continent. After Kolodny [23].

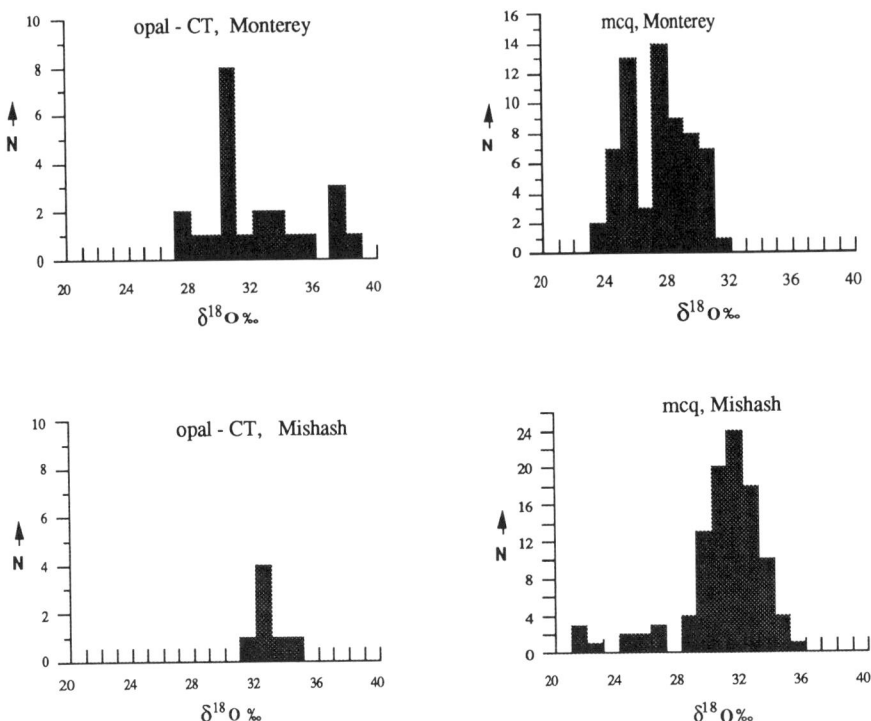

Figure 10. Histograms of $\delta^{18}O$ values for opal-CT and diagenetic microcrystalline quartz in cherts and porcelanites in the Mishash and Monterey formations. Data from Kolodny et al. [22], Behl [25], Pisciotto [55] Haimson [68], and Kastner [105].

contortions in cherty beds, appearance of chert spheroids, all of which seem to be caused by mechanical properties of chert [25] and are impressively similar in both formations. The abundance of what is usually considered as a diagenetic precursor of chert is also common to both formations. The good preservation of opal-CT in the almost 80 m.y. old Mishash Formation is particularly noteworthy, whereas the much younger Monterey Formation contains large amounts of unaltered opal-A diatomites.

As repeatedly noted above, in contrast to the depositional similarity, the diagenesis of the Monterey and the Mishash were sharply different. The differences are recorded in the total observed sediment thickness and in the inferred geothermal gradients (Table 1). The difference in diagenetic paths is striking in some sedimentological features: whereas phosphogenesis in the Monterey was mostly frozen in the initial pristine stage in laminated sediments, and with only rare reworking by "water dynamics" [100, phosphogenesis in the shallow water Mishash Formation has reached its ultimate expression, the formation of high grade phosphorite deposits due to sedimentary reworking in shelf environments. On the other hand the fate of the organic matter diagenesis has been more favorable in the Monterey Formation where organic compounds commonly reached maturation to petroleum [74], whereas the organic rich Ghareb shales overlying the Mishash were preserved as kerogen, never reaching the "oil window" [101,102]. The difference in diagenetic history also left its imprint on three isotopic criteria: a. $\delta^{13}C$. Practically all Mishash authigenic carbonates are depleted in ^{13}C [23] (Fig 4), probably reflecting the prevalence of sulfate reduction within the Mishash sediments [15]. Authigenic carbonates in the Monterey Formation are in some cases depleted in other cases enriched in ^{13}C [52,103,104]. This indicates that in the Monterey diagenesis proceeded beyond the sulfate reduction stage and fermentation resulted in methane formation; ^{12}C was preferentially removed into the methane leaving "heavy" carbonate behind. b. $\delta^{18}O_{Si}$. The isotopic composition of oxygen in silica phases ranges in both formations over a more or less similar range: $\delta^{18}O_{Si}$ = 22 to 35‰ (Fig. 10). The interpretation of these similar figures is however very different in both cases, and is dictated by our understanding (or misunderstanding?) of the respective geological conditions. In the case of the Monterey there is no reason to assume participation of fresh water in the diagenesis of its siliceous rocks, whereas there is every reason to expect a strong temperature gradient with increasing burial. Hence the low $\delta^{18}O$ values are interpreted here as reflecting elevated temperatures of diagenesis of the Monterey cherts and porcelanites. On the other hand one can hardly assume sufficiently deep burial of Mishash sediments to increase temperatures in the expected geothermal gradient. Hence one is left with the interpretation of fresh water as explaining the low $\delta^{18}O$ values. Fortunately the latter interpretation is supported by the boron concentration analyses [22]. One need not emphasize how far reaching this divergent interpretation of similar numbers is, in interpreting paleo-environments. c. $\delta^{18}O_P$. The isotopic composition of oxygen in the phosphate ion is remarkably uniform in all phosphates of the Mishash Formation. The range of $\delta^{18}O_P$ for the entire Mishash Formation in all of Israel is 3.5‰ (16.8 to 20.3‰., 115 samples from [30,39] (see Fig. 5). In fact this range covers also 20 phosphorite samples from the entire Mediterranean - Tethyan phosphogenetic province, from Turkey to Morocco [39]. Part of this variance is explained by some regional trends, such as lower $\delta^{18}O_P$ values of phosphorites in the Elat and Sinai regions. Compare this with a range of 5.9‰ for the Monterey (14.9 to 20.8‰, 19 samples, from [95]). The difference is more evident when comparing single localities between them or when comparing histograms of $\delta^{18}O_P$ values in the two localities (Fig. 6): whereas the Mishash shows a strongly peaked unimodal distribution with a mode at 18.7‰, a mean at 18.3‰ and a standard deviation of 1‰, $\delta^{18}O_P$ in the Monterey phosphates is flatly distributed, with no mode, a mean of 17.6 and a standard deviation of 2‰. Shemesh et al. [106] and Kastner et al. [95] explained this difference in $\delta^{18}O_P$ spread by the ability of Monterey phosphates to record the diagenetic temperature range of 16 to 68°C due to sufficient microorganism activity in the fermenting Monterey sediment, whereas the Mishash phosphates recorded principally only their depositional signal. This difference is further stressed by the large difference between $\delta^{18}O_P$ of fossil fish and of apatitic peloids in the Monterey formation, whereas $\delta^{18}O_P$ of fish debris and peloids is practically indistinguishable in the Mishash Formation.

Acknowledgments
A large part of this work was done while Y.K was a Visiting Allan Cox professor of Geology at Stanford University. R.E.G.'s work on the Monterey Formation has been supported over the years by grants from the National Science Foundation and the Petroleum Research Fund of the American Chemical Society (particularly grants PRF 18439-AC3 and PRF 25659-AC2); we gratefully acknowledge the donors of the PRF/ACS grants. We are also grateful to A. Starinsky, A.M. Abed and R. Behl for helpful discussions and to Y. Bartov, Z. Reiss and A. Iijima for reviews of this manuscript along with many useful suggestions. E. Portillo-Hegemier helped with the drafting, and R. Kunnanz-Petersen, M.-B. Harhen, E. Boring and C. Dunlap provided invaluable assistance in preparation of the manuscript.

REFERENCES

1. Z. Garfunkel. The pre-Quaternary geology of Israel. In: *The zoogeography of Israel*. Y. Yom-Tov and E. Tchernov (Eds). vol. pp. 7-34. Dr. W. Junk Publishers, Dordecht (1988).
2. Z. Lewy. Transgressions, regressions and relative sea level changes on the Cretaceous shelf of Israel and adjacent countries, a critical evaluation of Cretaceous global sea level correlations, *Paleoceanography* **5**, 619-637 (1990).
3. G. M. Lees. The chert beds of Palestine, *Proc. geol. Ass*. **39**, 445-462.(1928).
4. Y. Arkin and M. Hamaoui. The Judea Group (Upper Cretaceous) in central and southern Israel, *Isr. Geol. Surv. Bull*. **42**, 1-17 (1967).
5. E. Krenkel. Der Syrische Bogen, *Cent. Miner. Geol. Paleontol. Abh. B*,. **9, 10**, 274-281, 301-313. (1924).
6. Y. K. Bentor. Relations entre la tectonique et les depots de phosphates dans le Neguev israelien. In: *C.r Geol. Cong. Int. 19th sess*. Algiers. Sec. XI, pt.11, 93-101 (1953)

7. A. Flexer. Stratigraphy and facies development of the Mount Scopus Group (Senonian - Paleocene) in Israel and adjacent countries, *Isr. J. Earth-Sci.* **17**, 85-114 (1968).
8. A. Flexer. Late Cretaceous paleogeography of northern Israel and its significance for the Levant geology, *Peleogeogr. Paleoclimat. Paleoecol.* **10**, 293-316 (1971).
9. Z. Reiss. Assemblages from a Senonian high-productivity sea, *Rev. Paleobiol., Spec. Publ. "Benthos '86". fasc. 2*, 323-332 (1988).
10. Y. Bartov, Y. Eyal, Z. Garfunkel and G. Steinitz. Late Cretaceous and Tertiary stratigraphy and paleogeography of southern Israel, *Isr. J. Earth-Sci.* **21**, 69-97 (1972).
11. Y. Kolodny. Lithostratigraphy in the Mishash Formation, Northern Negev, *Isr. J. Earth-Sci.* **16**, 57-73 (1967).
12. Y. Nathan, Y. Shiloni, R. Roded, I. Gal and Y. Deutsch. The geochemistry of the northern and central Negev phosphorites (Southern Israel), *Isr. Geol. Surv. Bull.* **73**, 43 pp. (1979).
13. D. Soudry, Y. Nathan and R. Roded. The Ashosh-Haroz facies and their significance for the Mishash paleogeography and phosphorite accumulation in the Northern and Central Negev (Southern Israel), *Isr. J. Earth-Sci.* **34**, 211-220 (1985).
14. Y. K. Bentor and A. Vroman. *The geological map of the Negev, 1:100,000. Sheet Abde (Ovdat), 1st ed.* Tel Aviv (1951).
15. G. Steinitz. Enigmatic chert structures in the Senonian cherts of Israel, *Isr. Geol. Surv. Bull.* **75**, 1-46 (1981).
16. G. Steinitz. Paleogeography of the Menuha and Mishash Formations in the Eastern Ramon area, Southern Israel, *Isr. J. Earth-Sci.* **25**, 70-75 (1976).
17. Y. Kolodny, Y. Nathan and E. Sass. Porcellanite in the Mishash Formation, Negev, Southern Israel, *J. Sed. Pet.* **35**, 454-463 (1965).
18. J. B. Jones and E. R. Segnit. The nature of opal: I. Nomenclature and constituent phases, *J. geol. Soc. Australia* **18**, 56-68 (1971).
19. Y. Kolodny. Petrology of Siliceous Rocks in the Mishash Formation (Negev, Israel), *J. Sed. Pet.* **39**, 166-175 (1969).
20. H. R. Wenk and Y. Kolodny. Preferred orientation of quartz in a chert breccia., *Proc. Nat. Acad. Sci.* **59**, 1061-1066 (1968).
21. W. H. Berger and U. von Rad. Cretaceous and Cenozoic sediments from the Atlantic Ocean. In: *Initial Reports of the Deep Sea Drilling Project.* D. E. Hayes A. C. Pimm *et al.* (Eds). vol. XIV, pp. 787-954. U. S. Government Printing Office. Washington (1972).
22. Y. Kolodny, A. Tarablous and U. Frieslander. Participation of fresh water in chert diagenesis - evidence from stable isotopes and boron-track mapping, *Sedimentology* **27**, 305-316 (1980).
23. Y. Kolodny. Carbon isotopes and depositional environment of a high productivity sedimentary sequence - the case of the Mishash - Ghareb Formations, Israel, *Isr. J. Earth-Sci.* **29**, 147-156 (1980).
24. N. L. Taliaferro. Contraction phenomena in cherts, *Geol. Soc. Am. Bull.* **45**, 189-232 (1934).
25. R. J. Behl. *Chertification in the Monterey Formation of California and Deep-Sea Sediments of the West Pacific.* Unpublished Ph. D. Thesis. University of California, Santa Cruz (1992).
26. M. Kastner. Authigenic silicates in deep-sea sediments: formation and diagenesis. In: *The Sea, The Oceanic Lithosphere.* C. Emiliani (Ed.). vol. 7, pp. 915-980. John Wiley & Sons, New York (1981).
27. D. Soudry, S. Moshkovitz and A. Ehrlich. Occurrence of siliceous microfossils (diatoms, silicoflagellates and sponge spicules) in the Campanian Mishash Formation, southern Israel, *Eclogae geol. Helv.* **74**, 97-107 (1981).
28. S. Moshkovitz, A. Ehrlich and D. Soudry. Siliceous microfossils of Upper Cretaceous Mishash Formation, central Negev, Israel, *Cretaceous Res.* **4**, 173-194 (1983).
29. Y. Haas, Z. Reiss and G. Honig. Note on Senonian Radiolaria from Israel, *Isr. J. Earth-Sci.* **34**, 167-171 (1985).
30. A. Vengosh, Y. Kolodny and M. Tepperberg. Multi-phase oxygen isotopic analysis as a tracer of diagenesis: the example of the Mishash Formation, Cretaceous, Israel, *Chem. Geol. (Isotope Geosc.)* **65**, 235-253 (1987).
31. D. Soudry. Ultra-fine structures and genesis of the Campanian Negev high-grade phosphorites (southern Israel), *Sedimentology* **34**, 641-660 (1987).
32. E. Sass and Y. Kolodny. Isotope geochemistry and origin of carbonate concretions, *Chem. Geol.* **10**, 261-286 (1973).
33. H. Irwin, C. Curtis and M. Coleman. Isotopic evidence for source of diagenetic carbonates formed during burial of organic rich sediments, *Nature* **269**, 209-213 (1977).
34. T. F. Pedersen and S. E. Calvert. Anoxia vs. productivity: what controls the formation of organic-carbon-rich sediments and sedimentary rocks?, *AAPG Bull.* **74**, 454-466 (1990).
35. D. Soudry and Y. Champetier. Microbial processes in the Negev phosphorites (southern Israel), *Sedimentology* **30**, 411- 423 (1983).
36. D. Soudry. Intervention de schyzophytes dans la phosphomicritisation de debris osseux, *C.r. hebd. Seanc. Acad.Sci., Paris.* **288D**, 669-671 (1979).

37. Y. Avital, A. Starinsky and Y. Kolodny. Uranium geochemistry and fission-track mapping of phosphorites, Zef'a Field, Israel, *Econ. Geol.* **78,** 121-131(1983).
38. A. Shemesh, Y. Kolodny and B. Luz. Oxygen isotope variations in phosphate of biogenic apatites. II. Phosphorite rocks, *Earth Planet. Sci. Lett.* **64,** 405-416 (1983).
39. A. Shemesh and Y. Kolodny. Oxygen isotope variations in phosphorites from the southeastern Tethys, *Isr. J. Earth-Sci.* **37,** 1-15 (1988).
40. Y. Kolodny, B. Luz and O. Navon. Oxygen isotope variations in phosphate of biogenic apatites. I. Fish bone apatite -- rechecking the rules of the game, *Earth Planet. Sci. Lett.* **64,** 398-404 (1983).
41. Y. Kolodny and M. Raab. Oxygen isotopes in phosphatic fish remains from Israel: paleothermometry of tropical Cretaceous and Tertiary shelf waters, *Paleogeogr. Paleoclimatol. Paleoecol.* **64,** 59-67 (1988).
42. A. Longinelli and S. Nuti. Revised phosphate-water isotopic temperature scale, *Earth Planet. Sci. Lett.* **19,** 373-376 (1973).
43. P. Grandjean, H. Capetta and F. Albarede. The REE and εNd of 40-70 Ma old fish debris from the West-African platform, *Geophys. Res. Lett.* **15,** 389-392 (1988).
44. J.C. Ingle. Origin of Neogene diatomites around the North Pacific Rim. In: *The Monterey Formation and Related Siliceous Rocks of California.* R.E. Garrison and R.G. Douglas (Eds). pp. 159-179, Pacific Sec., Soc. Econ. Paleon. Min. Pub. (1981).
45. R.B. Dunbar, R.C. Marty and P.A. Baker. Cenozoic marine sedimentation in the Sechura and Pisco basins, Peru, *Palaeogeogr. Palaeoclimatol. Palaeoecol.* **77,** 235-262 (1990).
46. T. Atwater. Implications of plate tectonics for the Cenozoic tectonic evolution of western North America, *Geol. Soc. Amer. Bull.* **81,** 3513-3536 (1970).
47. M.C. Blake, R.H. Campbell, T.W. Dibblee, D.G. Howell, T.H. Nilsen, W.R. Normark, J.C. Vedder, and E.A. Silver. Neogene basin formation in relation to plate-tectonic evolution of San Andreas fault system, California, *Amer. Assoc. Petroleum Geologists Bull.* **62,** 344-372 (1978).
48. S.A. Graham. Tectonic controls on petroleum occurrence in central California. In: *Cenozoic Basin Development of Coastal California.* R.V. Ingersoll and W.G. Ernst, W.G. (Eds.). Ruby Volume VI, pp. 407-426. Prentice-Hall Inc., Inglewood Cliffs, N.J. (1987).
49. J.C. Ingle. Cenozoic depositional history of the northern continental borderland of southern California and the origin of associated diatomites. In: *Guide to the Monterey Formation in the California coastal area, Ventura to San Luis Obispo.* C.M. Isaacs (Ed). pp. 1-8, Pacific Sec., Amer. Assoc. Petroleum Geologists, 52 (1981).
50. W.R. Dickinson, R.A. Armin, N. Beckvar, T.C. Goodin, S.U. Janecke, R.A. Mark, R.D. Norris, G. Radel, and A.A. Wortman. Geohistory analysis of sediment accumulation and subsidence for selected California basins. In: *Cenozoic Basin Development of Coastal California.* R.V. Ingersoll and W.G. Ernst (Eds), Rubey Volume VI, pp. 7-23. Prentice-Hall, Inglewood Cliffs, N.J. (1987).
51. C.M. Isaacs. Abundance versus rates of accumulation in fine-grained strata of the Miocene Santa Barbara Basin, California, *Geo-Marine Letters* **5,** 25-30 (1985).
52. K.A. Pisciotto and R.E. Garrison. Lithofacies and depositional environments of the Monterey Formation, California, In: *The Monterey Formation and Related Siliceous Rocks of California.* R.E. Garrison and R.G. Douglas (Eds). pp 97-122, Pacific Sec., Soc. Econ. Paleon. Min. Pub. (1981)..
53. M.B. Lagoe. Subsurface facies analysis of the Saltos Shale Member, Monterey Formation (Miocene) and associated rocks, Cuyama Valley, California. In: *The Monterey Formation and Related Siliceous Rocks of California.* R.E. Garrison and R.G. Douglas (Eds) pp. 199-211, Pacific Sec., Soc. Econ. Paleon. Min. Pub. (1981).
54. M.B. Lagoe. Depositional environments in the Monterey Formation, Cuyama Basin, California, *Geol. Soc. Amer. Bull.* **96,** 1296-1312 (1985).
55. K.A. Pisciotto. *Basinal Sedimentary Facies and Diagenetic Aspects of the Monterey Shale, California.* Ph.D. Thesis, University of California, Santa Cruz (1978).
56. C.R. Canfield. Subsurface stratigraphy of Santa Maria Valley oil field and adjacent parts of Santa Maria Valley, California, *Amer. Assoc. Petroleum Geologists Bull.* **23,** 45-81 (1939).
57. C.M. Isaacs. *Diagenesis in the Monterey Formation examined laterally along the coast near Santa Barbara. California.* Ph.D. Thesis, Stanford University, California (1980).
58. J.A. Barron. Paleoceanographic and tectonic controls on deposition of the Monterey Formation and related siliceous rocks in California, *Palaeogeogr. Palaeoclimatol. Palaeecol.* **53,** 27-45 (1986).
59. C.M. Isaacs, C.M.. The Miocene Monterey Formation - depositional and diagenetic facies along the Santa Barbara, California coastal area, In: *Field notes on the Monterey Formation, Santa Barbara area.* C.M. Isaacs (Ed.). pp. 1-30, California Amer. Assoc. Petroleum Geol. Student Chapter Fieldtrip # 1, Guidebook (1987).
60. C.M. Isaacs and M.B. Lagoe. Mid-Tertiary biogeneous silica deposition in California - the pre-Monterey record. In: *Pacific Neogene Event Studies.* R. Tsuchi (Ed), pp. 29-42. Shizuoka University, Shizuoka (1987).
61. G. Keller and J.A. Barron. Paleoceanographic implications of Miocene deep-sea hiatuses, *Geol. Soc. Amer. Bull.* **94,** 590-613 (1983).
62. J.A. Barron and G. Keller. Paleotemperature oscillation in the middle and late Miocene of the northeastern Pacific, *Micropaleontology* **29,** 150-181 (1983).

63. M.N. Bramlette, The Monterey Formation of California and the origin of its siliceous rocks, *U.S. Geol. Survey Prof. Paper* **212**, 1-57 (1946).
64. K.J. Murata and J.R. Nakata. Cristobalitic stage in the diagenesis of diatomaceous shale, *Science* **184**, 567-568 (1974).
65. K.J. Murata and R.R. Larson. Diagenesis of Miocene siliceous shales, Temblor Range, California. *U.S. Geol. Survey Jour. Res.*, **3**, 553-566 (1975).
66. K.J. Murata, I. Freidman and J.D. Gleason. Oxygen isotope relations between diagenetic silica minerals in Monterey Shale, Temblor Range, California. *Amer. J. Sci.* **277**, 259-272 (1977).
67. K.A. Pisciotto. Diagenetic trends in the siliceous facies of the Monterey Shale in the Santa Maria region, California, *Sedimentology* **28**, 547-571 (1981).
68. M. Haimson. *Oxygen Isotope Studies of Silica in the Monterey Formation, California*. M.Sc. Thesis, Arizona State University (1982).
69. M.A. Keller and C.M. Isaacs. An evaluation of temperature scales for silica diagenesis in diatomaceous sequences including a new approach based on the Miocene Monterey Formation, California, *Geo-Marine Letters* **5**, 31-35 (1985).
70. C.M. Isaacs. Influence of rock composition on kinetics of silica phase changes in the Monterey Formation, Santa Barbara area, California, *Geology* **10**, 304-308 (1982).
71. G. Steinitz. Chert "dike" structures in Senonian chert beds, southern Negev, Israel, *J. Sed. Pet.* **40**, 1241-1245 (1970).
72. G.B. Hieshima. *Sedimentology of Miocene Monterey Formation Diatomites, California*. M.S. Thesis, University of Wisconsin, Madison (1987).
73. J.S. Hornafius. Facies analysis of the Monterey Formation in the northern Santa Barbara Channel, *Amer. Assoc. Petroleum Geologists Bull.* **75**, 894-909 (1991).
74. C.M. Isaacs and N.F. Petersen, N.F. Petroleum in the Miocene Monterey Formation, California. In: *Siliceous Sedimentary Rock-Hosted Ores and Petroleum*. J.R. Hein (Ed.). pp. 83-116, Van Nostrand Reinhold, New York (1987).
75. C.E. Savdra and D.J. Bottjer. Trace-fossil model for reconstruction of paleo-oxygenation in bottom waters, *Geology* **14**, 3-6 (1986).
76. F.M. Govean and R.E. Garrison. Significance of laminated and massive diatomites in the upper part of the Monterey Formation, California. In: *The Monterey Formation and Related Siliceous Rocks of California*. R.E. Garrison and R.G. Douglas (Eds). pp. 181-198, Pacific Sec., Soc. Econ. Paleon. Min. Pub. (1981).
77. T. MacKinnon. Origin of the Miocene Monterey Formation in California. In: *Oil in the California Monterey Formation*. T. MacKinnon (Ed.). pp. 1-10, 28th Int. Geol. Congress, Field Trip Guidebook T311 (1989).
78. K.M. Bohacs. Sequences stratigraphy of the Monterey Formation, Santa Barbara County: Integration of physical, chemical, and biofacies data from outcrop and subsurface. In: *Miocene and Oligocene Petroleum Reservoirs of the Santa Maria and Santa Barbara-Ventura Basins, California*. M.A. Keller and M.K. McGowen (Eds). pp. 139-201. Soc. Econ. Paleon. Min. Core Workshop No. 24 (1990).
79. J.R. Schwalbach and K.M. Bohacs (Eds). *Sequence Stratigraphy in Fine-grained rocks: examples from the Monterey Formation*. Pacific Sec., Soc. Econ. Paleon. Min. Pub., vol. 70, 1-80 (1992).
80. L.D. White. *Chronostratigraphic and Paleoceanographic Aspects of Selected Chert Intervals in the Miocene Monterey Formation California*. Ph.D. Thesis, University of California, Santa Cruz (1989).
81. R.E. Garrison. Neogene lithofacies and depositional sequences associated with upwelling regions along the eastern margin of the Pacific. In: *Pacific Neogene: Environment, Evolution and Events*. R. Tsuchi and J.C. Ingle (Eds). pp. 43-68, Univ. of Tokyo Press, Tokyo (1992).
82. K.A. Pisciotto. Review of secondary carbonates in the Monterey Formation, California. In: *The Monterey Formation and Related Siliceous Rocks of California*.. R.E. Garrison and R.G. Douglas (Eds). pp. 273-283, Pac. Sec., Soc. Econ. Paleon. Min. Pub. (1981).
83. K.A. Pisciotto and J.J. Mahoney. Isotopic survey of diagenetic carbonates, DSDP Leg 63. In: *Initial Reports of the Deep Sea Drilling Project*. R.S. Yeats and B.U. Haq (Eds). vol. 63, pp. 595-609. U.S. Govt. Printing Office, Washington (1981).
84. R.E. Garrison and S.A. Graham. Early diagenetic dolomites and the origin of dolomite-bearing breccias, lower Monterey Formation, Arroyo Seco, Monterey County, California. In: *Dolomites of the Monterey Formation and Other Organic-rich Units*. R.E. Garrison, M. Kastner, and D.H. Zenger (Eds). pp. 87-102, Pacific Sec., Soc. Econ. Paleon. Min. Pub. no. **41** (1984).
85. J. Kushnir and M. Kastner. Two forms of dolomite occurrences in the Monterey Formation, California: concretions and layers - a comparative mineralogical, geochemical and isotopic study. In: *Dolomites of the Monterey Formation and Other Organic-rich units*. R.E. Garrison, M. Kastner and D.H. Zenger (Eds), pp. 171-184, Pacific Sec., Soc. Econ. Paleon. Min. Pub. no. **41** (1984).
86. P.A. Baker and S.J. Burns. Occurrence and formation of dolomite in organic-rich continental margin sediments, *Amer. Assoc. Petroleum Geol. Bull.* **69**, 1917-1930 (1985).
87. I. Friedman and K.J. Murata. Origin of dolomite in Miocene Monterey Shale and related formations in the Temblor Range, California.,*Geochim. Cosmochim. Acta* **43**, 1357-1365 (1979).

88. R.E. Garrison, M. Kastner, and Y. Kolodny. Phosphorites and phosphatic rocks in the Monterey Formation and related Miocene units, coastal California. In: *Cenozoic Basin Development in Coastal California*. R.V. Ingersoll and W.G. Ernst (Eds). pp. 349-381. Prentice Hall, Englewood Cliffs, N.J. (1987)
89. R.E. Garrison, M. Kastner, and C.E. Reimers. Miocene phosphogenesis in California. In: *Phosphate Deposits of the World, vol. 3, Neogene to Modern Phosphorites*. W. C. Burnett and S.R. Riggs (Eds.). pp. 285-299. Cambridge Univ. Press (1990).
90. K.B. Föllmi and R.E. Garrison. Phosphatic sediments, ordinary or extraordinary deposits? The example of the Miocene Monterey Formation (California). In: *Controversies in Modern Geology*. D.W. Müller, J.A. McKenzie, and H. Weissert (Eds). pp. 55-84. Academic Press, London (1991).
91. K.B. Föllmi, R.E. Garrison, and K.A. Grimm. Stratification in phosphatic sediments: illustrations from the Neogene of central California. In: *Cycles and Events in Stratigraphy*. Einsele, G., Ricken, W., and Seilacher, A. (Eds). pp. 492-507. Springer-Verlag, Berlin (1991).
92. C.E. Reimers, M. Kastner, and R.E. Garrison. The role of bacterial mats in phosphate mineralization with particular reference to the Monterey Formation. In: *Phosphate Deposits of the World, Vol. 3, Neogene to Modern Phosphorites*. W.C. Burnett and S.R. Riggs (Eds). pp. 300-311, Ch. 24. Cambridge Univ. Press (1990).
93. A.E. Roberts and T.L. Vercoutere. Geology and petrology of the upper Miocene phosphate deposit near New Cuyama, Santa Barbara County, California, *U.S. Geological Survey Bull.* **B-1635**, 89 pp. (1985).
94. R.E. Garrison and M. Kastner. Phosphatic sediments and rocks recovered from the Peru margin during ODP Leg 112. In: *Proceedings of the Ocean Drilling Program, Scientific Results*. E. Suess, R. von Huene, R. and K. Emeis (Eds). vol. 112, pp. 111-134. College Station, Texas (1990).
95. M. Kastner, R.E. Garrison, Y. Kolodny, C.E. Reimers and A. Shemesh. Coupled changes of oxygen isotopes in PO_4^{3-} and CO_3^{2-} in apatite, with emphasis on the Monterey Formation, California. In: *Phosphate Deposits of the World, Vol. 3, Neogene to Modern Phosphorites*. W.C. Burnett and S.R. Riggs (Eds). pp. 312-324, Ch. 25. Cambridge Univ. Press (1990).
96. G. Steinitz. Evaporite-chert associations in Senonian bedded cherts, Israel, *Isr. J. Earth-Sci.* **26**, 55-63 (1977).
97. L. P. Knauth. A model for the origin of chert in limestone, *Geology*. **7**, 274-277. (1979).
98. Y. Kolodny and B. Luz. Oxygen isotopes in phosphates of fossil fish - Devonian to Recent. In: *Stable Isotope Geochemistry: A Tribute to Samuel Epstein*. H. P. Taylor Jr., J. R. O'Neil and I. R. Kaplan (Eds). vol. 3, pp. 105-119. The Geochemical Society, London(1992).
99. W.H. Berger and J.C. Herguera, Reading the Sedimentary Record of the Ocean's Productivity. in: *Primary Productivity and Biogeochemical Cycles in the Sea*. P.G. Falkowski and A.D. Woodhead (Eds). pp. 455-486, Plenum Press, New York and London
100. G. N. Baturin. Formation of phosphate sediments and water dynamics, *Oceanology* **11**, 373-376 (1971).
101. A. Bein and O. Amit. Depositional environments of the Senonian chert, phosphorite and oil shale sequence as deduced from their organic matter composition, *Sedimentology*. **29**, 81-90 (1982).
102. B. Spiro, L. Heller-Kallai and Z. Aizenshtat. Environment of deposition and diagenesis of some "oil shales" in Israel, *Abstr. 10th Int. Sedim. Congr.* **2**, 632-633 (1978).
103. K. J. Murata, I. Friedman and M. Cremer. Geochemistry of diagenetic dolomites in Miocene formations of California and Oregon. *U.S. Geol. Surv. Prof. Paper*. **724C**, 12 pp. (1972).
104. K. J. Murata, I. Friedman and B. M. Madsen. Isotopic composition of diagenetic carbonates in marine Miocene formations of California and Oregon. *U.S. Geol. Surv.,.Prof. Paper* **614-B**, 24 pp. (1969).
105. M. Kastner. Carbonate and silica diagenesis. In: *The Geochemical and Paleoenvironmental History of the Monterey Formation: Sediments and Hydrocarbons. Vol. 1, Data Synthesis and Text Volume*. pp. 178-207, Global Geochemistry Corp., Canoga Park, California (1985).
106 A. Shemesh, Y. Kolodny and B. Luz. Isotope geochemistry of oxygen and carbon in phosphate and carbonate of phosphorite francolite, *Geochim. Cosmochim. Acta* **52**, 2565-2572 (1988).

Are modern and ancient phosphorites really so different?

C. R. GLENN[1], M. A. ARTHUR[2], J. M. RESIG[1], W. C. BURNETT[3], W. E. DEAN[4] and R.A. JAHNKE[5]

[1]Dept. Geology and Geophysics, University of Hawaii, Honolulu, HI 96822, U.S.A., [2]Dept. of Geosciences, Pennsylvania State University, University Park, PA 16802, U.S.A., [3]Dept. of Oceanography, Florida State University, Tallahassee, FL, 32306-3048, U.S.A., [4]U.S. Geological Survey, MS 939, Federal Center, Denver, CO, 80225, U.S.A., and [5]Skidaway Institute of Oceanography, PO Box 13687, Savannah, GA, 31416, U.S.A.

ABSTRACT

Modern phosphorites and associated facies off the coast of Peru are compared with wide-spread Upper Cretaceous deposits in Egypt. New data suggest that the modern deposits are much larger than previously assumed. The modern deposits contain pelletal phosphorites and are dominated by phosphoritic hardgrounds. The ancient are dominated by pelletal phosphorite beds. Phosphate precipitation occurred at or directly below the seafloor in both settings. Precipitation at deeper burial depths appears to be limited by lattice poisoning by dissolved carbonate and/or the diffusive limits of dissolved fluoride. Bottom currents keep the sediment-water interface clear of detritus and thus help prolong phosphate growth at critical burial levels. Bottom currents also winnow dispersed *in situ* CFA pellets from muds and concentrate them into thick pelletal phosphorite beds. Extensive reworking in the ancient setting resulted in thick, amalgamated, pelletal-phosphorite sandwaves. Thick glauconite sands also occur in both settings and the grains of both are thought to precipitate along the intersection of anoxic and oxygenated bottom waters. Organic-carbon-rich muds and biosiliceous sediments reflect high oceanic fertility in both the modern and ancient deposits. The principal difference that does appear to exist between modern Peruvian and ancient Egyptian phosphorites is the source of P: although upwelled water is the main source of P off Peru, fluvial inputs of particulate-borne P may likely be more important in tropical cratonic settings such as those exemplified in Egypt.

Keywords: Phosphorus, phosphorites, phosphatic shales, biosiliceous sediments, organic-carbon-rich sediments, glauconites, upwelling, fluvial P flux, current winnowing, sedimentology, petrology, geochemistry, stable isotopes, pore-water geochemistry.

INTRODUCTION

It has been held that modern phosphorites are poor analogs for their ancient counterparts (e.g. [1,2]). Often cited differences include the size of the deposits and the amount of phosphorus they contain, the ultimate source of the phosphorus, differences in facies associations (i.e. carbonate and glauconite associations in ancient deposits and the apparent lack of such facies in modern settings), differences in mineral paragenesis, differences in sedimentation rates, formation below (Recent phosphorites) versus above (e.g. apatitic oolites of some ancient phosphorites) the sediment-water interface, and differences in morphologies (pellets in the ancient, nodules in the present). This paper addresses such similarities and differences by examining two apparently disparate occurrences: (1) Pleistocene and Recent phosphorites occurring off the coast of Peru on an active margin slope and (2) areally widespread Late Cretaceous (Campanian) phosphorites that were deposited on a broad, stable craton in northwest Africa. Our sample basis comes from several years work in Egypt [3], results from the R/V *Robert A. Conrad* expedition 23-06 to Peru in 1982 [4], ship-board and post-cruise results from the Ocean Drilling Program (ODP) Leg 112 to Peru, and recent results from submersible and shipboard operations off Peru aboard the R/V *Seward Johnson* in late 1992.

PHOSPHOGENESIS ON THE PERU SHELF AND SLOPE

Physical Setting and Phosphorite Distribution
Phosphatic facies of the Peru continental margin chiefly reside on the upper slope-outer shelf within biosiliceous and organic-carbon-rich muds that result from low continental fluvial discharge, sustained nutrient upwelling and high biological productivity [5-9]. The sediments are laminated to burrow-mottled [10,11] and numerous erosional unconformities in the sediments suggest intermittent slope failure and scouring by bottom currents over periods of thousands of years [8,12]. As discussed below, such variations between sedimentation and sediment reworking/winnowing have likely played a major role in concentrating authigenic phosphatic detritus into winnowed lag deposits, as well as serving as a mechanism for maintaining phosphatic phases at shallow burial levels optimum for continued phosphate precipitation from pore waters.

The character and distribution of the Peru margin sediments changes both across the shelf/slope and latitudinally in response to variations in shelf morphology, primary productivity, and bottom current distribution and intensity [6,7]. The water column overlying the shelf/slope sediments is characterized by a relatively thin (<25m deep) north- and offshore-directed Eckman surface layer underlain by mainly southerly (Poleward) flowing undercurrents that supply upwelled water from relatively shallow (ca. 400 m) depths [13-15]. Available current measurements [14,16] indicate that during normal upwelling conditions these undercurrents obtain velocities of up to 20 cm/s over the slope at about 5°S. Between about 5°S and 10°S these currents migrate shoreward over a relatively broad (30 km wide) shelf where they impinge upon and rework the bottom [8]. South of this, the shelf narrows and deepens, and the shelf-slope transition becomes more gentle. In this region (10°S-16°S) outer shelf organic carbon accumulation rates reach a maximum [8] where bottom current velocities diminish to about 5 cm/s [14] as the undercurrent again migrates seaward on to the upper slope. During El Niño events, however, the undercurrent velocities on the shelf may increase to an average of about 25 to 30 cm/s at 10°S over 8- to 64-day periods [16].

In our recent submersible studies, we have found that the shelf and slope at 12°S and 13.5°S can be largely divided into cross-shelf tracts of surface sediments dominated by either organic-carbon-rich muds covered with benthic sulfide-oxidizing bacterial mats (*Thioploca* sp.) or relatively mud-free zones of phosphatic crusts and proto-crusts underlain by stiff green-gray muds and glauconite sands at greater depth (Figs. 1 and 2). At 12°S, between water depths of about 100m and 200m, the outer shelf is dominated by muds blanketed by *Thioploca* sp. mats. Seaward of this, the seafloor is almost entirely covered with gravelly phosphorites, phosphatic hardgrounds or pelletal sands, one exception being a medial mud lens centered at about 300 m (Fig. 1). Along another cross-shelf/slope transect, at 13.5°S, we mapped similar phosphatic facies between ca. 175 m and 600 m and glauconitic sands between 800 m and 1050 m (Fig. 2). There, mud facies occur on the outer shelf (< 100m) and at about 600 m. At both localities highest bottom water velocities were found to occur in association with the phosphorite-glauconite belts (Figs. 1 and 2). Lowest velocities typified the tracks of phosphorite-poor, organic-carbon-rich mud facies along the middle to outer shelf. Dissolved oxygen contents decrease rapidly between surface waters and about 100 m and the core of the oxygen-minimum zone impinges slope sediments at about 500 m at 12°S and at about 400 m at 13.5°S. Below these depths, bottom waters again become oxygenated.

Phosphorite Petrology and Nomenclature
Phosphatic components on the Peru margin consist of ovule to round, sand- to silt-sized phosphatic grains or peloids (as elsewhere, these are termed "pellets", although a fecal origin is not implied) and larger accretionary masses such as crusts and nodules [10,17]. Large portions of the sea floor along the upper slope have been densely cemented into phosphatic hardgrounds that are pavements of welded masses containing all of these morphologies. Pellet grains are found randomly disseminated in the sediments or as concentrations in friable layers or beds. Phosphatic pellets are also a common component of some carbonate fluorapatite (CFA) nodules and crusts where they are cemented into these accretionary bodies

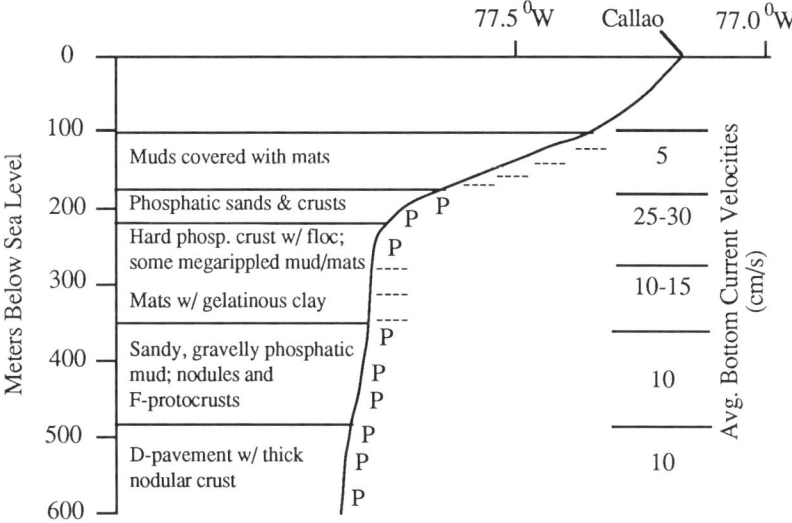

Figure 1. Sediment and phosphorite distribution together with measured bottom current velocities (cm/s) on the outer Peru shelf and upper slope at 12°S (results from R/V *Seward Johnson*, 1992 cruise; unpublished). "P" represents distribution of phosphorite. Nodular hardground pavements (D) and phosphatic sands occur in association with highest current velocities. Friable (F-phosphate) CFA-protocrusts (see text) occur between about 350 and 500 m. Bacterial mats are sulfide-oxidizing *Thioploca* sp.

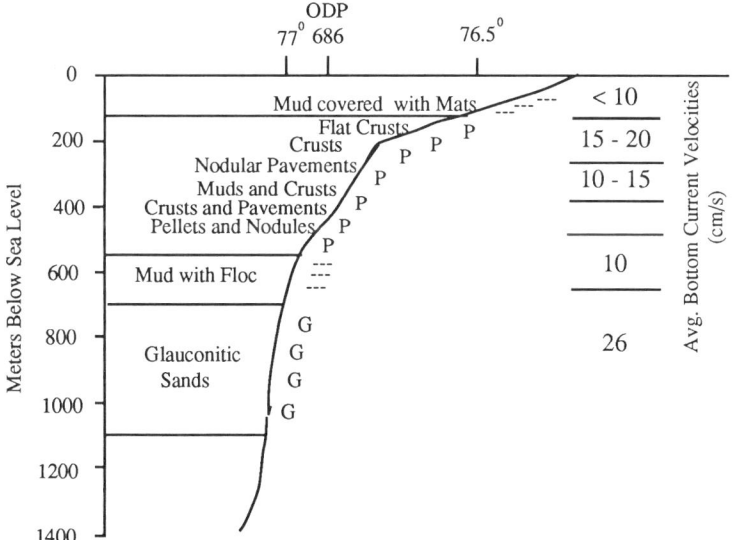

Figure 2. Sediment and phosphorite distribution together with measured bottom current velocities on the outer Peru shelf and upper slope at 13.5°S (results from R/V *Seward Johnson*, 1992 cruise; unpublished). Phosphorites (P) and peloidal glauconite sands (G) occur in association with highest current velocities. Bacterial mats are sulfide-oxidizing *Thioploca* sp.

as either randomly disseminated grains or as individually cemented sediment-grain layers. We use the terminology of Garrison and Kastner [10] and subdivide Peru margin phosphorites into F-phosphates, D-phosphates and P-phosphates:

1. **F-phosphates:** small nodules, peloids or laminae of *friable*, light colored CFA in diatomaceous muds. These were called collophane mudstones by Burnett [18]. Consistency varies from very friable and unconsolidated to more compact, though even the later can be scratched with a fingernail. Presumably, F-phosphates age or mature to eventually become D-phosphates.

2. **D-phosphates:** well lithified, often dark and dense nodules, gravels and hardgrounds of CFA.

3. **P-phosphates:** phosphoritic sands dominated by structureless and coated (micro-banded) phosphatic grains, with some admixtures of fish bones and teeth.

The main phosphatic mineral phase of all these phosphate types is CFA. Although an anisotropic mineral, it is frequently cryptocrystalline and thus due to aggregate polarization appears optically isotropic (i.e., pseudo-isotropic) under polarized light [17]. When the size of CFA crystallites reach a few microns, however, they begin to exhibit low orders of birefringence (first order grays to pale yellow). As discussed below, the variation in CFA crystallite size as observed in freshly deposited sediments may, in part, reflect variations in pore-water chemistry.

Phosphatic and Glauconitic Sands
Because of poor recovery when using dredge haul and grab core sampling techniques, the Peru Margin phosphorite province has been considered by many workers to be very different than most ancient occurrences because it did not produce sizable quantities of phosphate or glauconite peloids. Recent investigations, however, have demonstrated that beds of phosphatic and glauconite grains are actually quite common, both in the subsurface [4,10,19] as well as on the seafloor (e.g. Figs. 1-4).

Glauconite grains occur scattered in muds, within some phosphate nodules and crusts, admixed with phosphatic sands, and as relatively pure accumulations of glauconite beds. In most cases, glauconite grains are structureless, but in some instances they apparently show evidence for having precipitated within foraminiferal tests [19, 20]. The sands of glauconite beds are friable and well sorted (they contain little to no clays or diatoms) and contain various admixtures of foraminifers and phosphate pellets. The thickest and purist of sands observed by us occur as thick winnowed deposits blanketing the slope at water depths between 800 and 1100 m at 13.5°S (Figs. 2-3). Other winnowed, friable, normally-graded, and bioturbated Quaternary glauconitic sands are reported in the shallow subsurface at ODP Site 684 (Subunit IB; water depth 425 meters; 9°S) [10,19]. Both of these occurrences are beneath the core of the present oxygen-minimum zone (cf. [8,18,20]) and both sites have experienced significant current erosion. The Upper Quaternary sands at Site 684, for example, are cross-cut by erosional unconformities [10,19] and contain phosphatized foraminifers of Miocene age (this study). As based on available data, therefore, the glauconite sands appear to preferentially accumulate where bottom current activities and relatively high, and also where bottom waters are relatively oxygenated, i.e. along the lower edges of the oxygen minimum zone. We believe this spatial distribution is favored because: (1) glauconite contains iron in both reduced (Fe^{+2}) and oxidized (Fe^{+3}) states and this requires that its precipitation occur from mildly reducing solutions [3,17], and (2) the accumulation of the glauconite sands are a result of current winnowing of grains authigenically precipitated from interstitial or intraskeletal pore waters.

P-phosphate sands contain abundant CFA pellets and may also contain admixtures of siliciclastic grains, foraminifers, fish debris and glauconite. They occur as both relatively pure layers within otherwise quite-water diatomaceous sediment or as a sandy matrix surrounding D-phosphate nodules and gravels. In the subsurface, some of the beds are meters thick [19]. In some cases, they fill burrow traces and some of the beds are normally graded [10] indicating deposition from hydraulic flow. As in many ancient phosphorite deposits (cf.

Figure 3. Modern glauconite sand beds collected by box corer from the Peru margin at 13.28°S, water depth 931 m (R/V *Seward Johnson*, 1992 cruise). The upper half of this sand is dominated by black, coarse-sand-size glauconite pellets, abundant large benthic foraminifers and common worm tubes up to 10 cm long. The lower half is a clayey glauconite sand. The contact between the two is burrow mottled and marked by iron oxyhydroxide staining. Scale bar in cm.

[21-23]), Peru margin phosphatic grains have a variety of morphologies including structureless pellets, grains with multiple CFA coatings, phosphatized or originally phosphatic biogenic debris, intraclasts and irregular-shaped phosphatic "clumps" (Fig. 4). The latter are morphologically transitional between structureless pellets and interstitial cements. Subrounded to ovule-shaped cryptocrystalline structureless and coated grains dominate CFA grain abundances off of Peru. Whereas structureless peloids generally lack prominent internal morphology, coated grains typically contain a central phosphatic or siliciclastic nucleus grain surrounded by a concentric structure of light yellow brown to dark brown pseudo-isotropic lamina occasionally interlayered with thin (<10 µm) bands of relatively inclusion-free, first-order gray birefringent CFA (Fig. 4). Unlike typical carbonate ooids, the shape and internal arrangement of phosphatic coated grains are highly irregular, and their lamina typically thicken and thin, or pinch-out around the grain [17].

F-phosphate Nodules, Crusts and Softgrounds

F-phosphate nodules and crusts are relatively non-layered, semi-friable to well-indurated structures which may contain scattered glauconite and phosphate grains and well-preserved burrow structures. Relative to D-phosphates, F-phosphates are markedly homogenous. In the sediments, F-phosphates chiefly occur as nodules and small peloids, but we have also observed them directly on the seafloor where they occur as thin (<2 cm thick), light-brown to yellow crusts and nodules blanketing the seafloor (Figs. 5 and 6). We have termed the latter CFA proto-crusts or softgrounds, the implication being that we believe that these are the precursors of D-phosphate crusts and hardgrounds. They are very soft; they can be crushed and broken by hand and are easily cut with a knife, yet they are commonly bored. They are generally flat and more rigid on the upper surface and bulbous and softer on the underside. The upper surface may also be thinly coated with manganese. The nodules/crusts appear to grow mainly downward into the sediment. Figure 6 illustrates our interpretation of how the CFA proto-crusts evolve into mature, conglomeratic phosphatic hardgrounds.

Observations with the scanning electron microscope show the CFA cement of F-phosphates and recent seafloor CFA proto-crusts to be composed of very small globular and ellipsoidal (spindle) structures (Fig. 7; also see [10,24]). The spindle structures very commonly display simple (two spindles nucleating from a single point) and more complex (crosses of 4 spindles, rosettes of 6 or more spindles) growth structures that indicate outward growth from a single point. With time, the F-phosphates may become better ordered and more crystalline [25], eventually developing hexagonal crystallites [10,24], and these crystallographic changes may reflect CFA growth rates (discussed below). In addition, our preliminary observations suggest that the spindles making up the CFA proto-crusts directly collected from the seafloor tend to be smaller (usually <0.5 µm in length) than those collected from both F- and D-phosphates recovered at depth (usually ca. 5 µm in length). Taken together, we feel that the rosette structures and such variations in crystallinity argue against the notion that these microstructures represent phosphatized bacteria. Similar rosettes also occur in synthetically-precipitated iron oxyhydroxides [26].

D-Phosphate Nodules, Crusts and Hardgrounds
D-phosphate nodules and crusts are typically well-indurated, dark-brown to black bodies often composed of alternating, lenticular layers of phosphatically cemented size- and/or morphologically-sorted phosphatic, biogenic, and siliciclastic grains (Fig. 8). They develop through multiple stages of CFA cementation, sediment phosphatization, bioerosion

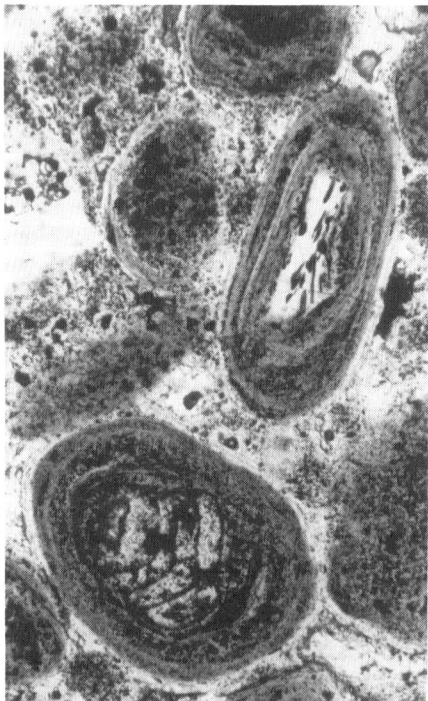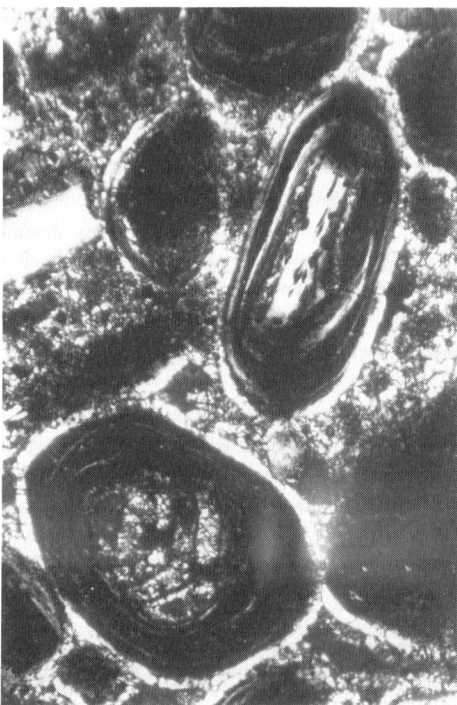

Figure 4. Phosphatic pellets from a modern Peru margin phosphorite nodule from 10.58°S, water depth ca. 450 m. Left is plain polarized, right is cross polarized. The lower-most grain is a CFA coated and replaced benthic foraminifer. The ovule-shaped grain in the center right is a CFA coated fish bone. The grain at center left is a coated CFA grain with a structureless CFA nucleus and an eroded outer rim. Note the internal and fringing bands of birefringent CFA cements. Interparticle pore space is filled with both birefringent and pseudo-isotropic CFA cements. Vertical dimension of photo is 0.5 mm.

Figure 5.
Looking down on a very friable, light-olive colored CFA proto-crust on gelatinous mud in a box core (BC 33) recovered from the Peru margin at 13.5°S (R/V Seaward Johnson, 1992 cruise). The crust is bored and rounded on its edges. It was likely broken up by the coring procedure. Water depth 373 m.

(burrowing and boring) and current erosion and break up, and re-phosphatization (Fig. 6). The end result is the development of true phosphatic hardgrounds composed of many generations of welded D-phosphate nodules and crusts that now cover large tracts of the Peru margin upper slope (Figs. 1, 2, 8-10).

Individual sediment layers coating individual D-nodules often each record the sequential precipitation of: (1) birefringent pore-lining and grain-fringing CFA cements, followed by (2) infilling of remaining pore space by more equant and pseudo-isotropic interstitial CFA cements, and finally (3) capping of each sediment-cement generation with a thin, completely pseudo-isotropic and relatively inclusion-free microlaminated CFA crust [10,17]. The growth of such layered nodules is therefore punctuated, with each additional sediment layer being bound by three generations of phosphatic cement. Interstitial CFA cements (Figs. 4 and 8) range from pseudo-isotropic to more birefringent varieties transitional with fringe cements, and cements binding individual grain-layers are usually distinct from one another in terms of their relative birefringence, crystallinity, and microinclusion densities. Increases in interstitial cement pseudo-isotropism and diminishing crystallite size are usually accompanied by increases in the degree of phosphatization and a progressive darkening of the groundmass. Older D-phosphates (pre-Quaternary) tend to be dominated by dark interstitial CFA cements [10]. Within individual nodules, the older sediment generations always have interstitial cements, whereas fringing cements only occur in the younger layers [10,17,27].

Associated Authigenic Phases and Mineral Paragenesis
Diagenetic accessory minerals associated with the Peru margin phosphorites include glauconite, pyrite and dolomite, and their occurrence in the phosphatic nodules and peloids supplies information about various stages of organic matter degradation and mineral paragenesis characterizing Peru margin sediments during CFA precipitation. The distribution of these minerals generally reflects the depth distribution of suboxic to anoxic bacterial

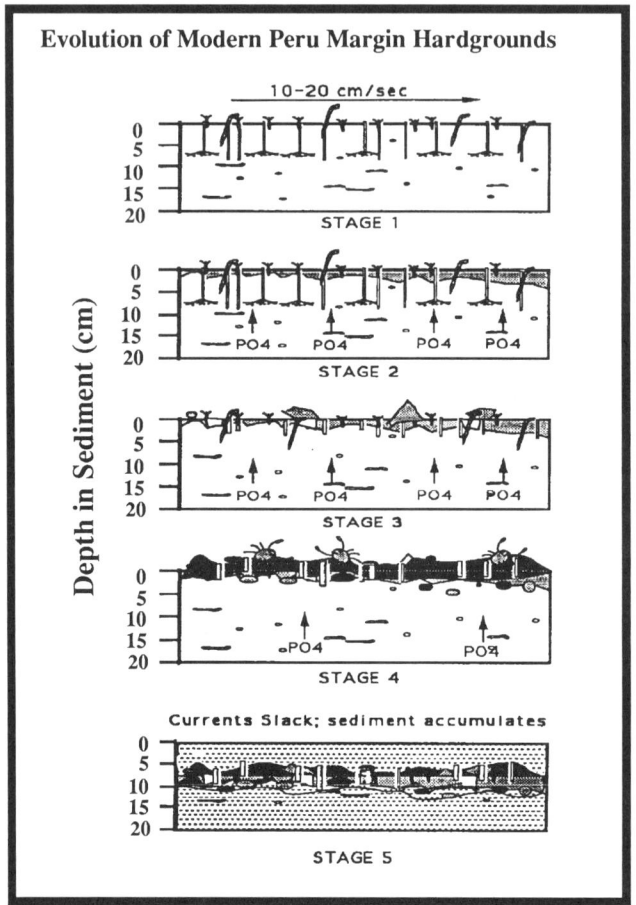

Figure 6. Evolution of Peru margin CFA proto-crusts and sequential transformation to phosphatic hardgrounds.

STAGE 1: Clay containing relatively high concentrations of Corg (~2%) is deposited at moderate rates under low-oxygen concentrations in bottom waters followed by a hiatus; polychaetes and other worms adapted to low-oxygen conditions burrow; many open vertical tubes; currents keep bottom swept clear of fine-grained sediment.

STAGE 2: Slowing or cessation of deposition allows phosphatization of the upper several centimeters of sediment. A stiff but friable phosphatized claystone (CFA proto-crust) is the initial product. This continues to be burrowed and bored.

STAGE 3: The crust lithifies slowly and increases in thickness mainly by downward cementation into the sediment. The crust, still friable, is periodically broken up by the activity of organisms as well as by bottom-current flow. Individual clasts become somewhat rounded by abrasion. Boring continues, as does cementation of broken pieces.

STAGE 4: The crust solidifies by progressive phosphatization as currents continue to sweep the bottom. Smaller nodules may continue to grow in the sediment, but overall rate of growth slows. Organisms tolerant of low- oxygen conditions encrust and inhabit the lower and upper surfaces; boring and cementation continues. Nodules grown in the sediment and phosphatic and dolomitic gravel created by fragmentation are accreted.

STAGE 5: As currents slacken, sediment begins to accumulate and bury crusts; crusts may be partly exhumed and reburied a number of times. In some cases, full crusts do not form and only a phosphatic gravel lag deposit occurs.

Figure 7. Scanning electron micrograph of a F-phosphate composed entirely of CFA spindles and rosettes. The rosette in the center has grown as six spindles nucleating from a single point. A smaller seventh spindle is nucleating off the front arm. The other spindles of the groundmass fused into a stubby mass as they competed for growth space. Sample collected from 7.22 meters below the seafloor at ODP site 679B (112-679B-1H-05, 22-24). Black scale bar is 1μm long.

populations in marine pore waters [17].

Because glauconite contains iron in both reduced and oxidized states (Fe^{2+}:Fe^{3+} approximately 1:7), its development is probably very early, taking place within suboxic sediments after partial reduction of ferric iron. There, soluble reduced iron may be reoxidized and incorporated along with small amounts of Fe^{2+} during the glauconitization process. Within organic-rich sediments such as those off Peru (and in Egypt; see below), this suboxic zone may occur within pore waters located directly below the sediment-water interface (Table 1). In the lateral dimension, it may occur along the intersection of oxygenated and oxygen-deficient bottom waters [3,28]. Glauconite precipitation may also occur with reducing microvolumes in foraminiferal tests [20]. Within Peru CFA nodules, both CFA and pyrite replace, and thus generally post-date glauconite formation [17].

Pyrite is an important indicator of sulfate reduction processes [29,30] and is a ubiquitous phase in the Peru margin phosphorites. Its formation is favored by available reactive iron and an abundance of highly reactive marine organic compounds in these sediments. Within Peru CFA nodules, it occurs as both a common mineral inclusion in all CFA grain and cement morphologies as well as a replacement (chiefly after glauconite and dolomite) and pore-filling cement (Fig. 8c). Precipitation of pyrite and CFA appears to be nearly coincident, with pyrite precipitation continuing beyond that of CFA where it may replace the CFA, or infill remaining pore space after partial interstitial CFA cementation [17].

Nodules composed of well-ordered dolomite form the cores of many of the D-phosphate nodules and crusts recovered off of Peru (Fig. 8). Carbon isotopic ratios of these dolomites

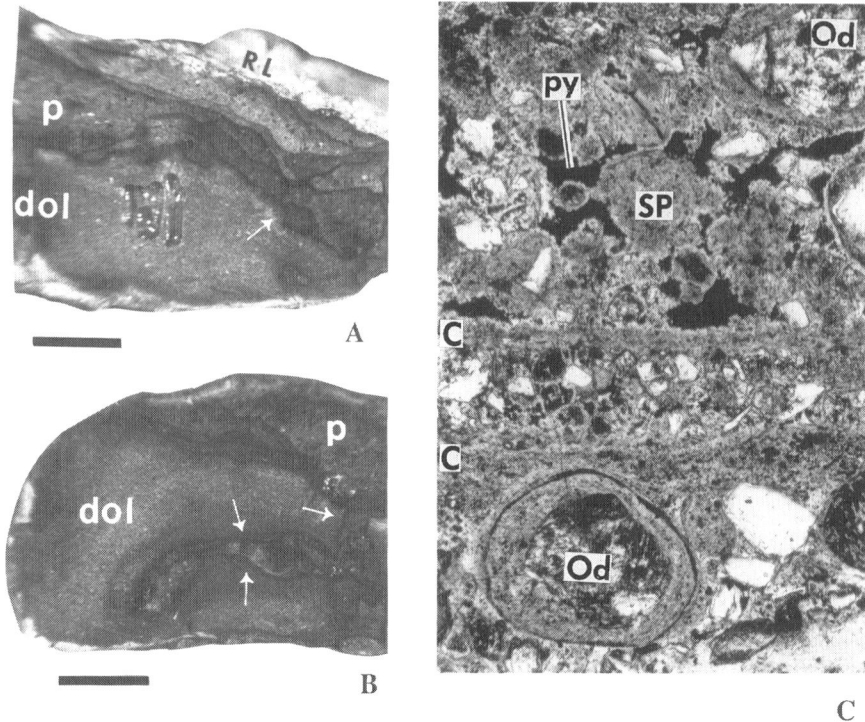

Figure 8. (a) and (b) D-phosphate nodules from the Peru margin (12°S; water depth 204 m) with dolomicritic cores (dol); scale bars 1 cm. The outer portion of both nodules (p) are covered with multiple generations of phosphatic sediment and cement. The outermost, lighter-colored layer in (a) is a friable F-phosphate layer and underlying layers become progressively darker as they get older towards the nodule's core. The dolomitic core in (b) has been extensively bored and the boring has been infilled with phosphatic pellets, sediment, and cement (arrows). (c) Three layers of sediment cemented to D-phosphate nodule by interstitial CFA and pyrite cements. Long dimension of photo is 1.3 mm. Od = coated pellets with siliciclastic nuclei; SP = structureless pellet; Py = interstitial pyrite cement; C = microlaminated coatings separating the layers. Samples collected during R/V *Conrad* expedition

(see below) indicate precipitation within the zone of microbial sulfate reduction and later methanogenesis. Subsequent biological boring and encrustation of these dolomicrites by multiple generations of phosphatic and pyritic cements (Fig. 8), however, indicates their later exhumation and exposure at the sea floor followed by reburial. In addition, Garrison and Kastner [10] found that F-peloids and micro-nodules in Leg 112 cores often occur enclosed within and partly replaced by fine crystalline dolomite, which is evidence for phosphatization preceding dolomitization. Most dolomites thus appear to post date an initial phase of CFA precipitation.

Stable Carbon Isotopes of Modern CFA
Carbon isotope and lattice-bound CO_2 data from CFA helps constrain where and when phosphorite formation occurs. The primary assumption is that as carbonate fluorapatite phases grow, they do so in isotopic and chemical equilibrium with the total dissolved carbon (TDC) in the waters from which they precipitate (see discussions in [27,31]). If CFA precipitates directly from seawater at the sediment/water interface, then the carbon isotopic composition of incorporated CO_2 should be characteristic of bottom water (today, about 0 per mil [‰], relative to PDB). If, on the other hand, CFA precipitates within the sediments, the

Figure 9. Hard, well-indurated modern Peru margin conglomeratic D-type CFA hardground collected from the seafloor by submersible at 250 meters water depth, 13.5°S (JSL 3352; R/V *Seward Johnson*, 1992 cruise). Scale bars at upper left in cm.

Figure 10. Close-up photo of the hardground shown in Figure 9. The rubble is composed of rounded as well as broken and abraded phosphate nodules and fragments of crusts that have been welded together by CFA cements and progressive phosphatization. The surface is pitted with borings. These structures are morphologically identical to the famous phosphatic "pebbles" of the Miocene Florida Land Pebble beds.

$\delta^{13}C$ values of lattice-bound CO_2 should mimic those of the pore-water TDC and display relatively extreme negative to positive values characteristic of suboxic to anoxic organic matter degradation processes (Table 1).

Carbon isotope ratios (relative to PDB) for the structural carbonate of Recent Peru margin CFA and dolomite are shown in Fig. 11 and values for total dissolved carbon of modern Peru margin pore waters are shown in Fig. 12. $\delta^{13}C$ values obtained from CFA nodules and CFA grains are about the same, ranging from about -5 to 0‰ and -3 to 0‰, respectively. In comparison with pore-water $\delta^{13}C$ values from these sediments (e.g. Fig. 12), the range of CFA $\delta^{13}C$ suggests that all CFA morphologies formed at or near the sediment-water interface in largely suboxic pore waters, perhaps to the early phases of sulfate reduction. Shallow depths of precipitation are also suggested by dissolved interstitial phosphorus and fluoride profiles, which probably indicate uptake by CFA precipitation within a few to perhaps a few tens of centimeters of the sediment-water interface (Fig. 12; [37,38]). The dolomitic cores of CFA nodules recovered from 12°S by grab sampling and dredge hauls, however, display both more strongly negative CFA-$\delta^{13}C$ values (to about -11‰), which suggest precipitation in the zone of sulfate reduction, and more positive values (to about +6‰) indicative of precipitation at greater depth in the sediment within the zone of methanogenesis (Fig. 11). Although ages for the dolomitic (interior) portions of the nodules encrusted by CFA have not been determined, it is apparent that they must have formed first, perhaps meters below the sediment-water interface, and then later have been exhumed and exposed to the seafloor (where they were bored) prior to shallow reburial and encrustation by CFA and pyrite.

The petrographic and isotopic considerations discussed above indicate that Peru margin CFA nucleates within a few to at most a few tens of centimeters of the sediment-water interface. Figure 12 illustrates that although dissolved phosphate is provided to the Peru margin pore waters during organic matter oxidation and sulfate reduction deeper in the sediment column, the locus of phosphate precipitation appears more strongly associated with an interfacial phosphate maxima that occurs close to the sediment water interface [38]. Similar interfacial

Table 1.
Change in isotopic composition of marine pore waters accompanying successive stages of microbial oxidation of organic mater (CH_2O). The reactions reflect downward decreasing metabolic free energy yields [32-36].

TDC-$\delta^{13}C$	Environment	Diagenetic Zone	Reaction
FROM: +/- 0.5‰ (bottom water)	oxic	aerobic oxidation	$CH_2O + O_2 \rightarrow$ $CO_2 + H_2O \rightarrow HCO_3^- + H^+$
~ ~ ~	suboxic	manganese reduction	$CH_2O + 3CO_2 + H_2O + 2MnO_2$ $\rightarrow 2Mn^{++} + 4HCO_3^-$
~ ~ ~		nitrate reduction	$5CH_2O + 4NO_3^- \rightarrow$ $2N_2 + 4HCO_3^- + CO_2 + 3H_2O$
~ ~		ferric iron reduction	$CH_2O + 7CO_2 + 4Fe(OH)_3 \rightarrow$ $4Fe^{++} + 8HCO_3^- + 3H_2O$
~ TO: -25‰	anoxic	sulfate reduction	$2CH_2O + SO_4^{--} \rightarrow$ $H_2S + 2HCO_3^-$
~ ~ TO: +25‰	anoxic	methanogenesis	$2CH_2O \rightarrow CH_4 + CO_2$

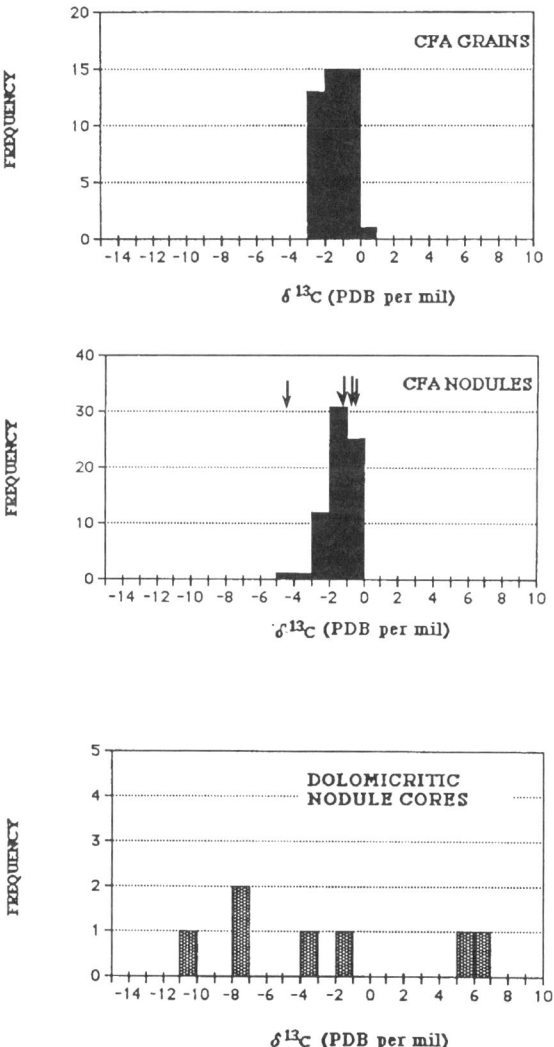

Figure 11. Carbon isotopic composition of Peru margin CFA grains and nodules and dolomite nodule cores. Arrows in center panel indicate values for recent F-phosphate CFA proto-crusts recovered from the seafloor; other data from Glenn at al. [31].

phosphate spikes are also found in association with the formation of Recent phosphorite along the Mexican continental margin [39] and off the eastern coast of Australia [40,41]. These interfacial phosphate spikes may be produced by suboxic organic degradation processes at the interface (Table 1) [34,37,38], the release of adsorbed phosphorus from ferric oxyhydroxides upon encountering reducing conditions in pore waters [27,38,40,42-45], or by the dissolution of fish debris (hydroxyapatite)[37,38,46] (however, see [47]). It has also been suggested that these spikes may be in some way related to the metabolic activity of sulfur-oxidizing bacterial mats (*Thioploca sp.*) commonly present on the seafloor in this region. In addition to explaining the origin of this interfacial phosphate spike, an additional intriguing problem that remains is explaining why CFA precipitation does not occur at greater depths in the sediment pile. We return to this question in our discussions below.

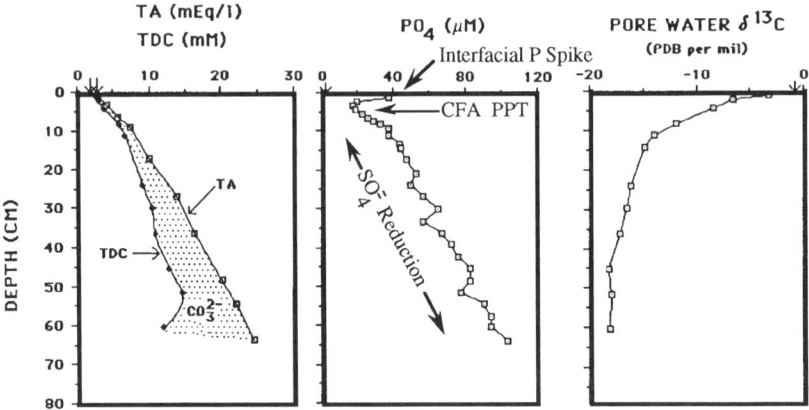

Figure 12. Peru margin pore water profiles of total alkalinity (TA), total dissolved carbon (TDC), dissolved phosphate and the carbon isotopic composition of the TDC in Peru margin sediments. Arrows on concentration axes indicate bottom water concentrations. The difference between TA and TDC (stippled) approximates the concentration of dissolved carbonate (mM/l). CFA precipitation (CFA PPT) occurs within a few centimeters of the sediment-water interface in association with the interfacial phosphate spike visible in the center diagram. After Glenn et al. [31].

Currents, Winnowing and Glacial-Interglacial Change
We believe that current winnowing and erosion strongly influence, if not regulate, phosphorite formation on the Peru shelf. Modern phosphorite crusts on the seafloor appear to be concentrated where bottom current velocities reach a maximum (Figs. 1 and 2) and, in the subsurface, the phosphorites are commonly associated with unconformities [8,19]. Radiometric data of Burnett et al. [48] suggest that recently buried phosphorite grains of the Peru Shelf grow quickly on time scales of a few years and that some of the near-surface pellet beds contain mixtures of older and younger pellets. The P-phosphorite beds are often graded, poorly sorted (contain clays), occur within a poorly sorted sediment [49], and are often bioturbated at the base [10]. Garrison and Kastner [10] suggested that F-phosphates tend to occur within laminated intervals which reflect deposition under oxygen-deficient conditions, whereas D-phosphates are more typically associated with bioturbated intervals representative of more oxygenated conditions. Wefer et al. [52] showed that interglacial episodes correlate with increases in organic carbon in Leg 112 sediments.

Figure 13 illustrates the stratigraphic succession for the upper 14 meters of Pleistocene-Recent sediments at ODP Site 679. This site was drilled on the upper slope at 11°S in a water depth of 450 m. We illustrate this section because it is unique in that it well represents the interplay between phosphorite genesis and current activity during Quaternary times. The section is subdivided into a number of repetitive sedimentary cycles. Each cycle: (1) begins at its base with homogenous sediment that (2) passes upwards through sediment containing an increase in benthic foraminifers and (3) finally ends at its top with a layer of phosphatic nodules residing in a foraminifer-rich sand. The sand layers directly below the phosphorite layers are commonly bioturbated. There are at least 9 such cycles (with minor variations) within this portion of the record at site 679B. The upwards increasing foraminifer abundance is largely due to an increase in what we have termed the poleward undercurrent assemblage (PUA). This assemblage is described and illustrated by Resig [50,51] as the Quaternary Lower-Bathyal Current Biofacies. It represents a collection of robust species concentrated at

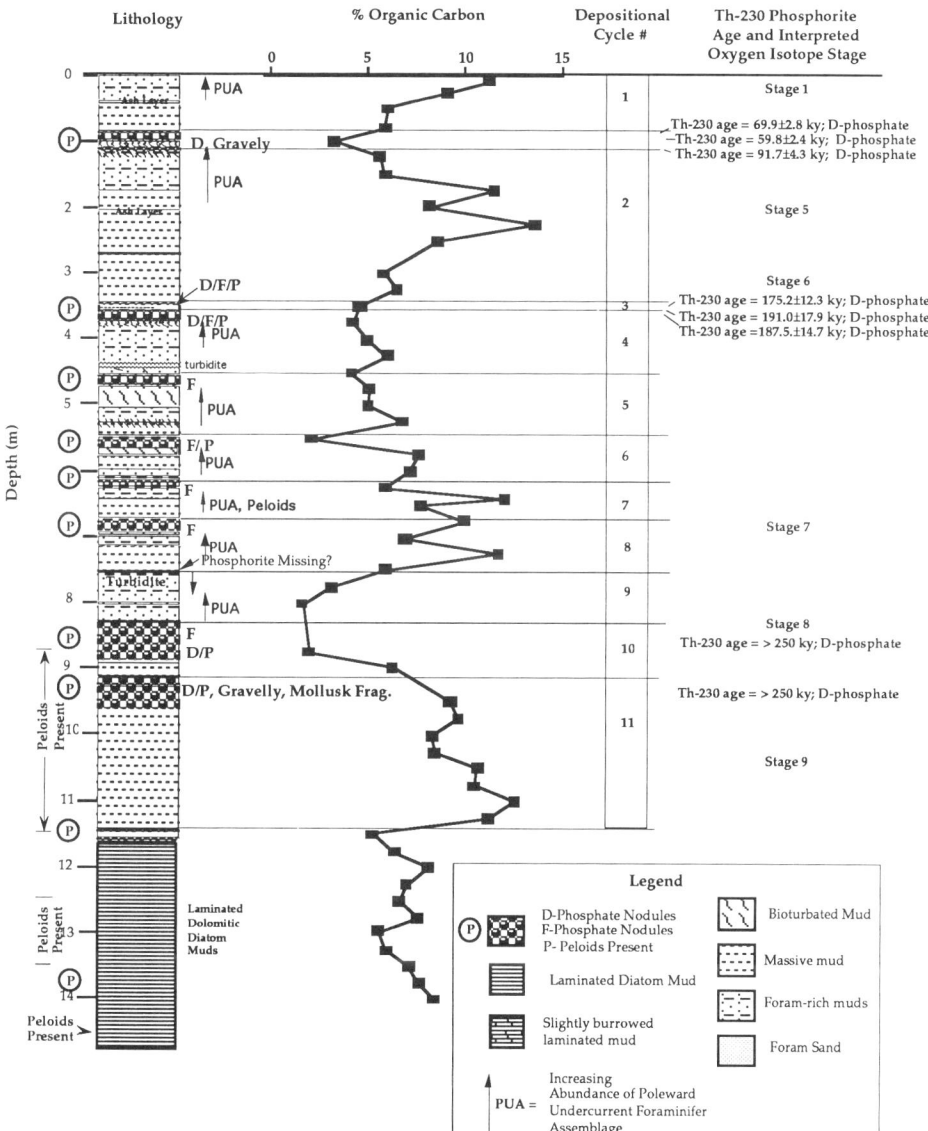

Figure 13. Detailed stratigraphy, ages and organic carbon contents of the upper 14 meters at ODP Hole 679B 11°S; water depth 450 m). The positions of the oxygen isotope stages should be regarded as preliminary.

times into foraminiferal sands by current action and it occurs at water depths characterizing the Peru Poleward Undercurrent. Thus, it appears that the phosphorites at this site were deposited at times of increased bottom-current activity. ^{230}Th ages for two horizons of these nodules show that the ages of individual nodules are mixed (Fig. 13), indicating some erosion and redeposition. Further, a relative decrease in organic carbon content and the presence of bioturbation beneath most of the phosphorite horizons indicates that these depositional phases were marked by a relative increase in bottom-water oxygenation. The oxygen isotope stages indicated in Figure 13 are preliminary.

We believe that the concentration and cyclic nature of phosphorite development at Site 679 is linked to the movement and spatial variability of the poleward undercurrent. F-phosphates are likely precursors for D-phosphates and, on scales of millimeters at a time, D-phosphate growth takes place as a series of punctuated steps involving discrete episodes of sedimentation and subsequent stages of CFA cementation. Phosphorite growth rates on the Peru shelf are relatively rapid, but they are still much slower then ambient sedimentation rates (centimeters per thousand years) [25,53-55]. Individual pellets may form on time scales of a few years, but larger masses such as crusts and nodules grow at rates of millimeters per thousand years. Importantly, because pore water, carbon isotope and petrologic data all indicate that phosphate precipitation is confined to within a few centimeters of the sediment-water interface, the discrepancy between CFA growth rates and bulk sedimentation rates requires some mechanism for sustaining the length of time that phosphorite grains and particles may reside within this zone of phosphogenesis. At Site 679, at least, this mechanism appears to be current winnowing as well as current erosion. During current winnowing, sedimentation rates are reduced as the phosphorites are swept clean of most fine-grained detritus. During current erosion, previously precipitated phosphorites and other authigenic phases may be exhumed and reexposed to the seafloor where phosphogenesis may be re-initiated.

ANCIENT PHOSPHORITES ON THE EGYPTIAN CRATON: THE DUWI GROUP

Physical Setting and Stratigraphy
Phosphatic strata of the Duwi group occur as thin, widespread shallow marine deposits of Late Cretaceous age that crop out in a generally east-west trending belt spanning the middle latitudes of Egypt [3]. These strata are of economic worth and form a portion of an extensive Middle East-North African phosphogenic province. In total, the rocks of this province contain in excess of 70 billion metric tons of phosphate rock. The phosphate resources in Egypt alone are estimated to exceed 3 billion metric tons [56].

The Duwi group phosphorites were deposited accompanying a major marine transgression of shallow epeiric seas that flanked the Tethyan trough to the north. To the north (seaward) of the phosphorite belt, Upper Cretaceous sediments are dominated by fine-grained carbonate and sandy facies. To the south (landward), phosphatic rocks occur interstratified with terrigenous clastics and sedimentary iron ores. Rocks of the Duwi group lie above and intercalate with marginal-marine to shallow-marine shales of the Variegated Shales (upper Nubia Formation), some of which are part of the Duwi group [28]. Overlying the Duwi group are deeper-water marine marls and chalks of Maastrichtian age.

In Egypt's Eastern Desert (Red Sea Coast) the principal phosphorite-bearing strata are known as the Duwi Formation [57], in the Nile Valley region as the Sibâîya Phosphate Formation [57-58] and in the Western Desert as the Phosphate Formation [59]. Glenn [28] collectively referred to these three formations as Duwi group strata. The stratigraphic and facies relationships between these formations is illustrated in Figure 14. The correlations between sections are tied through sequence stratigraphic analysis [cf. 60-62] and are based on interpretations of depositional environments, relative paleowater depths between facies and available age data [3, 28]. The bounding surfaces separating the systems tracts are thought to represent time-synchronous stratal boundaries [61]. Figure 14 thus illustrates the Duwi group as a series of systems tracts which represent a linkage of contemporaneous depositional packages. Two main depositional realms are inferred from the lithofacies assemblages: (1) a deeper-water hemipelagic environment (TST-1 to HST, Fig. 14) accompanying maximum transgression dominated by deposition of phosphorites, organic carbon-rich shales and biosiliceous sediments (porcelanites), which, after maximum flooding, shoals upwards into (2) a progradational stage accompanying sea level fall (HST and LPW, Fig. 14) during which oyster banks with brackish back-reef sediments (Red Sea Coast) and deltaic sediments (Nile Valley) dominated eastern portions of the phosphorite belt, while greensands were reworked seaward in areas to the west. Following this, the cycle repeated with the deposition of organic-carbon-rich phosphatic shales and marls and associated phosphorites in the earliest Maastrichtian (TST-2).

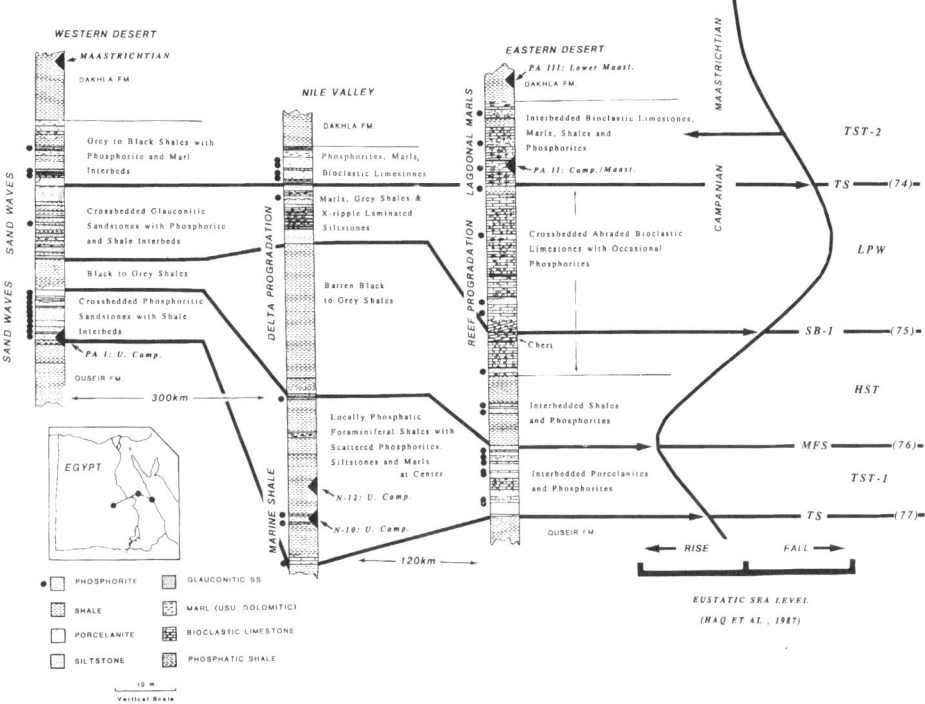

Figure 14. Duwi Group stratigraphy and lithofacies correlations between the Western Desert (Phosphate Formation), Nile Valley (Sibâîya Phosphate Formation), and Eastern Desert (Duwi Formation) localities. Inset shows location of type sections illustrated. The tie to eustatic sea level variations [60] is based on age data and sequence stratigraphic interpretations [3, 59]. Surfaces: SB = sequence boundary, MFS = maximum flooding surface, TS = transgressive surface (first flooding surface above maximum progradation). Systems tracts: LPW = lowstand progradational wedge, TST = lower transgressive systems tract (1 & 2 refer to this study), HST = highstand systems tract. From [3].

Importance of Associated Laminated, Organically-Derived Sediments
As is common to many phosphorite sequences, the phosphatic rocks of the Duwi group occur in association with laminated organic-carbon-rich shales, cherts and porcelanites (i.e. biosiliceous sediments [3]) (Fig. 14). These rocks are correlative with similar Campanian-age facies in Israel (see Kolodny and Garrison, this volume), in Jordan (see Abed, this volume) and in Syria (see Al Maleh and Mouty, this volume) and signify the influence of biologically-productive surface waters. Because these rocks are typically finely laminated and not bioturbated, they also indicate deposition beneath anoxic or oxygen-poor bottom waters. The porcelanites are best developed in the Eastern Desert where they reach a thickness of 10 m at the base of the Duwi Formation. They are scarce in the Nile Valley region and absent in the Western Desert. They are now composed of well-ordered microcrystalline quartz and clays, yet retain some traces of siliceous diatoms and sponge spicules. They also contain scattered *in situ* phosphate pellets.

Marine shales are widespread through the Duwi Group and tend to predominate near the base and top of the group in association with the principal phosphorites (see below). Pyrolysis data [3] indicate a marine algal to mixed terrigenous/marine algal source for the organic matter of these rocks and the organic carbon content of unweathered shales ranges from several percent to greater than 20%. Like the porcelanites, the shales locally may be highly phosphatic (1% to 12% P_2O_5), containing scattered ovules randomly disseminated in the groundmass (Fig. 15). Compactional distortion of the shale laminae surrounding these grains indicates that they were hard and thus formed prior to significant burial. Precipitation of these pellets from sediment pore waters is suggested by shale laminations or shell debris passing through the pellets, pellet inclusion of clays, dolomite, and pyrite, and by the pellets' isotopic composition (discussed below). Importantly, these pellets are very similar to those found in associated phosphorites and appear to represent the *in situ* source grains from which, through current winnowing, the intercalated phosphorites were mechanically derived.

Figure 15. *In situ* phosphate grains in laminated organic-carbon-rich (TOC = 13.9%) black shale from the Eastern Desert. Such grains are likely the starting point from which the Egyptian phosphorites were mechanically derived. Coin diameter 2.4 cm.

Phosphorite Sedimentology
The principal phosphatic components of the main Duwi phosphorite beds are diagenetic, silt- to granule-sized pellets and skeletal grains (phosphatized or primary fish debris). The pellet grains are structureless pellets or of intraclast pellet types. Coated grains, such as those which are common on the modern Peru slope, do not occur in the Egyptian deposits. Internally structureless pellets usually predominate and often contain clots of organic matter, pyrite inclusions and occasional siliciclastics.

Individual phosphorite beds may range from a few millimeters to tens of centimeters thick, and thin, lenticular beds are found interbedded with all associated lithofacies shown in Figure 14. They are, however, most commonly associated with organic-carbon-rich shales. Despite this association, extensive bioturbation and reworking of the beds is common and, as a result, internal stratification is often lacking (Fig. 16). Where not extensively bioturbated, the beds may display shallow basal scour, faint internal layering, clay intercalations or drapes, faint fining- or coarsening-upwards of phosphatic grains, or crossbedding. They typically have very low organic carbon contents (<1%), low hydrogen indices and high oxygen indices [3] indicating a well-oxygenated depositional environment. The thickest accumulations occur along the base of the sequence at Abu Tartur, near the top of the sequence in the Nile Valley, and both along the base and along the top of the sequence along the Eastern Desert. Coarse-grained phosphorite lags are typical in more southerly areas and along the top of the Duwi group throughout the region. More massive units typically are composed of multiple accumulations of thinner individual beds that have developed by repeated episodes of sediment reworking and bed amalgamation (Fig. 17), and in most instances are interbedded with either fine-grained porcelanites, shales, or occasionally marls (Fig. 14).

Western Desert Phosphorite Sandwaves
The thickest accumulation of phosphorite rocks in Egypt occur at the base of the sequence in the Western Desert (Fig. 14) where the beds locally combine to form a single seam averaging, yet often exceeding, 4 m thick. Exploratory bore hole information indicates that these deposits underlie much of the 1200 km² area now occupied by the Abu Tartur Plateau [56,63]. These striking pelletal beds likely accumulated as northeast-southwest trending

Figure 16. Silicified pelletal phosphorite from the Duwi group. Note massive texture and large bored rip-up clast of biosiliceous sediment near the top of the middle bed.

offshore phosphatic sandwaves [3,28,64]. The beds display large-scale, low-angle crossbedding, are amalgamated into larger bodies that thicken from 1.5 m to more than 10 m in thickness over distances of several hundred meters in "ridge and swale" topography (Fig. 18). They are similar in size and morphology to modern tidal- and/or storm-dominated sandwaves [3,28]. Bore-hole data [65] indicate that these sand bodies are up to several thousand meters long [3]. Although cross-shelf tidal currents may have played a significant role in maintaining these bedforms, the appearance of burrowed clay drapes on crossbed sets and dominantly unidirectional crossbedding directions in them suggests a rather punctuated development and storm reworking [66]. As with the other phosphorite occurrences in Egypt, the shales intercalated and laterally associated with the Abu Tartur phosphorites are locally phosphatic, containing randomly disseminated phosphate pellets and skeletal debris. It is probable that tidally-dominated current winnowing and erosion on the western Egyptian shelf acted to separate the phosphatic grains from such fine-grained phosphatic sediments and further concentrated and molded these accumulations into the giant sand waves observed.

Western Desert Crossbedded Greensands
Also occurring at Abu Tartur in the Western Desert are thick accumulations of tabular and trough crossbedded glauconitic greensands. These occur near the upper middle part of the stratigraphic section and developed where soft glauconitic peloids were reworked and concentrated into thick units during a relative sea level fall (Fig. 14). They are temporally equivalent to seaward prograding oyster reef complexes in the Eastern Desert area and to prograding deltaic muds and sands in the Nile Valley region (Fig. 14). They are medium- to coarse-grained, friable to well cemented, mixed with varying proportions of angular to subangular siliciclastic grains, pyritized phosphate pellets and fish debris, and contain both interstitial and peloidal siderite cements. Phosphate nodules up to a few centimeters in

Figure 17. Growth and amalgamation of two pelletal phosphorite beds (to left of hammer handle) into one (right of handle) at the expense of intervening shales. Western Desert.

Figure 18. Pelletal phosphorite sandwaves (between arrows) in the Western Desert. Note swale depression between second arrows left and wave crest between arrows on the far right. At the crest (arrows far right), the wave is about 7 m thick. The swales between the waves are filled with black phosphatic shales.

diameter also occur in these sands. As with the underlying phosphorites, their crossbedding indicates substantial current energies. Periodic hardground development in these sands is indicated by pholad and other borings into sideritic and goethitic cements. Like the underlying phosphorites, these glauconites are likely reworked diagenetic precipitates. They do not possess obvious carbonate replacement or infilling textures [e.g. 20,68], but do show high aluminum contents and good correlations between increases in iron and potassium, both of which suggest glauconitization of a precursory illitic phase [69; cf. 70-71].

CARBONATE CONTENTS AND CARBON ISOTOPE GEOCHEMISTRY: MODERN AND ANCIENT

The carbon isotopic composition and percentage of carbonate (measured by X-ray diffraction and conventionally expressed as CO_2 [72]) extracted from individual *in situ* phosphate pellets in the Egyptian shales is shown in Figure 19. The $\delta^{13}C$ values range from about +3 to -9‰ and thus suggest precipitation from pore waters characterized by bacterial sulfate reduction (Table 1). The linear relationship between $\delta^{13}C$ and $\%CO_2$ for these samples suggests that as the pore waters of the shales became more negative (with shallow burial) there was a progressive increase in the substitution of CO_3 for PO_4 in pelletal CFA. This is consistent with the inference that bicarbonate and carbonate ion concentrations increase with burial depth in anoxic pore waters off Peru (Fig. 12, Table 1). Figure 19 also shows a similar relationship for one *in situ* F-phosphate crust recovered from the Peru margin. The slope of the line for the Peru samples is lower, however, as the Peru phosphorites yield CFA-carbonate $\delta^{13}C$ values no more negative than -5‰ (Fig. 12). In contrast, the $\delta^{13}C$-$\%CO_2$ relationship for individual phosphate pellets from Egyptian phosphorites is different in that it shows much scatter (Fig. 20). We consider this scatter to be the result of winnowing and concentration of CFA grains which precipitated in a variety of pore-water environments. Each of these environments may have been characterized by differences in local pore water carbonate-ion buffering due to carbonate dissolution, organic matter degradation rates, fluoride gradients in pore waters [37,38], or variations imposed by relative grain size and/or multiple stages of precipitation (as also indicated by intraclasts, compound pellets, etc.).

Figure 21 illustrates how CFA-$\delta^{13}C$ may vary with rates of CFA growth. Over a period of 1.38 ky, this crust grew upward in the Peru margin muds towards the sediment-water interface. The upwards increase in CFA-$\delta^{13}C$ in this crust is in agreement with the concept of decreasing pore water ^{13}C with burial depth. Growth rates in this nodule increased as it grew closer towards the sediment-water interface. These growth rate changes also correlate with changes in CFA crystallinity and chemical composition. Kim and Burnett [25] showed that the crystallinity (XRD peak height and sharpness) and phosphorus content of the crust increases downward, and Lamboy [24] demonstrated that whereas the upper part of the crust contains spindle-shaped CFA crystallites (e.g. Fig. 7), the lower part of the crust contains hexagonal crystallites. It appears, therefore, that slower CFA growth rates may favor increased carbonate substitution and better crystallographic ordering in CFA.

Changes in CFA growth rates may be the result of several factors. The first is that as the nodule accreted towards the sediment-water interface it grew more rapidly because its upper surface came progressively closer to the interstitial dissolved phosphate maxima in these sediments (see above) and because the diffusive distance for fluoride also decreased. The second is that there may be a lower depth limit to which the mineral can grow and that this is controlled by the dissolved carbonate activity of the pore waters [17,31]. Experimental results, for example, indicate that increases in concentration of carbonate in solution increase CFA solubility [39], increased bicarbonate to phosphate ratios in solution reduce apatite replacement rates (after calcite) [73], and increases in carbonate substitution in apatites limits the size to which the apatite crystallites can grow [74-76]. Such "lattice poisoning" by carbonate may explain why Peru phosphorites appear to precipitate only near the sediment-water interface, despite the ample supply of dissolved phosphate that may occur at depth in these sediments as a result of sulfate reduction processes. A third, interrelated explanation for the variance in CFA growth rates may be related to variations in sedimentation rate. Van

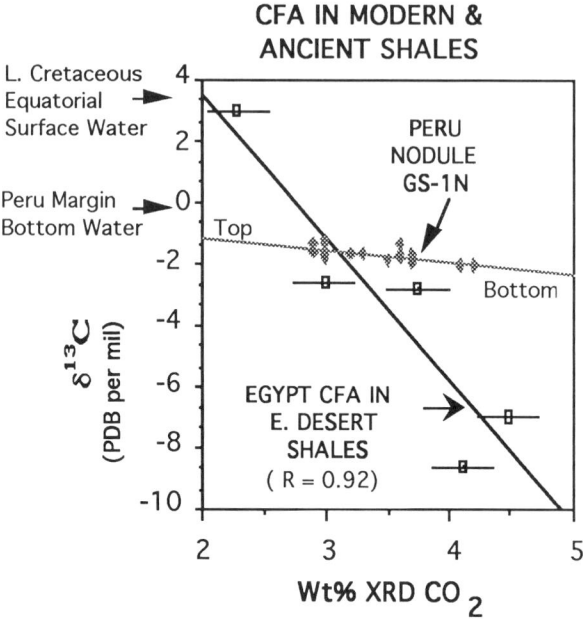

Figure 19. $\delta^{13}C$ versus weight percent CO_2 for individual CFA pellets from Egyptian phosphatic shales as compared with bottom to top slices from a Peru margin F-type phosphate nodule. The modern CFA precipitated within the suboxic diagenetic zone. The Egyptian grains formed in association with sulfate reduction processes. Data from [3,31].

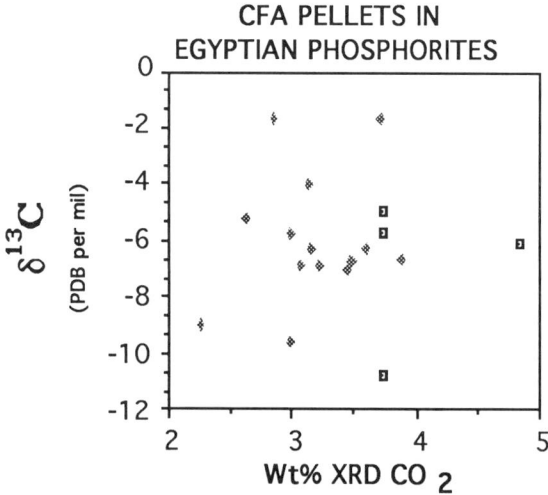

Figure 20. CFA $\delta^{13}C$ versus weight percent CO_2 for individual CFA pellets extracted from Egyptian phosphorites. These relations imply precipitation from a variety of pore water environments and latter mixing and reconcentration of the grains by current winnowing. Squares = Eastern Desert samples; Diamonds = Western Desert samples. Data from [3].

Cappellen and Berner [47] deduced that there is an optimum sedimentation rate for effective CFA precipitation on the basis of numerical modeling. They suggested that when sedimentation rates are too high the precipitating mineral phase is mechanically diluted. When too low, all reactive P is regenerated so close to the sediment-water interface that upward diffusion prevents pore water dissolved P from reaching its saturation value. As illustrated by our discussions above, such an optimum sedimentation rate may be controlled by the strength of bottom currents.

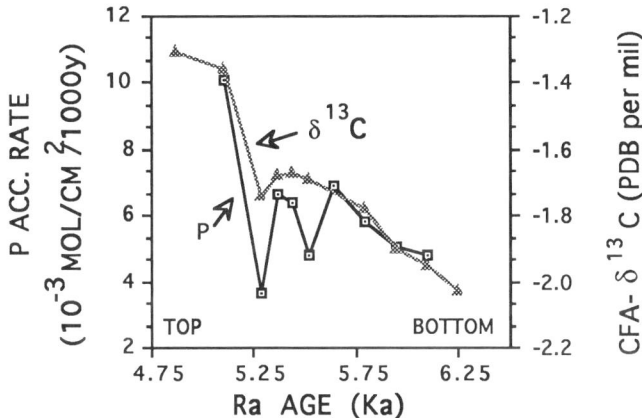

Figure 21. CFA δ^{13}C and CFA growth rate for a F-phosphate crust (GS-1N) from the Peru margin. This nodule grew from its bottom towards its top over a period of 1.38 ky. Slower rates of growth correlate with more negative CFA δ^{13}C values, increases in phosphorus and carbonate contents, and improved crystallographic ordering (see text). Compare with Figure 19. After Glenn et al. [31].

DISCUSSION AND CONCLUSIONS

The above discussions may seem rather superficial in that we have only compared two phosphorite deposits with which we are most familiar. However, we believe that these comparisons are instructive because they illustrate that modern phosphorite and associated facies development on the Peru slope and shelf is really very akin to the development of similar deposits in the ancient record. We summarize our comparisons in Table 2. We believe that the size, lateral extent and "reserves" of the modern facies are really quite large, in spite of their recent age (Table 2). Their morphologies are also similar to those of the ancient record. Pellet beds occur on the modern Peru slope, just as they do in the Late Cretaceous of North Africa, the Middle East, and elsewhere. Hardgrounds, crusts and nodules appear to be more prevalent on the Peruvian shelf than they do in the ancient record of Egypt. Yet, phosphatic hardgrounds are really not all that uncommon elsewhere in the record. Cook et al. [77], for example, recently noted that for many years phosphatic hardgrounds were the "Cinderellas of the phosphorite world!" Notable exceptions were the Miocene (microsphorite) hardgrounds and their "land-pebble" derivatives in the southeast U.S. [23], hardgrounds in Late Cretaceous chalks of England [78,79] and France [80,81], and Cambrian "phoscrete" hardgrounds and coated grains in Australia [82]. Many other phosphatic hardgrounds may have been overlooked in settings where units are well lithified and show little differential weathering. Off Peru, the hardgrounds develop through progressive phosphatization, starting out as a thin, very soft and friable crust (F-phosphate CFA proto-crusts) that is easily broken by current erosion and bioerosion. Currents help to sweep these crusts clear of detritus. With time, these crusts harden and darken and continue to incorporate granule to cobble sized broken clasts of microsphorite. The end result is a hard, thick, extensively-bored CFA pavement which resists the swing of a hammer.

Table 2.
Comparison of salient features between phosphoritic sediments of Egypt and offshore Peru

Parameter	Egypt (Campanian, Late Cretaceous)	Peru (Pleistocene-Present)
Depositional Setting	Epicontinental Seaway	Open Ocean Shelf and Slope
Types of Phosphorite	Pelletal (minor nodular/hardground)	Hardground/Nodular/Pelletal
Source of Phosphorus	Fluvial and/or Oceanic Upwelling	Oceanic Upwelling
Hinterland Climate	Humid	Arid
Currents	Tidal/Cross Shelf?	Poleward Undercurrent
Current Velocity	ca. >50 cm/s (sand waves)	up to 30 cm/s (measurements)
Water Depth	Neritic	Upper to Middle Bathyal (200-1100m)
Mean Sedimentation Rate (m/My)	20 (100m/my.)	14-20 (Site 679B, top 4 meters)
Primary Producer	Diatoms	Diatoms
Associated Diagenetic Phases	Glauconite, Siderite Dolomite, Pyrite	Pyrite, Dolomite, Glauconite, Siderite?
Associated Depositional Facies	Shales, Porcelanites and Cherts, Glauconitic Sands, (Oyster Reefs)	Organic-rich Diatomaceous Muds and Mudstones; Glauconitic Sands
Shale Organic Carbon	1% - 22%	1% - 23%
Hydrogen Index, Shales (mgHC/gOC)	ca. 300-800	ca. 400 - 600
$\delta^{13}C$ of CFA	+3‰ to -11‰	+1‰ to -5‰
Where Phosphorite Precipitates	Sulfate Reduction Zone (10's-100's cm in seds.?)	Suboxic Diagenetic Zone (<15 cm)
Size of P Deposits (10^6 Met. Tonnes Elemental P)	Egypt: 327 for 5 m.y.[a]; Egypt + Israel + Jordan + Syria + Turkey: 661 for 5 m.y. episode[a]	140 for 15 cm modern deposit, 12°S-13.5°S, 20 km swath[b] 746 for 15 cm, 7°S-15°S, 20 km swath[c]
Modern Amazon R. P-Flux (10^6 Met. Tonnes P/1000y)	1,000[d]	

[a] From resource data of [56] using an average of 22% P_2O_5 for phosphorite rock.
[b] Assumes 15 cm thick layer of phosphatic hardgrounds (P_2O_5 = 25%) and phosphatic sands (P_2O_5 = 25%) occupying a 20 km swath between 12°S and 13.5°S; dry bulk density of 2.55g/cc calculated as average between a phosphatic sand with 30% porosity and a well lithified phosphatic crust.
[c] Same assumptions as b (above) except assumes 20 km wide swath area between 7°S and 15°S.
[d] See text for sources of data

Sedimentological and stable isotopic evidence clearly shows that the phosphorites of both the modern and Egyptian setting formed from solutions undergoing various stages of bacterial decay of organic matter. Similar $\delta^{13}C$ values have also been reported for other phosphorite occurrences [e.g. 83-88]. In the modern setting, it appears that CFA precipitation is limited to the suboxic zone and occurs in association with a pore water phosphorus spike that occurs directly below the sediment-water interface. The carbon isotope values for the Egyptian occurrences imply precipitation at deeper depths in the sediment column in association with sulfate reduction. Both occurrences (where unaffected by winnowing and mixing of isotopic signatures) show evidence for a systematic correlation between the amount of carbonate for phosphate substitution that occurs in the CFA lattice and the $\delta^{13}C$ values of that carbonate. Carbonate substitution in CFA is known to have an upper limit of about 6 weight percent, and several lines of evidence suggest that carbonate poisoning may limit the precipitation of CFA. Another possibility is that the lower burial depth of CFA precipitation may be controlled by the relative diffusion rate of fluoride, which is a function of sedimentation rate and porosity. In any case, it appears that the higher bottom water temperatures on the Egyptian shelf (ca. 25°C [69,89]) may have facilitated saturation with respect to CFA at deeper burial depths by lowering its relative solubility [90,91].

The depositional settings of Egypt and Peru have both similarities and major differences. Indicators of elevated primary productivity and biologic fertility (high organic carbon contents and hydrogen indices of muds, abundant fish bones and teeth, remnants of diatoms) are common to both occurrences, yet the general depositional settings of the two deposits may be very different. The modern setting has an arid climate, receives little contribution of sediment from the adjacent landmass and is dominated by active upwelling. However, it is not readily clear what principal source provided and transported phosphorus in the Egyptian setting. Entrainment of upwelled water from the Tethyan margin 400 km to 500 km to the north is a potential source and oxygen isotope data suggest a possible cooling of ocean waters in association with phosphorite deposition in Israel [89]. However, it is also possible that fluvial inputs of phosphorus may have been important in this setting. Glenn and Arthur [3] suggested that the main source of P for the Egyptian phosphorites may have been P sorbed on clays and iron compounds that was released from these particulates upon encountering reducing bottom waters. Today, for example, the Amazon River, a relatively pristine tropical drainage system, delivers approximately 10^{11} gPyr^{-1} to relatively high salinity fringes (25 ppt) of the Amazon shelf [92]. The majority (77%) of Amazon P transport occurs as inorganic P (sorbed and some detrital P) associated with the river's suspended load [93]. Berner et al. [94] suggest that an appreciable portion of this P is sorbed on iron compounds and that it constitutes about half of the total P sedimented on the Amazon delta. Extrapolated over 5 m.y. (i.e. the time it took the Egyptian phosphorites to accumulate), the flux of fine-fraction associated P exiting the Amazon drainage system at 'Obidos [93], could provide the Amazon shelf with more than 3.8 trillion metric tonnes of P (Table 2)! Glauconite also occurs on the Amazon shelf in association with vertically extensive iron reduction zones in muds supersaturated with respect to authigenic vivianite (iron phosphate) and siderite (iron carbonate) [95].

Late Cretaceous climate was warm and encroachment and evaporation of the shallow Campanian sea into tropical paleolatitudes of 1-10°N suggests a warm and humid climate in which lateritic weathering profiles were probably well developed in hinterlands to the south. Reworked, phosphatic and sideritic glauconite greensands developed in the Western Desert region at the same time as maximum delta progradation in the Nile Valley region (Fig 14), and pyrolysis data for organic carbon indicate a mixed marine-terrestrial signal for Western Desert intercalated shales [3]. Deltaic progradation also has been suggested for the lower Dakhla Shales at Abu Tartur [96], and palynological studies of the Dakhla Shales in both the Eastern and Western Deserts suggest warm and humid subtropical to tropical conditions with estuarine shorelines lined by hundreds to thousands of miles of mangrove swamps and elevated hinterlands dominated by heavy rainfall to the south [96,97]. From the underlying phosphorites at Abu Tartur, periodically brackish conditions have been suggested from microflora data [97], and enrichments in rare-earth elements and iron have been interpreted as indicating strong terrigenous influences and possible influx of solutions derived from chemical weathering processes [98]. These factors, taken together with our analogy to the modern Amazon system, suggest high rates of chemical weathering in hinterland areas as well as pronounced fluvial discharge of iron and phosphorus to the Egyptian sea. The coupling of shelf current winnowing and near-shore sediment trapping during rapid sea level rise may have provided a major mechanism of sequence condensation which helps explain the apparent conflict between low overall sedimentation rates and major tropical riverine inputs of nutrients, iron and fine-grained siliciclastic detritus [27,28]. The redox coupling of sorbed Fe-P may also explain the common association between phosphorites, glauconites and organic-carbon-rich shales [3]. While we have not ruled out the potential of upwelled nutrients in the Egyptian setting, we think it is clear that more studies are needed to resolve the main source of phosphorus for many ancient phosphorite deposits.

The glauconites at Abu Tartur may be similar to those encountered off Peru in that they both are well sorted sands that lack admixtures of clays. As such, they represent winnowed sands deposited by high energy conditions within an otherwise quiet-water hemipelagic environment. They also appear to have accumulated where pore waters or bottom waters were suboxic (mildly reducing) and contained both ferric and ferrous iron. Although reworked, the glauconitic sands off Peru appear to be most highly concentrated along the lower margin of the present day oxygen-minimum zone, and it has been suggested [3,28] that the glauconite of the Abu Tartur greensands originally precipitated at the intersection of anoxic and oxygenated bottom waters. As with pelletal phosphorite beds, the thicker glauconite beds of both Egypt and offshore Peru thus appear to require a two step process involving precipitation from pore solutions and later reworking and winnowing by bottom currents. Further study is required to firmly establish where the glauconite grains precipitate initially.

Phosphorites have long been recognized to occur within condensed sections. In the Egyptian setting, the phosphorites and fine-grained organically-derived sediments developed at times of regional transgression and maximum flooding accompanying two third-order changes of sea level over a time interval of about 5 Ma. These intervals served to trap the locus of coarser siliciclastic detritus in more landward portions of the Egyptian shelf. Other, higher-frequency variations in sea level, such as those affecting Quaternary phosphorite development off Peru (e.g. Fig. 13), may have also aided the Egyptian phosphorite "machine" by lowering wave base and thus increasing current winnowing and bed amalgamation during relative sea level falls. However, generating higher-frequency sea level variations on a ice-free world is problematic and storm and/or tidal current reworking of the Egyptian shelf is likely more probable. Whatever the mechanism, our comparison between the Egyptian and Peruvian deposits demonstrates that current activities have had a major impact on both their depositional histories. Current winnowing and reworking of phosphatic sediment appears to be important in both because it serves to reduce sedimentation rates and thus maintain CFA near the sediment-water interface. This may be a necessary prerequisite when considering the relatively slow growth rates of nodules, crusts and hardgrounds. Millimeter-sized pellets, however, may grow over a time span of a few years and this may explain the predominance of phosphatic pellets over nodules in the geologic record [48]. Current winnowing of fines is also important because it serves to remove CFA from fine-grained sediment and concentrate it into lag deposits of potential economic worth.

Acknowledgments
Financial support for this work was provided by grants from the National Science Foundation to CRG (OCE-9201305), MAA (OCE-9014801), WCB (OCE-9214493), and RAJ (OCE-9011829), a post cruise grant to JMR from the Ocean Drilling Program, and financial assistance from the U.S. Geological Survey. Richard Knight and Thomas Marchitto aided in the collection of the Leg 112 foraminifer and oxygen isotope data and Rajan Sivaramakrishnan completed the isotope analysis of recently collected Peru margin CFA proto-crusts. We thank R.E. Garrison and A. Iijima for their reviews of the manuscript. We extend our gratitude and appreciation to our fellow scientists and the crews of the *JOIDES Resolution* and the R/V *Seward Johnson*. This is a contribution to IGCP Project 325 'Palaeogeography of Authigenic Minerals and Phosphorites', and is University of Hawaii

School of Ocean and Earth Science Contribution Number 3253.

REFERENCES

1. Y.K. Bentor. Phosphorites - the unsolved problems. *Soc. Econ. Petrol. Mineral. Sp. Publ.* **29**, 3-18 (1980).
2. Y.K. Bentor. Modern phosphorites-not a sure guide for the interpretation of ancient deposits. In: Rep. Mar. Phosphat. Workshop. p. 29. W.C. Burnett and R.P. Sheldon (Eds) Honolulu, (1979).
3. C.R. Glenn and M.A. Arthur. Anatomy and origin of a Cretaceous phosphorite-greensand giant, Egypt. *Sedimentology*, **37**, 123-154 (1990).
4. W.C. Burnett and P.N. Froelich (Eds). *The Origin of Marine Phosphorite -- Results of the R/V Robert D. Conrad Cruise 23-06 to the Peru Shelf.* Marine Geology **80**, III-VI (1988).
5. V.R. Rosato. *Peruvian deep-sea sediments: Evidence for continental accretion.* Ms. Thesis, Oregon State University, Corvallis (1974).
6. L.A. Krissek. and K.F. Scheidegger. Environmental controls on sediment texture and composition in low oxygen zones off Peru and Oregon. In: *Coastal Upwelling, Its Sediment Record, Part B: Sedimentary Records of Ancient Coastal Upwelling.* J. Thiede and E. Suess (Eds), pp. 163-180, Plenum Press, New York (1983).
7. K.F. Scheidegger and L.A. Krissek.. Zooplankton and nekton: Natural barriers to the seaward transport of suspended terrigenous particles off Peru. In: *Coastal Upwelling, Its Sediment Record, Part B: Sedimentary Records of Ancient Coastal Upwelling.* J. Thiede and E. Suess (Eds), pp. 163-180, Plenum Press, New York (1983).
8. C.E. Reimers and E. Suess. Spatial and temporal patterns of organic matter accumulation on the Peru continental margin. In: *Coastal Upwelling - Its Sedimentary Record, Part B: Sedimentary Records of Ancient Coastal Upwelling.* E. Suess and J. Thiede (Eds). pp. 311-346. Plenum Press, New York (1983).
9. C.E. Reimers and E. Suess. Late Quaternary fluctuations in the cycling of organic matter off central Peru: A proto-kerogen record. In: *Coastal Upwelling - Its Sedimentary Record, Part B: Sedimentary Records of Ancient Coastal Upwelling.* E. Suess and J. Thiede (Eds). pp. 497-526. Plenum Press, New York (1983).
10. R.E. Garrison and M. Kastner. Phosphatic sediments and rocks recovered from the Peru margin during ODP Leg 112. In: *Proc. ODP, Sci. Results.* E. Suess, von Huene et al., (Eds). **112**, pp. 111-134, College Station, TX (Ocean Drilling Program) (1990).
11. H. Schrader and R. Sorknes. Peruvian coastal upwelling: late Quaternary productivity changes revealed by diatoms. *Marine Geology* **97**, 233-249 (1991).
12. T.J. DeVries and Pearcy. Fish debris in sediments of the upwelling zone off central Peru: a late Quaternary record. *Deep-Sea Res.* **28**, 87-109 (1982).
13. K. Wyrtki. The horizontal and vertical field of motion in the Peru Current. *Bull. Scripps Inst. Oceanogr.* **8**, 313-346 (1963).
14. C. Brockmann, E. Fahrbach, A. Huyer and R.L. Smith. The poleward undercurrent along the Peru coast: 5 to 15°S, *Deep-Sea Res..* **27A**, 847-856 (1980).
15. L.A. Codispoti, R.C. Dugdale and H.J. Minas. A comparison of the nutrient regimes off Northwest Africa, Peru, and Baja California. *Rapp. P. -v. Reun. Cons. int. Explor. Mer.* **1980**, 184-201 (1982).
16. R.L. Smith. Peru coastal currents during El Nino: 1976 to 1982, *Science* **221**, 1397-1399 (1983).
17. C.R. Glenn and M.A. Arthur. Petrology and major element geochemistry of Peru margin phosphorites and associated diagenetic minerals: authigenesis in modern organic-rich sediments. *Marine Geology*, **80**, 231-267 (1988).
18. W.C. Burnett, H.H. Veeh and A. Soutar. U-series, oceanographic and sedimentary evidence in support of recent formation of phosphate nodules off Peru. *Soc. Econ. Paleontol. Mineral. Spec. Publ.* **29**, 61-71 (1980).
19. E. Suess, R. von Huene, et al. (eds.).*Initial Reports of the Ocean Drilling Program* **Vol. 112**. College Station, Texas (1988).
20. G.S. Odin and A. Matter. De glauconiarum origine. *Sedimentology* **28**, 611-641 (1981).
21. C.P. Mabie and H.D. Hess, H.D. Petrographic study and classification of western phosphate ores. *U.S. Bureau of Mines Report of Invest.*, **6468**, 1-95 (1964).
22. P.J. Cook. Sedimentary phosphate deposits. In: *Handbook of Strata-bound and Stratiform Ore Deposits, II, Regional Studies and Specific Deposits, Vol. 2.* K.H. Wolf (Ed.). pp. 505-535, Elsevier, Amsterdam (1976).
23. S.R. Riggs. Petrology of the Tertiary phosphorite system of Florida. *Econ. Geol.* **74**, 195-220 (1979).
24. M. Lamboy. Microstructures of a phosphatic crust from the Peruvian continetal margin: phosphatized bacteria and associated phenomena, *Oceanologica Acta* **13**, 439-451 (1990).
25. K.H. Kim and W.C. Burnett. Uranium-series growth history of a quaternary phosphatic crust from the Peruvian continental margin. *Chemical Geology* **58**, 227-244 (1986).
26. R. Giovanoli and G. Arrhenius. Structural chemistry of marine maganese and iron minerals and synthetic model compounds. In: *The Manganese Nodule Belt of the Pacific Ocean.* P. Halback, G. Friedrich and U. von Stackelberg (Eds). pp. 20-37. Ferdinand Enke Verlag, Stuttgart (1988).

27. C.R. Glenn. Pore water, petrologic and stable carbon isotopic data bearing on the origin of modern Peru margin phosphorites and associated authigenic phases. In: *Phosphate Deposits of the World: Volume 3, Genesis of Neogene to Recent Phosphorites*. W.C. Burnett and S.R. Riggs (Eds). pp. 46-61, Cambridge University Press (1990).
28. C.R. Glenn. Deposition of the Duwi, Sibâîya and Phosphate Formations, Egypt: phosphogenesis and glauconitization in a Late Cretaceous epeiric sea. In: *Phosphorite Research and Development*. A.J.G. Notholt, and I. Jarvis (Eds). Geological Society Spec. Publ. No. 52, pp. 205-222, The Geological Society, Bath (1990).
29. R.A. Berner. Sedimentary pyrite formation: an update, *Geochim. Cosmochim. Acta* **48**, 605-615 (1984).
30. R.A. Berner. Sulfate reduction, organic matter decomposition and pyrite formation, *Phil. Trans. R. Soc. London (A)* **315**, 25-38 (1985).
31. C.R. Glenn, M.A. Arthur, H.W. Yeh and W.C. Burnett. Carbon isotopic composition and lattice-bound carbonate of Peru-Chile margin phosphorites, *Marine Geology* **80**, 287-307 (1988).
32. C.D. Curtis. Sedimentary geochemistry: environments and processes dominated by involvement of an aqueous phase, *Philosophical Transactions of the Royal Society of London (A)* **286**, 353-72 (1977).
33. H. Irwin C.D. Curtis,, and M. Coleman. Isotopic evidence for source of diagenetic carbonates formed during burial of organic-rich sediments, *Nature* **269**, 209-13 (1977).
34. P.N. Froelich, G.P. Klinkhammer, M.L. Bender N.A. Luedtke, G.R. Heath, D. Cullen, P. Dauphin, D. Hammond, B. Hartman and V. Maynard. Early oxidation of organic matter in pelagic sediments of the eastern equitorial Atlantic: suboxic diagenesis, *Geochim. Cosmochim. Acta* **43**, 1075-90 (1979).
35. R.A. Berner, R.A. *Early Diagenesis, A Theoretical Approach*. Princeton University Press, Princeton (1980).
36. M.L. Coleman. Geochemistry of diagenetic non-silicate minerals: kinetic considerations, *Philosophical Transactions of the Royal Society of London (A)* 315, 39-56 (1985)
37. P.N. Froelich, K.H. Kim, R. Jahnke, W.C. Burnett, A. Soutar and M. Deakin. Pore water fluoride in Peru continental margin sediment: uptake from seawater, *Geochim. Cosmochim. Acta* **47**, 1605-1612 (1983).
38. P.N. Froelich, M.A. Arthur, W.C. Burnett, M. Deakin, V. Hensley, R. Jahnke, L. Kaul, K.H. Kim, K. Roe, A. Soutar and C. Vathakanon. Early diagenesis of organic matter in Peru continental margin sediments: phosphorite precipitation. *Marine Geology* **80**, 309-343 (1988).
39. R.A. Jahnke. The synthesis and solubility of carbonate fluorapatite, *Amer. J.Sci.* **284**, 58-78 (1984).
40. D.T. Heggie, G.W. Skyring, D.W. O'Brien, C. Reimers, A. Herczeg, D.J.W. Moriarty, W.C. Burnett and A.R. Milnes. Organic carbon cycling and modern phosphorite formation on the East Australian continental margin: an overview. In: *Phosphorite Research and Development*. A.J.G. Notholt, and I. Jarvis (Eds). Geological Society Spec. Publ. No. 52, pp. 87-117. The Geological Society, Bath (1990).
41. P.J. Cook and G.W. O'Brien. Neogene to Holocene phosphorites of Australia. In: *Phosphorite Deposits of the World*. W.C. Burnett and S.R. Riggs (Eds). pp. 98-121, Cambridge University Press, Cambridge (1990).
42. R.A. Berner. Phosphate removal from sea-water by absorption on volcanogenic ferric oxides, *Earth Planet. Sci. Letters* **18**, 77-86 (1973).
43. M.D. Krom and R.A. Berner. Adsorption of phosphate in anoxic marine sediments. *Limnology and Oceanography* **25**, 797-806 (1980).
44. M.D. Krom and R.A. Berner. The diagenesis of phosphorus in a nearshore marine sediment, *Geochim. Cosmochim. Acta* **45**, 207-16 (1981)
45. G.J. de Lange. Early diagenetic reactions in interbedded pelagic and turbidic sediments in the Nares Abyssal Plain (western North Atlantic): Consequences for the composition of sediment and interstitial water. *Geochim. Cosmochim. Acta* **50**, 2543-2561 (1986).
46. E. Suess. Phosphate regeneration from sediments of the Peru continental margin by dissolution of fish debris: *Geochim. Cosmochim. Acta* **45**, 577-88 (1981).
47. P. Van Cappellen and R.A. Berner. A mathematical model for the early diagenesis of phosphorus and fluorine in marine sediments: Apatite Precipitation. Amer. J. Sci. **288**, 289-333 (1988).
48. W.C. Burnett, K.B. Baker, P.A. Chin, W. McCabe and R. Ditchburn. Uranium-series and AMS ^{14}C studies of modern phosphatic pellets from Peru shelf muds, *Marine Geology* **80**, 215-230 (1988).
49. K.B. Baker and W.C. Burnett. Distribution, texture and composition of modern phosphate pellets in Peru shelf muds, *Marine Geology* **80**, 195-213 (1988).
50. J. Resig. Biogeography of benthic foraminifera of the northern Nazca Plate and adjacent continental margin. *Geol. Soc. Am. Mem.* **154**, 619-666 (1981).
51. J. Resig. Benthic foraminiferal stratigraphy and paleoenvironments off Peru, Ocean Drilling Program Leg 112. In: *Proc. ODP, Sci. Results*. E. Suess, R. von Huene, et al. (Eds). vol. 112, pp. 263-296, College Station, TX (Ocean Drilling Program) (1990).
52. G. Wefer, P. Heinze and E. Suess. Stratigraphy and sedimentation rates from oxygen isotope composition, organic carbon content, and grain-size distribution at the Peru upwelling region: Holes 680B and 686B. In: *Proc. ODP, Sci. Results*. E. Suess, R. von Huene, et al.(Eds). vol. 112, pp. 355-366, College Station, TX (Ocean Drilling Program) (1990).
55. K.H. Kim and W.C. Burnett. 226Ra in phosphate nodules from the Peru/Chile seafloor, *Geochim. Cosmochim. Acta* **49**, 1073-1081 (1985).

54. K.H. Kim and W.C. Burnett. Uranium-series growth history of a quaternary phosphatic crust from the Peruvian continental margin, *Chemical Geology* **58**, 227-244 (1986).
55. K.H. Kim and W.C. Burnett. Accumulation and biological mixing of Peru margin sediments. *Marine Geology* **80**, 181-194 (1988).
56. A.J.G. Notholt. Phosphorite resources in the Mediterranean (Tethyan) Phosphogenic Province: A progress report, *Sci. Géol. Mém.* **77**, 9-21 (1985).
57. M.I. Youssef. Upper Cretaceous rocks in Kosseir area. *Bull. Inst. Desert. Egypt*, **T. VII**, 35-53 (1957).
58. Z.R. El-Nagger. Stratigraphy and planktonic foraminifera of the Upper Cretaceous-Lower Tertiary succession in the Esna-Idfu region, Nile Valley, Egypt, *Bull. Br. Mus. Nat. Hist., Supp. 2* (1966).
59. G.H. Awad, and M.G. Ghobrial. Zonal stratigraphy of Kharga Oasis, *Geol. Surv. Egypt* **34**, 1-77 (1966).
60. B.U. Haq, J. Hardenbol and P.R. Vail. Chronology of fluctuating sea levels since the Triassic, *Science* **235**, 1156-1167 (1987).
61. P.R. Vail. Seismic stratigraphic interpretation procedure. In: *Atlas of Seismic Stratigraphy , Vol. 1*. A.W. Bally (Ed.). pp. 1-10, Amer. Assoc. Petrol. Geol. Studies in Geol., Tulsa **27** (1987).
62. J.C. Wagoner, R.M. Mitchum Jr., H.W. Posamentier and P.R. Vail. Key definitions of sequence stratigraphy. In: *Atlas of Seismic Stratigraphy , Vol. 1*. A.W. Bally (Ed.). pp. 11-14, Amer. Assoc. Petrol. Geol. Studies in Geol., Tulsa, **27**, (1987).
63. A.S. Wassef. On the results of geological investigations and ore reserves calculations of Abu Tartur phosphorite deposit. *Annals Geol. Surv. Egypt* **7**, 1-60 (1977).
64. R.E. Garrison, C.R. Glenn, P.D. Snavely and S.E.A. Mansour. Sedimentology and origin of Upper Cretaceous phosphorite deposits at Abu Tartur, Western Desert, Egypt *Annals Geol. Surv. Egypt* **9**, 261-281 (1979).
65. M.H. Hermina. Preliminary evaluation of Maghrabi-Liffiya phosphorites, Abu Tartur area, Western Desert, Egypt, *Ann. Geol. Surv. Egypt* **3**, 39-74 (1973).
66. P.K. Bose, G. Ghosh, S. Shome and Bardhan. Evidence of superimposition of storm waves on tidal currents in rocks from the Tithonian-Neocomian Umia Member, Kutch, India, *Sedimentary Geology* **54**, 321-329 (1988).
68. C.R. Glenn, J.D. Kronen, Jr., P.A. Symonds, W. Wei, and D. Kroon. High resolution sequence stratigraphy, condensed sections and flooding events off the Great Barrier Reef: 0-1.5 Ma. In: *Proc. ODP, Sci. Results*, P.J. Davies, J.A. Mckenzie, A. Palmer-Julson (Eds.) **133**, College Station, TX (Ocean Drilling Program) (in press).
69. C.R. Glenn. *Phosphorus fluxes, phosphorite sedimentation and associated diagenesis in oxygen-deficient basins: The modern Black Sea and Peru margin and the Upper Cretaceous of Egypt.* Univ. Rhode Island, PhD Dissert., University Microfilms (1987).
70. V. Berg-Madsen. High-alumina glaucony from the Middle Cambrian of Öland and Bornholm, southern Baltoscandia, *Jour. Sed. Petrol.* **51**, 875-893 (1983).
71. B.D. Bornhold and P. Giresse. Gluconitic sediments on the continental shelf off Vancouver Island, British Columbia, Canada. *Jour. Sed. Petrol.* **55**, 653-664 (1985).
72. R.A. Gulbrandsen. Relation of carbon dioxide content of apatite of the Phosphoria Formation to regional facies. *U.S. Geol. Surv. Prof. Paper*, **700-B**, B9-B13 (1970).
73. L.L. Ames. The genesis of carbonate-apatite, *Econ. Geol.* **54**, 829-41 (1959).
74. R.Z. LeGeros, O.R. Trautz, J.P. LeGeros, and E. Klein. Apatite crystallites: Effects of carbonate on morphology, *Science* **155**, 1409-1411 (1967).
75. LeGeros, R.Z., and LeGeros, J.P., 1984. Phosphate minerals in human tissue. In: *Phosphate Minerals*. J.O. Nriagu and P.B. Moore (Eds). pp. 351-385 Springer-Verlag, Berlin-Heidelberg (1985).
76. G.H. McClellan, and J.R. Lehr Crystal chemical investigation of natural apatites, *Amer. Mineralogist* **54**, 1374-91 (1969).
77. P.J. Cook, J.H. Shergold, W.C. Burnett and S.R. Riggs. Phosphorite research: a historical overview. In: *Phosphorite Research and Development*. A.J.G. Notholt and I. Jarvis (Eds). pp. 1-22. Geol. Soc. Sp. Publ. No. 52. Geol. Soc., Bath (1990).
78. R.E. Garrison, W.J. Kennedy and T.J. Palmer. Early lithification and hardgrounds in Upper Albian and Cenomanian calcarenites, Southwest England, *Cretaceous Res.* **8**, 103-140 (1987).
79. W.J. Kennedy and R.E. Garrison. Morphology and genesis of nodular chalks and hardgrounds in the Upper Creataceous of southern England, *Sedimentology* **22**, 311-386 (1975).
80. I. Jarvis. Geochemistry of phosphatic chalks and hardgrounds from the Santonian to Early Campanian (Cretaceous) of northern France, *J. Geol. Soc. London.* **137**, 705-732 (1980).
81. I. Jarvis. The initiation of phosphatic chalk sedimentation - the Senonian (Cretaceous) of the Anglo-Paris Basin. In: *Marine Phosphorites*. Y.K. Bentor (Ed.). pp. 167-192, Soc. Econ. Paleont. Mineral. Sp. Publ. 29, Tulsa (1980).
82. P.N. Southgate. Cambrian phoscrete profiles, coated grains, and microbial processes in phosphogenesis: Georgina Basin, Australia. *Jour. Sed. Pet.* **56**, 429-441 (1986).
83. D.Z. Piper and Y. Kolodny. The stable isotopic composition of a phosphorite deposit: $\delta^{13}C$, $\delta^{34}S$ and $\delta^{18}O$, *Deep-Sea Res.* **34**, 897-911 (1987).
84. K.S. Al-Bassam. Carbon and oxygen composition of some sedimentary apatites from Iraq, *Econ. Geol.* **75**,

1231-1233 (1980).
85. R.A. Benmore, M.L. Coleman, and J.M. McArthur. Origin of sedimentary francolite from its sulfur and carbon isotope composition, *Nature*, **302**, 516-518 (1983).
86. Y. Kolodny and I.R. Kaplan. Carbon and oxygen isotopes in apatite-CO_2 and co-existing calcite from sedimentary phosphorite. *Jour. Sed. Petrol.* **40**, 954-959 (1970)
87. J.M. McArthur, M.L. Coleman, and J.M. Bremner. Carbon and oxygen isotopic composition of structural carbonate in sedimentary francolite, *J. Geol. Soc. London* **137**, 669-673 (1980).
88. J.M. McArthur, R.A. Benmore, M.L. Coleman, C. Soldi, H.-W. Yeh, and G.W. O'Brien. Stable isotopic characterization of francolite formation, *Earth Planet. Sci. Lett.* **77**, 20-34 (1986).
89. A. Shemesh and Y. Kolodny. Oxygen isotopes in phosphorites from the Southeastern Tethys, *Isr. J. Earth Sci.* **37**, 1-15 (1988).
90. E.L.Atlas. *Phosphate equilibria in seawater and interstitial waters*. Ph.D. Thesis. Oregon State University, Corvallis (1975).
91. E.L. Atlas, & R.M. Pytkowicz. Solubility behavior of apatites in seawater, *Limnol. Oceanogr.* **22**, 290-300 (1977).
92. L.E. Fox, S.L. Sager and S.C. Wofsy. The chemical control of soluble phosphorus in the Amazon estuary. *Geochim. Cosmochim. Acta* **50**, 783-794 (1986).
93. J.E. Richy and R.L. Victoria. C, N, and P export dynamics in the Amazon River. In: *Interactions of C, N, P and S Biogeochemical Cycles and Global Change*. R. Wollast, F.T. Mackenzie and L. Chou (Eds), pp. 123-139, Springer-Verlag, Berlin (1993).
94. R.A. Berner, K.C. Ruttenburg, E.D. Ingall and J.-L. Rao. The nature of phosphorus burial in modern marine sediments. In: *Interactions of C, N, P and S Biogeochemical Cycles and Global Change*. R. Wollast, F.T. Mackenzie and L. Chou (Eds), pp. 365-378, Springer-Verlag, Berlin (1993).
95. R.C. Aller, J.E. Mackin, and R.T. Cox Jr. Diagenesis of Fe and S in Amazon inner shelf muds: apparent dominance of Fe reduction and implications for the genesis of ironstones: *Continental Shelf Res.* **6**, 263-289 (1986).
96. E. Schrank. Organic-geochemical and palynological studies of a Dahkla Shale profile (Late Cretaceous) in southeast Egypt: Part A: Succession of microfloras and depositional environment, *Berliner geowiss. Abh. (A)* **50**, 189-207 (1984).
97. E. Schrank. Organic-walled microfossils and sedimentary facies in the Abu Tartur phosphates (Late Cretaceous, Egypt), *Berliner geowiss. Abh. (A)* **50**, 177-187 (1984).
98. K. Germann, W.-D. Bock, and T. Schröter. Properties and origin of Upper Cretaceous Campanian phosphorites in Egypt, *Sci. Géol. Mém.* **77**, 23-33 (1985).

Phosphogenesis and the controls on phosphorus accumulation in continental margin sediments

G. FILIPPELLI[1] and M. DELANEY[2]
[1,2] *Earth Sciences Department and Institute of Marine Sciences, University of California, Santa Cruz, CA 95064, USA*

Abstract-- Phosphorus (P) is a limiting element for oceanic productivity over geologically-long time scales; therefore, it is important to understand the history of P accumulation in oceanic sediments through time. To elucidate the controls on P accumulation and phosphogenesis (the in-situ formation of P-rich minerals in marine sediments), we examined P accumulation on two scales: a Miocene low-oxygen basin of the Monterey Formation that displays intense phosphogenesis, and three phosphorite giant deposits (large accumulations of P-rich sedimentary rocks) spanning about 250 m.y.
Phosphorus accumulation rates (in units of μmol $P \cdot cm^{-2} \cdot y^{-1}$) were much higher in phosphatic shale strata (1.5 - 29) than in dolomitic or siliceous strata (0.1 - 3.1) of the Shell Beach section of the Monterey Formation. Based on the geochemistry, sedimentology, and oceanographic setting of these rocks, the factors controlling phosphogenesis appear to be the flux of organic matter, low oxygen bottom water concentrations, and porosity and permeability of the sediments.
We calculated P accumulation and burial rates for several well-studied phosphorite giants, and found that these parameters were comparable to those for the modern Peru phosphogenic margin. Though phosphorite giant deposition probably requires favorable sedimentological, tectonic, and/or oceanographic conditions, it is not a geochemically-anomalous phenomenon.

Keywords: Phosphorus accumulation, phosphogenesis, phosphorite giants

Introduction

Phosphorus (P) is probably the most important limiting element for oceanic productivity over long time scales because it lacks a volatile form to facilitate transfer between the ocean, atmosphere, and terrestrial biosphere [1], and its only significant source to the ocean is continental weathering [2]. Thus, the availability of P in ancient oceans may have played a role in modulating paleoclimates through the link between productivity and atmospheric composition [3-6]. To this end, determining the burial flux of P is important for understanding the long-term record of P input to the ocean [3].
The oceanic mass balance of dissolved phosphate, with an oceanic residence time of about 80,000 years, is dominated by one source, river input, and by removal in various P-bearing forms in marine sediments [7,8]. Deep ocean basins areally dominate the oceanic system, but continental margins are the most important sink for phosphorus; Baturin [9] calculates that approximately 80% of total P accumulates in continental margin sediments. A major portion of the P buried in continental margin sediments is in the form of carbonate fluorapatite (CFA) and is the result of phosphogenesis (the in-situ formation of authigenic CFA in marine sediments). If sediments that have undergone phosphogenesis are reworked, the fine and uncemented fraction of sediment particles can be winnowed away, leaving the relatively unreactive and cemented phosphate minerals as a lag deposit. These P-rich lag deposits are termed phosphorites, and they can have very enriched P

concentrations (greater than 5 wt% P_2O_5; [10,11]) compared to average continental margin sediments. It is important to separate P accumulation during phosphogenesis from that occurring during phosphorite formation, because the first process results in net removal of P from the marine environment, while the second is simply the concentration of P within sedimentary horizons (Fig. 1). Therefore, understanding the relationship between P burial, phosphogenesis, and phosphorite formation in continental margin sediments is an important step to understanding the dynamics of the oceanic P mass balance.

In this paper, we examine P accumulation and the controls on phosphogenesis on two scales: a Miocene low-oxygen basin of the Monterey Formation that displays intense phosphogenesis, and three phosphorite giant deposits (large accumulations of P-rich sedimentary rocks) spanning about 250 m.y.

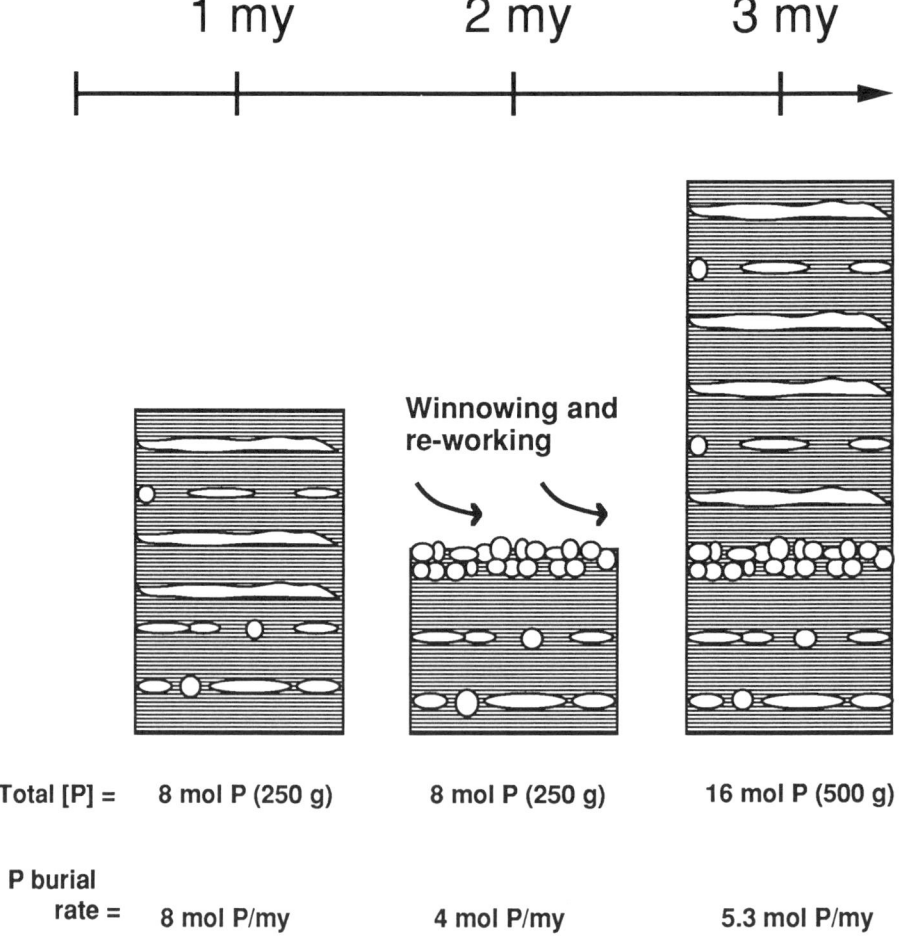

Figure 1. Schematic diagram of phosphorus concentrations and net phosphorus burial rates during phosphogenesis and phosphorite formation, at three time intervals. Note that, though total phosphorus concentration increases at the 3 m.y. time interval, the net phosphorus accumulation rate of the section is below the value for the time interval preceding phosphorite formation (1 m.y.).

PHOSPHORUS ACCUMULATION IN THE MIOCENE MONTEREY FORMATION AT SHELL BEACH, PISMO BASIN, CALIFORNIA

The Shell Beach section of the Monterey Formation is located approximately 15 km south of the city of San Luis Obispo in central California. The section lies within the Pismo basin, one of the many small Neogene extensional basins that developed in the California borderland region during the late Oligocene to early Miocene [12] and was a depocenter during most of the Miocene and Pliocene [13]. The Shell Beach section is divisible into two distinct facies: a lower Phosphatic Facies and an upper Siliceous Facies. The lower Phosphatic Facies conformably overlies the middle Miocene [14] Obispo Tuff and its contact with the upper Siliceous Facies is marked by a hiatus spanning a time of approximately 1 m.y. [15]. It consists of calcareous-rich phosphatic mudrocks, marlstones, and shales [dark, laminated, and organic-rich with thin (0.5 - 10 mm), white, discontinuous laminae and micronodules of CFA] interbedded with dolomites. The upper Siliceous Facies consists of interbedded porcellanite, porcellanous shale, and siliceous mudstone, dolomite layers, and laminated phosphatic shale strata. It is unconformably overlain by the late Miocene to Pliocene Pismo Formation, a clastic-rich marine unit. The Phosphatic Facies was deposited from 15.2 Ma (0 meters) to the 14.3 Ma base of the phosphatic hardgrounds (75 meters), with approximate rock abundances of 73% phosphatic shale strata and 27% dolomitic strata by thickness. The portion of the Siliceous Facies studied here was from the top of the phosphatic hardgrounds at 13.25 Ma (75 meters) to 12.76 Ma (150 meters), with approximate rock abundances of 20% phosphatic shale strata and 80% siliceous strata by thickness (age assignments from [15]). Estimated linear sedimentation rates are 88 m/my for the lower interval and 154 m/my for the upper interval of the section studied here [15].

The phosphatic shales of the Shell Beach section show clear evidence of intense phosphogenesis, and therefore the Pismo Basin is an excellent example of the Miocene low oxygen basins that probably had a significant role in removing P from the Miocene oceanic reservoir [16]. Furthermore, the Shell Beach section is characterized by interbedded phosphatic and non-phosphatic (siliceous and dolomitic) strata. Because of this variation in lithologies, the Shell Beach section is ideal for examining variability in P accumulation and the geochemical and sedimentological controls on phosphogenesis.

Phosphorus concentrations varied widely between the various lithologies at Shell Beach (Table 1; Fig. 2a). Phosphorus concentrations in phosphatic shales ranged from 0.36 to 3.9 wt%, and in siliceous and dolomitic strata ranged from 0.02 to 0.42 wt%. Thus, P concentrations are much higher in the shale strata, which were also visibly enriched in phosphatic minerals. The phosphatic hardground at 75 m, which separates the lower and upper facies, has a P concentration of 4.5 wt%. These P concentrations, normalized to wt% P_2O_5, are 1.9 to 9.0 for phosphatic shale strata, and 10.4 for the phosphatic hardgrounds. As phosphorite deposits are defined as containing greater than 5 wt% P_2O_5, some of the phosphatic shale strata can be classified as phosphorites. However, phosphorites are generally thought to result from the reworking of phosphatic-rich sediments and the concentration of P within certain horizons ([17] and references therein). The phosphatic shales show no obvious signs of reworking, and the P-rich laminae of the shales appear to have formed by the in-situ precipitation of authigenic CFA. The hardground, on the other hand, shows rounded and bored phosphate nodules as well as multiple cementation events, evidence for long exposure and reworking at the sediment-water interface and concentration of dissolution-resistant P-rich grains and clasts to form a phosphorite horizon.

Phosphorus accumulation rates were calculated from P concentrations, the average post-compaction sedimentation rate of each facies [15], and average sediment density [18].

Table 1.

Phosphorus concentrations and calculated accumulation rates from Shell Beach, California.

DESCRIPTION	HEIGHT (m)	AGE[a] (Ma)	PHOSPHORUS wt%	std. dev.	acc. rate[b]
siliceous	137	12.85	0.020	0.001	0.149
siliceous	135	12.86	0.053	0.001	0.399
siliceous	132	12.88	0.066	0.001	0.490
siliceous	130	12.89	0.020	0.002	0.151
siliceous	121	12.95	0.420	0.033	3.14
ph. shale	118	12.97	0.944	0.039	7.04
ph. shale	116	12.98	2.97	0.249	22.4
ph. shale	115	12.99	2.51	0.060	18.7
ph. shale	110	13.02	1.18	0.025	8.78
ph. shale	110	13.02	3.80	0.117	28.3
ph. shale	109	13.03	3.10	0.016	23.1
siliceous	108	13.04	0.247	0.008	1.83
ph. shale	105	13.06	3.35	0.042	25.0
siliceous	104	13.06	0.349	0.025	2.61
ph. shale	86	13.18	1.81	0.040	13.5
ph. shale	78	13.23	3.91	0.078	29.1
hardground[c]	75	--	8.32	0.205	--
ph. shale	74	14.31	2.81	0.043	12.0
ph. shale	57	14.50	2.10	0.059	8.95
dolomite	53	14.55	0.076	0.002	0.323
ph. shale	52	14.56	0.529	0.027	2.26
dolomite	40	14.70	0.070	0.006	0.298
ph. shale	36	14.74	1.68	0.057	7.15
ph. shale	25	14.87	2.20	0.161	9.36
ph. shale	24	14.88	2.05	0.021	8.72
ph. shale	21	14.91	1.76	0.069	7.48
ph. shale	20	14.92	1.48	0.023	6.29
ph. shale	10	15.04	1.50	0.093	6.38
ph. shale	9	15.05	1.09	0.080	4.64
dolomite	7	15.07	0.062	0.009	0.264
ph. shale	5	15.09	0.910	0.123	3.88
dolomite	4	15.10	0.132	0.002	0.56
ph. shale	3	15.12	0.535	0.015	2.28
ph. shale	2	15.13	0.354	0.012	1.51
tuff	1	15.14	0.025	0.003	0.108

a. Age determinations assuming linear sedimentation rates between age control points [15].
b. P accumulation rate (μmol P·cm^{-2}·y^{-1}), calculated by the equation

$$\text{P accumulation rate} = \frac{(\text{wt\% P}) \cdot d \cdot s}{100 \cdot 30.97 \cdot 10^6}$$

where d is the dry bulk density (we use 1.5 g/cm^3 [18]), s is the post-compaction sedimentation rate (cm/my), and the denominator is the product of 100 (correction from wt% value in numerator), 30.97 (the gram formula weight of phosphorus), and 10^6 (to convert result from my to y).
c. The 0.5 meter phosphatic hardground records substantial and repeated reworking; thus no age or accumulation rate values are calculated.

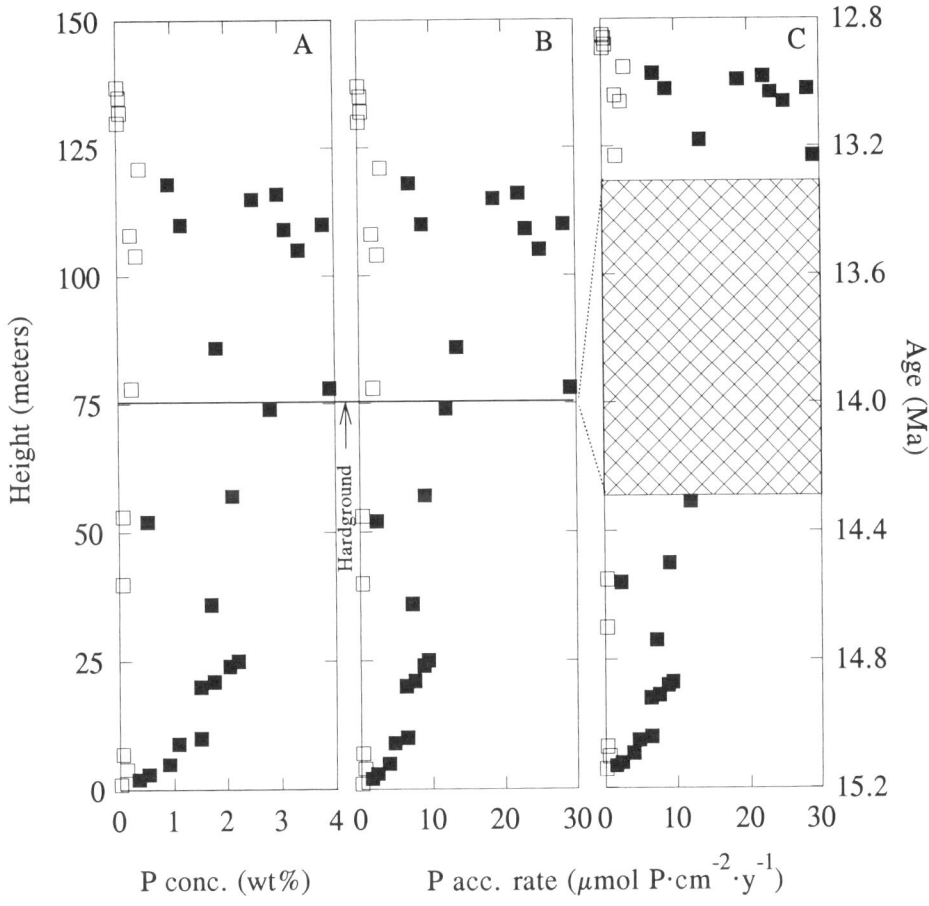

Figure 2. Phosphorus concentrations and accumulation rates in the Miocene Monterey Formation at Shell Beach, California (see Table 1). A. Phosphorus concentration (wt% P); B. Phosphorus accumulation rate versus stratigraphic height, and C. Phosphorus accumulation rate (μmol·cm^{-2}·y^{-1}) versus age. Solid squares are values in phosphatic shale strata and hollow squares are values in siliceous or dolomitic strata.

Density and differential sediment compaction depend on factors such as initial porosity, grain density, permeability, and cementation affinity; these factors are exceedingly difficult to model in sections with such variable lithology and intense authigenic mineralization, and thus we use an average density value for all lithologies as a best estimate.

Phosphorus accumulation rates (reported as μmol P·cm^{-2}·y^{-1} for comparison to modern measurements) were much higher in phosphatic shale strata (1.5 to 29) than in non-phosphatic strata (0.1 to 3.1; Table 1; Fig. 2b). Furthermore, P accumulation rates were generally higher in phosphatic shale strata of the upper Siliceous Facies (3.1 to 29) compared to the lower Phosphatic Facies (1.5 to 14).

Mean P accumulation rates, calculated using linear sedimentation rates [15], showed clear differences between lithologies and facies. We report means with errors calculated as $\pm 2\sigma/\sqrt{n}$ propagated through the calculations to allow comparison of errors from different sized sample sets. The mean P accumulation rate in phosphatic shales was 6.2 ± 3.0 in the Phosphatic Facies, where phosphatic shales comprise 73% of total strata by thickness, and 19.5 ± 7.7 in the Siliceous Facies, where they comprise only 20% of the total thickness. The mean P accumulation rate for dolomites was 0.4 ± 0.1 in the Phosphatic Facies (27% of total by thickness) and for siliceous strata was 1.3 ± 1.2 in the Siliceous Facies (80% of total by thickness).

A calculation of total mean P accumulation rates, weighted by average wt% P and fraction of total thickness of each lithology, yields an interesting result: the total mean P accumulation rate in the Phosphatic Facies (4.6 ± 2.2) is similar to that in the Siliceous Facies (4.9 ± 1.8). This result sheds light on an apparent anomaly. The Siliceous Facies is regarded as recording an increase in oceanic upwelling and productivity, yet the Phosphatic Facies, despite lower mean sedimentation rates, is visibly enriched in P-containing strata. Our geochemical results show that higher P accumulation rates in phosphatic shale strata of the Siliceous Facies offset the low relative abundance of phosphatic shale strata compared to the Phosphatic Facies. Though the onset of siliceous deposition certainly marked a change of depositional styles from mainly phosphatic shale to mainly biosiliceous deposition, the assumption that P deposition is more significant in the Phosphatic Facies than the Siliceous Facies needs to be reevaluated, or at least applied with caution.

Controls on P accumulation and phosphogenesis at Shell Beach

Our geochemical analyses reveal a substantial difference between wt% P in phosphatic shales and non-phosphatic strata. This difference is also observed visually on the outcrop level, with phosphatic shales displaying numerous white laminae and micronodules of authigenic CFA, contrasted with the lack of visible CFA in non-phosphatic strata. Though we have quantified the magnitude and frequency of P accumulation rate variability, the causes of this variability are less clear. We therefore discuss the process of phosphogenesis in light of our P accumulation rate results in order to understand the controls on P variability at Shell Beach.

Though the subject of intense scrutiny over the past few decades, phosphogenesis remains a complicated and poorly constrained process. Phosphogenesis has long been linked to sites of oceanic upwelling, with the breakdown of organic matter in areas of high organic matter flux postulated as the source of limiting phosphate [10,19,20]. Possible mechanisms driving phosphogenesis in near-surface sediments include inorganic precipitation from interstitial water [2,21-23], replacement of pre-existing carbonates [24-26], microbially-mediated precipitation of CFA [27-30], sedimentary redox cycling with or without microbial mediation [31-37], and possibly a combination of many mechanisms [38]. Geochemical modeling indicates that, in addition to a viable source or sources of necessary ions, the flux of these ions as modified by sedimentary physical properties also acts to control the distribution and mineralogical purity of CFA layers [39,40]. These models successfully replicate the thickness and spacing of modern CFA layers in sediments (e.g., [40,41]), using factors such as porosity, permeability, P fluxes from the breakdown of organic matter, and diffusion models to simulate sedimentary geochemical conditions. The layered aspect of CFA horizons has also been suggested to result from event deposition and the trapping of P sources (e.g., microbial mats) in the subsurface [42,43]; this theory is difficult to substantiate at the Shell Beach section, however, where there is no visual or petrographic evidence of sedimentary reworking. Therefore, the control of phosphogenesis by physical properties (e.g., sediment porosity, permeability, diffusion of phosphate) is probably most important within the phosphatic shale strata at Shell Beach, where CFA laminae and peloids are distributed in discrete, closely spaced horizons.

Field and geochemical evidence indicates that significant phosphogenesis occurred only within the shale intervals at Shell Beach; therefore, there must be some inhibition to phosphogenesis within non-phosphatic strata. Two controls on the presence or absence of phosphogenesis within particular strata or lithologies are geochemical and sedimentary. We discuss geochemical (organic matter regeneration, interstitial water phosphate concentrations and low-oxygen conditions) and sedimentological (permeability and phosphate residence time in interstitial water) factors controlling phosphogenesis and P variability at Shell Beach.

The flux of organic matter, and therefore P, to the seafloor is probably the most important factor controlling phosphogenesis within shale strata. Without adequate supply of P, interstitial water phosphate concentrations are generally not high enough to promote phosphogenesis. This control is used to explain the distribution of modern phosphatic sediments in upwelling areas off Peru and Chile [21,34,44], and the inferred relationship of phosphorites, black shales, and chert to deposition beneath ancient upwelling systems [17,22,45-47]. Phosphatic shale strata of the Monterey Formation commonly have high organic carbon (mean 13 wt%; [18]), and the presence of phosphatic strata, black shales, and porcelanites and cherts appears to support the interpretation that the Shell Beach section was deposited under conditions of high surface productivity and low terrigenous dilution, as well as the contention that the high flux of organic matter with associated P to the sediments led to intense phosphogenesis.

Low-oxygen bottom water and sedimentary conditions are also important for phosphogenesis at Shell Beach. The black, laminated shales indicate deposition within an oxygen minimum zone, with a rapid depletion of oxygen by oxic respiration due to high organic matter rain rates. Interstitial water phosphate concentrations in the modern Santa Barbara Basin are highest where bottom water dissolved oxygen is lowest [48], and most modern sites of phosphogenesis are located within oxygen minimum zones [21,34,41,49]. The lack of bioturbation and bioirrigation within the organic-rich muds allow the interstitial water concentration of dissolved phosphate to increase to levels sufficient to induce phosphogenesis: extensive bioturbation leads to high fluxes of dissolved phosphate out of the sediment and into the overlying water column, reducing interstitial water phosphate concentrations and thus inhibiting phosphogenesis. Though phosphogenesis within suboxic/anoxic sediments has been thought to occur by inorganic processes, recent evidence indicates that this might be a microbially-mediated process [27,34,50]. One example of this is the subsurface trapping of phosphate beneath microbial mats colonizing the H_2S/O_2 interface [30,51]. In this scenario P is released during organic matter breakdown and diffuses upward until reaching the microbial mat, which acts as a seal to chemical diffusion, greatly reducing diffusion coefficients and resulting in elevated interstitial water phosphate concentrations and CFA precipitation. A confirmation of microbial mediation of phosphogenesis at Shell Beach awaits fine-scale electron microscopy to determine the presence or absence of phosphatized microbial remains.

The porosity and permeability of sediments is critical for phosphogenesis. Laboratory studies, combined with field data from modern phosphatic environments, of rates of phosphogenesis indicate that CFA mineralizes from seawater (with elevated phosphate and fluorine concentrations) in 10 to 100 years, after undergoing a few metastable solid transitions [34,41,52]. Phosphogenesis therefore requires a supply of phosphate high enough to produce supersaturation (with respect to CFA) within a sedimentary horizon, and an interstitial water phosphate residence time long enough to overcome kinetic obstacles to CFA precipitation. This process could take place in the uppermost few centimeters at sedimentation rates appropriate for the Shell Beach section. The sedimentary conditions favorable to phosphogenesis appear to be satisfied in fine-grained organic-rich mudstones [39-41], where there is a high flux of P to the sediments (typically 0.5 - 15 μmol P·cm^{-2}·y^{-1}; [39]), low overall sedimentation rates so that P is not "diluted" by detritus [53,54](typically 10 - 500 m/my; [39]), and low porewater flushing rates (yielding mean diffusional fluxes of dissolved phosphate typically about 80 cm^2/y; [39]).

It is difficult to estimate the sedimentary diffusional flux of P, but the P accumulation rate (about 11 μmol P·cm^{-2}·y^{-1}) and the uncompacted sedimentation rate (400 - 500 m/my) at Shell Beach indicate "ideal" conditions for phosphogenesis within shale strata. Though detailed studies of P diffusion in non-phosphatic sediments have not been performed, there are clear sedimentological differences which may prohibit phosphogenesis. Siliceous diatomaceous oozes, for example, have relatively higher initial porosities and permeabilities than muds. This leads to high interstitial water flushing rates and short phosphate residence times, thereby inhibiting CFA precipitation (even though initial P accumulation rates may be comparable to those during mudstone deposition).

PHOSPHORITE GIANT DEPOSITS

Phosphorite giants are large accumulations of P-rich marine sedimentary rocks, with authigenically-precipitated apatite as the major P-bearing phase. It is widely thought that the accumulation of P in ancient phosphorite giants was inherently different than modern P accumulation, with oceanic or geochemical conditions atypical of the modern ocean [45,55-59]. The relative rarity and episodicity of phosphorite giant deposition [11,57,60] has in part reinforced this view. Uncertainties about authigenic apatite formation in modern sediments have also contributed to this view. It was thought that apatite appears not to be forming at present [61,62], although later evidence confirmed that it is [31,41,63,64]. A primary line of evidence justifying the belief about the atypical nature of phosphorite giant formation has been the comparison of their P inventories to the modern oceanic reservoir of dissolved phosphate (e.g., [45,58,65]). The relatively large rock resources of P in individual examples have led to suggestions that such factors as oceanic P concentrations or oceanic productivity must have differed drastically during phosphorite formation [57,59]. We suggest that this comparison is misleading; a more appropriate comparison is of the burial rates of P in ancient phosphorite deposits to P burial rates in oceanic sediments in areas of modern phosphogenesis [16]. This comparison will show that oceanic geochemistry has become a "red herring" in the search for explanations of phosphorite giant accumulation.

Phosphorite giants: The evidence for geochemical modesty
We chose three well-studied phosphorite giants: (1) the Permian Phosphoria Formation of the western United States, (2) the Upper Cretaceous phosphorites of the Middle East, and (3) the Miocene phosphorites of the Hawthorn and Pungo River Formation (southeastern United States) and the Monterey Formation along coastal California. These phosphorite giants were chosen because they are well-studied deposits that span about 250 million years, and thus can best reflect possible changes in oceanographic behavior suggested by other workers to be involved in phosphorite giant deposition. The Permian Phosphoria Formation, and especially its phosphorite-rich Meade Peak Member, has been the basis of comparisons of phosphorite giant P resources versus the oceanic P reservoir. The Upper Cretaceous phosphorites of the Middle East were deposited along the margins of the Tethyan seaway and are notable economic deposits of phosphorite. The Miocene phosphorites were deposited during major oceanographic changes (including the eventual development of modern circulation patterns) and have been linked to major changes in ocean chemistry (e.g., the "Monterey Hypothesis," [4,65]).

We compiled published information about P resources, areal extents, durations of deposition, and, when available, P accumulation rates, for these phosphorite deposits (Table 2). We calculated P burial rates (in mol P/yr) as the quotient of total P resources and estimated duration of deposition and P accumulation rates (in μmol P·cm^{-2}·yr^{-1}) as the quotient of P burial rate and areal extent of deposition for each example (except for the

Miocene Monterey Formation, which we explain below).
Because of the different types of data available, the estimates for the Miocene deposits were derived in several ways. Total P resources, areal extent, and duration of phosphatic deposition for the economic phosphorite deposits of the Hawthorn and Pungo River Formations are from Riggs and Sheldon [66], who compiled data from many published sources. From these values we calculated average P burial and P accumulation rates as in the other examples (Table 2). Because of the pristine (i.e., not reworked by physical processes) nature of the phosphatic intervals of the Monterey Formation that we present here [43,67], these deposits are not economic and there are no P resource estimates. However, because they have not been reworked, they provide an ideal opportunity to calculate directly their P accumulation rates from measured P concentrations and mass accumulation rates.
Phosphorus accumulation rates, based on measurements of total sedimentary P and estimated deposition time for the Cuyama (Upper Phosphatic Mudstone Member, Santa Margarita Formation), and Santa Barbara (Carbonaceous Marl Member, Naples Beach section) basins are from Garrison et al. [67], and for the Shell Beach section they are from the earlier portion of this paper. We estimated the areal extent of these basins from Miocene paleobasin reconstructions (see [67] for an example) and calculated total P burial rates for each basin (Table 2).
The total Monterey Formation P burial rate was based on the estimated P accumulation rate range for phosphogenic regions and an estimated areal extent of phosphatic strata deposition equivalent to 50% of the Monterey Formation marginal basin area. This may be an overestimation of the extent of phosphatic strata deposition, but probably by no more than a factor of two to four. The estimated range of Miocene P burial rates for our examples ($80 - 180 \times 10^8$ mol P/yr) is based on the sum of the estimates for the Monterey Formation and the Hawthorn and Pungo River Formations, with a 40% probable range from the average value (130×10^8 mol P/yr) to incorporate the estimate errors discussed above.

Phosphorus comparison: modern phosphogenesis vs. ancient phosphorites
The geochemistry and sedimentology of the modern Peru-Chile margin, acknowledged as a site of modern phosphogenesis, has been studied extensively (e.g., [34,64,68]). Total P accumulation rates measured in box cores taken from six sites from the Peru margin range from 0.8 to 6.3 μmol P·cm^{-2}·yr^{-1} [23,34]. On the basis of a depositional area of 22×10^{14} cm^2 [69], the estimated P burial for the Peru margin is $18 - 140 \times 10^8$ mol P/yr. The P accumulation rates and burial rates for these modern phosphogenic margins are comparable to those estimated for ancient phosphorite deposits in this study (Fig. 3).
The precise dating of the Miocene Monterey Formation has yielded P accumulation rates which can be compared directly to those of the modern phosphogenic Peru margin. This makes the Monterey Formation an excellent model for evaluating whether, on an area-normalized basis, P accumulation rates were markedly different in the past. Phosphorus accumulation rates for the Monterey Formation range from 1.7 to 9 μmol P·cm^{-2}·yr^{-1} (Table 2), which compare to rates of 0.8 to 6.3 μmol P·cm^{-2}·yr^{-1} for the Peru margin (Fig. 3). The range in P accumulation rates in Table 2 for the Miocene phosphorite total is the range in P accumulation rates in individual basins; apparently, P accumulation rates are greater in some depositional environments than in others.
The uncertainties for phosphorite deposits with reported P resources, areal extents, and/or durations of deposition are dependent on the accuracy of each of those estimates. For the Miocene Hawthorn and Pungo River Formations and the Permian Phosphoria Formation, the area and depositional duration are well constrained from extensive stratigraphic work in these economic deposits, and the major source of uncertainty is the estimates of total authigenic P. These estimates are minimum estimates of the total authigenic P, because

they do not include non-economic strata. However, if total authigenic P is actually twice the value reported, the resulting P accumulation rate for the Hawthorn and Pungo River Formations combined would be 0.4 μmol P·cm^{-2}·yr^{-1}, and for the Permian Phosphoria Formation would be 2 - 6 μmol P·cm^{-2}·yr^{-1}. Neither of these P accumulation rates are anomalous compared to the Miocene Monterey Formation or the modern Peru margin. For the Middle East phosphorite deposits, on the other hand, the P resources and areal extents of these deposits are well known, but their duration of deposition is not well constrained. We chose 1 and 10 m.y. to examine end-member cases.

Phosphorus burial rates, derived from P accumulation rates integrated over the area of deposition, are an important indicator of total P removal. Thus, a location with very high P accumulation rates and a small depositional area can sometimes have a P burial rate comparable to a location with a low P accumulation rate but a very large depositional area. This is the case for the modern Peru and Namibian margins, with a combined area of less than 1% of the total continental margin area [8] but a P accumulation rate that is many times the average (at least 15 times higher; [70]). This results in P burial rates for these modern phosphogenic margins that are equivalent to about 10% of the modern oceanic P output flux [8], highlighting the importance of these areally insignificant phosphogenic areas in the modern oceanic P balance.

Comparing modern P burial rates to those calculated for phosphorite deposits reveals the similarities between these examples (Fig. 3). For the Miocene example, the total areal extent is about 220 x 10^{14} cm^2, or about ten times that of the modern Peru margin, whereas P burial rates are 80 - 180 x 10^8 mol P/yr, approximately equal to those of the modern example. Though approximately equal in depositional area, the Upper Cretaceous phosphorites of the Middle East are relatively insignificant as P burial sites compared to the modern example.

The P-rich Meade Peak Member of the Phosphoria Formation has a range of P accumulation rates slightly higher than the other examples (Table 2), though not necessarily anomalously so. Because of its limited depositional extent, its P burial rate has a range similar to that of other examples (Table 2). The total Phosphoria Formation areal extent and P burial rate are comparable to those of the modern Peru margin.

Phosphorus accumulation and burial rates for the modern Peru margin are based on a limited number of box cores, representing a "snapshot" of geologic time. The ancient phosphorite deposits, on the other hand, are an end product of phosphogenesis and phosphorite formation and burial. It is difficult to evaluate the extent of preservation of the products of recent phosphogenesis; the deposition of phosphorite-rich sediments is not continuous through time, even for the same phosphogenic margin. This distinction leads to important questions about the role of tectonics, sea level, sedimentation types and rates, and sedimentary geochemistry in the longevity of phosphogenesis and preservation and condensation of phosphatic sediments into phosphorite deposits. However, the modern P burial and accumulation rates from the Peru margin provide a reasonable comparison to ancient phosphorite deposits based on drilling results, which reveal that Peru margin phosphogenesis and phosphorite deposition have occurred since the middle Miocene [79]. Although phosphorites were deposited in some other areas during the periods that we describe, we are not attempting to calculate global P fluxes, but instead are concentrating on well-quantified ancient examples and comparing them to a well-studied modern example, the Peru margin. It may be impossible to know whether the modern phosphogenic mode is exactly like that of the past, but the comparisons presented in this study provide compelling evidence that, in terms of average P accumulation and burial rates, the modern environment is comparable to the ancient.

Table 2. Phosphorus resources, accumulation and burial rates of phosphorite deposits.

	Estimated deposition time (m.y.)	Total P resources[*] ($\times 10^{12}$ mol)	Areal extent ($\times 10^{14}$ cm^2)	P accum. rate (μmol·cm^{-2}·yr^{-1})	P burial rate (10^8 mol /yr)
Miocene					
Monterey Fm.[+]					
Shell Beach	2	-	1.5	5	7.5
Cuyama Basin	7	-	2	3.4	6.8
Santa Barb. Basin	7	-	10	1.7	17
Estimated total			20	4	80
S.E. United States[§]					
Hawthorn and Pungo River Fms.	3	14 000	200	0.2	47
Estimated total				0.2—9	80—180
Upper Cretaceous					
Middle East[#]					
Egypt	1—10	9.8	12	0.001—0.01	0.01—0.1
Israel	1—10	3.4	4	0.001—0.01	0.003—0.03
Jordan	1—10	5.7	2	0.003—0.03	0.006—0.06
Morocco	1—10	220	5	0.04—0.4	0.2—2.2
Saudi Arabia	1—10	11	0.1	0.1—1.1	0.01—0.1
Syria	1—10	2.1	0.6	0.004—0.04	0.002—0.02
Estimated total		250	24	0.15—1.5	0.3—2.5
Permian					
Phosphoria Fm.[**]					
Meade Peak	2—6	4 700	2	3.9—12	8—24
Estimated total	2—6	23 000	40	1.0—2.9	38—120

[*] The gram formula weight of P, 30.97, was used to convert grams to moles.
[+] Monterey Formation: Shell Beach, depositional duration of the studied section--[15]; Cuyama and Santa Barbara Basins--[67]. Areal extents were calculated from Miocene basin reconstructions (see [67], for example). Estimated total P accumulation rate range represents the range of values reported, and estimated total P burial rate range is based on an average value of 130 x 10^8 mol/yr and probable range of about 40%.
[§] [66].
[#] Middle East total P resources and areal extent data: Egypt--[71]; Israel--[72]; Jordan--[73]; Morocco--[74]; Saudi Arabia--[75]; Syria--[76]. Deposition times of phosphatic intervals are not reported; the range of 1-10 m.y. is given as reasonable end-member estimates based on the temporal extent of phosphatic deposition in other phosphorite deposits.
[**] Meade Peak data from [58]; total Phosphoria Formation data from [77] and [78].

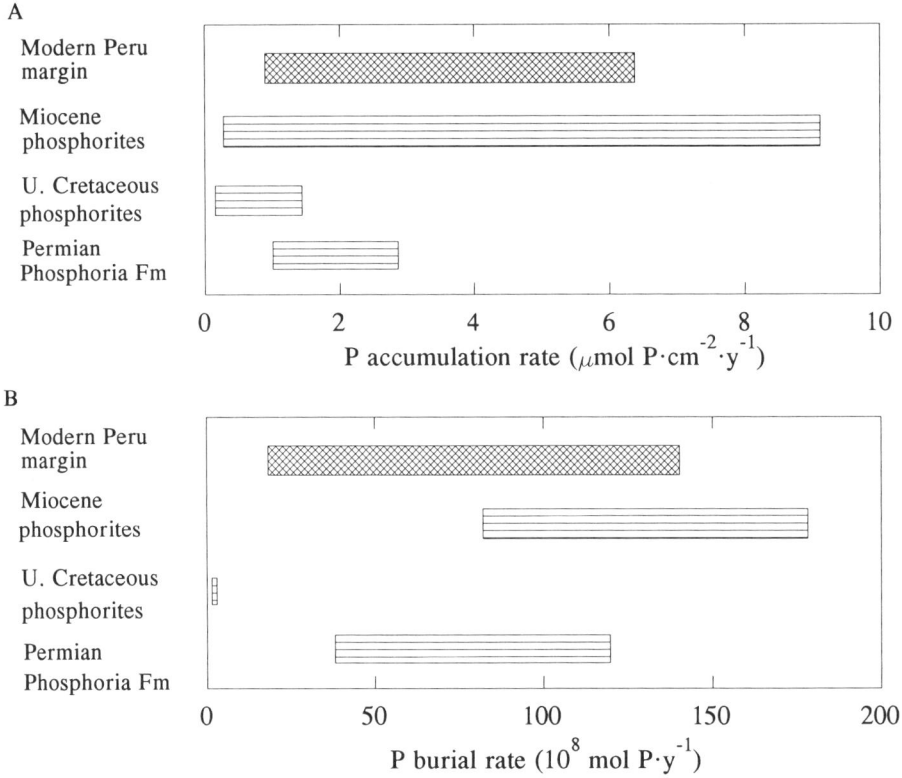

Figure 3. Phosphorus accumulation rates (A) and phosphorus burial rates (B) for the modern (phosphogenic) Peru margin and three ancient phosphorite deposits (data and ranges from Table 2).

CONCLUSIONS

Phosphorus accumulation rates are much higher in shale strata than in dolomitic or siliceous strata in the Miocene Monterey Formation at Shell Beach, California. This is probably due to sedimentary conditions (high dissolved phosphate concentrations and long interstitial water residence times) that promote phosphogenesis, trapping and preserving P within organic-rich muds but not within calcareous or biosiliceous sediments. Because P preservation is strongly controlled by lithology, episodic deposition of phosphatic sediments in a section do not necessarily record high productivity events in surface waters. Thus, the driving force behind the variability of P at Shell Beach is the variable lithology and not changing P fluxes to the sediments.

The three phosphorite giant examples (Permian Phosphoria Formation, Upper Cretaceous of the Middle East, and Miocene) had P burial and accumulation rates comparable to the modern phosphogenic Peru margin. Thus, major phosphorite deposition does not require oceanic geochemical conditions (e.g., oceanic P concentrations or oceanic productivity) drastically different from modern conditions. A slight reorganization of P output fluxes, possibly caused by changing sites of upwelling, sea-level changes, or increasing preservation or burial of P, could result in marginal areas with increased rates of phosphogenesis; favorable sedimentological and/or tectonic conditions could then lead to the concentration of sedimentary P into important ancient phosphorite deposits.

Acknowledgements

We thank the organizers of the Symposium 1-3-20 Volume (Symposium Volume on Siliceous, Phosphatic, and Glauconitic Sediments of the Tertiary and Mesozoic) of the 29th IGC for the opportunity to present this paper; the help and advice of A. Iijima is gratefully acknowledged. We also thank R.E. Garrison for support, advice, and review of this paper, J. Compton, D. Hodell, P. Froelich, K. Ruttenberg, and R. Behl for valuable discussions, and R. Franks and the UC Santa Cruz IMS Marine Analytical Laboratory for technical support. Support for the senior author as a National Defense Science and Engineering Graduate Fellow is graciously acknowledged.

REFERENCES

1. W.S. Broecker. Ocean chemistry during glacial time, *Geochim. Cosmochim. Acta* **46**, 1689-1705 (1982).
2. P.N. Froelich. Kinetic control of dissolved phosphate in natural rivers and estuaries: a primer on the phosphate buffer mechanism, *Limnology Oceanography* **33(4, part 2)**, 649-668 (1988).
3. W.S. Broecker and T.-H. Peng. *Tracers in the Sea.* Lamont-Doherty Geological Observatory, Columbia University, New York (1982).
4. E. Vincent and W.H. Berger. Carbon dioxide and polar cooling in the Miocene: the Monterey hypothesis. In: *The Carbon Cycle and Atmospheric CO_2: Natural Variations Archaen to Present.* E.T. Sundquist and W.S. Broecker (Eds). Amer. Geophys. Union Mono. **32**, 455-468 (1985).
5. A.C. Mix. Influence of productivity variations on long-term atmospheric CO_2, *Nature* **337**, 541-544 (1989).
6. H.D. Holland. Origins of breathable air, *Nature* **347**, 17 (1990).
7. P.N. Froelich, M.L. Bender, N.A. Luedtke, G.R. Heath and T. DeVries. The marine phosphorus cycle, *Amer. Jour. Sci.* **282**, 474-511 (1982).
8. P.N. Froelich. Interactions of the marine phosphorus and carbon cycles. In: *The interaction of global biogeochemical cycles.* B. Moore and M.N. Dastoor (Eds). California Institute of Technology, Pasadena, California, NASA-JPL Publication **84-21**, 141-176 (1984).
9. G.N. Baturin. Disseminated phosphorus in oceanic sediments - a review, *Mar. Geol.* **84**, 95-104 (1988).
10. V.E. McKelvey. Phosphate deposits, *Bull. U.S. Geol. Survey* **1252(D)**, 1-21 (1967).
11. P.J. Cook. Spatial and temporal controls on the formation of phosphate deposits - a review, In: *Phosphate Minerals.* J.O. Nriagu and P.B. Moore (Eds). pp. 242-272. Springer-Verlag, Berlin (1984).
12. M.C. Blake, Jr., R.H. Campbell, T.W. Dibblee, D.G. Howell, T. Nilsen, W.R. Normark, J.G. Vedder and E.A. Silver. Neogene basin formation in relation to plate-tectonic evolution of the San Andreas fault system, California, *Amer. Ass. Pet. Geol. Bull.* **62**, 344-372 (1978).
13. C.A. Hall. Geologic map of the Morro Bay South and Point San Luis Quadrangles, San Luis Obispo County, California. *U.S. Geological Survey Miscellaneous Field Studies Map, MF 511.* (1973).
14. D.L. Turner. Potassium-argon dating of Pacific coast Miocene foraminiferal stages, *Geol. Soc. Amer. Spec. Pap.* **124**, 91-129 (1970).
15. S.K. Omarzai. Monterey Formation of California at Shell Beach (Pismo Basin): Its lithofacies, paleomagnetism, age, and origin. In: *Sequence Stratigraphy in Fine-Grained Rocks: Example from the Monterey Formation.* J.R. Schwalbach and K.M. Bohacs (Eds). Pacific Sect. Soc. Sediment. Geol. (SEPM) Spec. Pub. **70**, 47-65 (1992).
16. G.M. Filippelli and M.L. Delaney. Similar phosphorus fluxes in ancient phosphorite deposits and a modern phosphogenic environment, *Geology* **20**, 709-712 (1992).
17. Y. Kolodny. Phosphorites. In: *The Sea.* C. Emiliani (Ed.). vol. 7, pp. 981-1023. Wiley, New York (1981).
18. C.M. Isaacs. Hemipelagic deposits in a Miocene basin, California: Toward a model of lithologic variation and sequence. In: *Fine-Grained Sediments: Deep Water Processes and Environments.* D.A.V.

Stow and D.J.M. Piper (Eds). Geol. Soc. Lon. Spec. Pub. **15**, 481-496 (1984).
19. A.V. Kazakov. The phosphorite facies and the genesis of phosphorites, *Scien. Inst. Fertil. Insecto-Fung. Transact.* **142** 95-113 (1937).
20. R.P. Sheldon. Paleolatitudinal and paleogeographic distribution of phosphorite, *Prof. Pap. U.S. Geol. Surv.* **501(C)**, 106-113 (1964).
21. W.C. Burnett. Geochemistry and origin of phosphorite deposits from off Peru and Chile, *Bull. Geol. Soc. Amer.* **88**, 813-823 (1977).
22. G.N. Baturin. Phosphorites on the sea floor. Origin, composition, and distribution. In: *Development in Sedimentology*, vol. 33. Elsevier, Amsterdam (1982).
23. P.N. Froelich, K.-H. Kim, R. Jahnke, W.C. Burnett, A. Soutar and M. Deakin. Pore water fluoride in Peru continental margin sediments: uptake from seawater, *Geochim. Cosmochim. Acta.* **47**, 1605-1612 (1983).
24. L.L. Ames, Jr. The genesis of carbonate apatites, *Econ. Geol.* **54**, 829-841 (1959).
25. W.J. Kennedy and R.E. Garrison. Morphology and genesis of nodular chalks and hardgrounds in the Upper Cretaceous of southern England, *Sediment.* **22**, 311-386 (1975).
26. J.M. McArthur, M.L. Coleman and J.M. Bremner. Carbon and oxygen isotopic composition of structural carbonate in sedimentary francolite, *Jour. Geol. Soc. Lon.* **137**, 669-673 (1980).
27. G.W. O'Brien, J.R. Harris, A.R. Milnes and H.H. Veeh. Bacterial origin of East Australian continental margin phosphorites, *Nature* **294**, 442-444 (1981).
28. J. Lucas and L. Prévôt. The synthesis of apatite by bacterial activity: Mechanism. In: *Phosphorites.* J. Lucas and L. Prévôt (Eds). Mém. Sci. Géol. **77**, 83-92 (1985).
29. L. Prévôt and J. Lucas. Microstructure of apatite replacing carbonate in synthesized and natural samples, *Jour. Sed. Pet.* **56**, 153-159 (1986).
30. C.E. Reimers, M. Kastner and R.E. Garrison. The role of bacterial mats in phosphate mineralization with particular reference to the Monterey Formation. In: *Phosphate Deposits of the World, Volume 3: Neogene to Modern Phosphorites.* W.C. Burnett and S.R. Riggs (Eds). pp. 300-311. Cambridge University Press, Cambridge (1990).
31. G.W. O'Brien and H.H. Veeh. Holocene phosphorite on the East Australian continental margin. *Nature.* **288**, 690-692 (1980).
32. G.F. Birch, J. Thomson, J.M. McArthur and W.C. Burnett. Pleistocene phosphorites off the west coast of South Africa, *Nature* **302**, 601-603 (1983).
33. G. Schaffer. Phosphate pumps and shuttles in the Black Sea. *Nature.* **321**, 515-517 (1986).
34. P.N. Froelich, M.A. Arthur, W.C. Burnett, M. Deakin, V. Hensley, R. Jahnke, L. Kaul, R.-H. Kim, K. Roe, A. Soutar and C. Vathakanon. Early diagenesis of organic matter in Peru continental margin sediments: Phosphorite precipitation, *Marine Geology* **80**, 309-343 (1988).
35. S.R. Riggs, S.W. Snyder, G.W. O'Brien, P.J. Cook and D.T. Heggie. Sedimentology of the Neogene to Modern glauconite-goethite-phosphate system: East Australian continental margin between 29° and 30° south latitude. In: *Apatite et Phosphorites.* J. Lucas, P.J. Cook and L. Prévôt (Eds). Bull. Sci. Géol. **42**, 185-204 (1989).
36. G.W. O'Brien, A.R. Milnes, H.H Veeh, D.T. Heggie, S.R. Riggs, D.J. Cullen, J.F. Marshall and P.J. Cook. Sedimentation dynamics and redox iron-cycling: Controlling factors for the apatite-glauconite association in the east Australian continental margin. In: *Phosphate Research and Development.* A.J.G. Notholt and I. Jarvis (Eds). Geol. Soc. Lon. Spec. Pub. **52**, 61-86 (1990).
37. D.T. Heggie, G.W Skyring, G.W. O'Brien, C. Reimers, A. Herczeg, D.J. Moriarty, W.C. Burnett and A.R. Milnes. Organic carbon cycling and modern phosphorite formation on the east Australian continental margin: An overview. In: *Phosphate Research and Development.* A.J.G. Notholt and I. Jarvis (Eds). Geol. Soc. Lon. Spec. Pub. **52**, 87-117 (1990).
38. I. Jarvis. Sedimentology, geochemistry and origin of phosphatic chalks: The Upper Cretaceous deposits of NW Europe, *Sediment.* **39**, 55-97 (1992).
39. P. van Cappellen and R.A. Berner. A mathematical model for the early diagenesis of phosphorus and fluorine in marine sediments: Apatite precipitation, *Amer. Jour. Sci.* **288**, 289-333 (1988).
40. J.D. Schuffert. A numerical model of multi-layered authigenic francolite formation, *Eos, Trans. Amer.*

Geophys. Union **72**, 20 (1991).
41. R.A. Jahnke, S.R. Emerson, K.K. Roe and W.C. Burnett. The present day formation of apatite in Mexican continental margin sediments, *Geochim. Cosmochim. Acta* **47**, 259-266 (1983).
42. K.B. Föllmi. Condensation and phosphogenesis: Example of the Helvetic mid-Cretaceous (northern Tethyan margin). In: *Phosphorite Research and Development*. A.J.G. Notholt and I. Jarvis (Eds). Geol. Soc. Lon. Spec. Pub. **52**, 237-252 (1990).
43. K.B. Föllmi and R.E. Garrison. Phosphatic sediments, ordinary or extraordinary deposits? The example of the Miocene Monterey Formation (California). In: *Controversies in Modern Geology*. D.W. Muller, J.A. McKenzie and H. Weisser (Eds). pp. 56-84. Academic Press, London (1991).
44. W.C. Burnett and H.H. Veeh. Uranium-series disequilibrium studies on phosphorite nodules from the west coast of South America, *Geochim. Cosmochim. Acta.* **41**, 755-764 (1977).
45. Y.K. Bentor. Phosphorites - the unsolved problems. In: *Marine phosphorites - Geochemistry, occurrence, genesis*. Y.K. Bentor (Ed.). Special Publication no. 29, pp. 3-18. Society of Economic Paleontologists and Mineralogists, Tulsa (1980).
46. S.E. Calvert and N.B. Price. Geochemistry of the Namibian shelf sediments. In: *Coastal Upwelling - Its Sediment Record, part A (NATO conference series IV, Marine Sciences, 10 A)*. E. Suess and J. Thiede (Eds). pp. 337-375. Plenum Press, New York (1983).
47. M. Kastner, K. Mertz, D. Hollander and R. Garrison. The association of dolomitite-phosphorite-chert: Causes and possible diagenetic sequences. In: *Dolomites of the Monterey Formation and Other Organic-Rich Units*. R.E. Garrison, M. Kastner and D.H. Zenger (Eds). Pacific Sect. Soc. Econ. Paleont. Mineral. **41**, 75-86 (1984).
48. E. Sholkovitz. Interstitial water chemistry of the Santa Barbara Basin sediments, *Geochim. Cosmochim. Acta* **37**, 2043-2073 (1973).
49. N.B. Price and S.E. Calvert. The geochemistry of phosphorites from the Namibian Shelf, *Chem. Geol.* **23**, 151-170 (1978).
50. R.E. Garrison, M. Kastner and Y. Kolodny. Phosphorites and phosphatic rocks in the Monterey Formation and related Miocene units, coastal California. In: *Cenozoic Basin Development in Coastal California*. R.V. Ingersoll and W.G. Ernst (Eds). Rubey Volume VI, pp. 348-381. Prentice-Hall, Englewood Cliffs (1987).
51. K.P. Krajewski and W.E. Krumbein. Microbial mats and phosphate mineralization in an ancient organic carbon-rich shelf environment: Triassic black shale sequence in Svalbard, Arctic Ocean, *Sediment*. (in press).
52. R.A. Gulbrandsen, C.E. Roberson and S.T. Neil. Time and crystallization of apatite in seawater, *Geochim. Cosmochim. Acta* **48**, 213-218 (1984).
53. F. Manheim, G.T. Rowe and D. Jipa. Marine phosphorite formation off Peru, *Jour. Sed. Pet.* **45**, 243-251 (1975).
54. E. Suess. Phosphate regeneration from sediments of the Peru continental margin by dissolution of fish debris, *Geochim. Cosmochim. Acta* **45** 577-588 (1981).
55. A.G. Fischer and M.A. Arthur. Secular variations in the pelagic realm. In: *Deepwater Carbonate Environments*. H.E. Cook and P. Enos (Eds). Soc. Econ. Paleon. Mineral. Spec. Pub. **25**, 19-50 (1977).
56. Y.K. Bentor. Modern phosphorites - not a sure guide for the interpretation of ancient deposits. In: *Report of the Marine Phosphatic Sediments Workshop: Honolulu, Hawaii* W.C. Burnett and R.P. Sheldon (Eds), pp. 29, Univ. Hawaii (1979).
57. R.P. Sheldon. Episodicity of phosphate deposition and deep ocean circulation: A hypothesis. In: *Marine phosphorites - Geochemistry, occurrence, genesis*. Y.K. Bentor (Ed.). Special Publication no. 29, pp. 239-247. Society of Economic Paleontologists and Mineralogists, Tulsa (1980).
58. R.P. Sheldon. Ancient marine phosphorites, *Ann. Rev. Earth Planet. Sci.* **9**, 251-284 (1981).
59. S.R. Riggs. Paleoceanographic model of Neogene phosphorite deposition, U.S. Atlantic continental margin, *Science* **223**, 123-131 (1984).
60. P.J. Cook and M.W. McElhinny. A reevaluation of the spatial and temporal distribution of sedimentary phosphate deposits in the light of plate tectonics, *Econ. Geol.* **74**, 315-330 (1979).

61. Y. Kolodny. Are marine phosphorites forming today? *Nature* **224**, 1017-1019 (1969).
62. Y. Kolodny and I.R. Kaplan. Uranium isotopes in sea floor phosphorites, *Geochim. Cosmochim. Acta.* **34**, 3-24 (1970).
63. G.N. Baturin. K.I. Merkulova and P.I. Chalov. Radiometric evidence for recent formation of phosphatic nodules in marine shelf sediments, *Mar. Geol.* **13**, 37-41 (1972).
64. H.H. Veeh, W.C. Burnett and A. Soutar. Contemporary phosphorites on the continental margin of Peru, *Science* **181**, 844-845 (1973).
65. J.S. Compton, S.W. Snyder and D.A. Hodell. Phosphogenesis and weathering of shelf sediments from the southeastern United States: Implications for Miocene $\delta^{13}C$ excursions and global cooling, *Geology* **18**, 1227-1230 (1990).
65. M.A. Arthur and H.C. Jenkyns. Phosphorites and paleoceanography, *Oceanol. Acta*, **Spec. Vol.**, 83-96 (1981).
66. S.R. Riggs and R.P. Sheldon. Paleoceanographic and paleoclimatic controls of the temporal and geographic distribution of Upper Cenozoic continental margin phosphorites. In: *Phosphate Deposits of the World, Volume 3: Neogene to Modern Phosphorites.* W.C. Burnett and S.R. Riggs (Eds). pp. 207-222. Cambridge University Press, Cambridge (1990).
67. R.E. Garrison, M. Kastner C.E. Reimers. Miocene phosphogenesis in California. In: *Phosphate Deposits of the World, Volume 3: Neogene to Modern Phosphorites.* W.C. Burnett and S.R. Riggs (Eds). pp. 285-299. Cambridge University Press, Cambridge (1990).
68. W.C. Burnett, M.J. Beers and K.K. Roe. Growth rates of phosphate nodules from the continental margin of Peru, *Science* **215**, 1616-1618 (1982).
69. D.J. DeMaster. The supply and accumulation of silica in the marine environment, *Geochim. Cosmochim. Acta* **45**, 1715-1732 (1981).
70. K.C. Ruttenberg. *Diagenesis and burial of phosphorus in marine sediments: implications for the marine phosphorus budget.* Ph.D. dissertation, Yale University, Massachusetts (1990).
71. B. Issawi. A review of Egyptian Late Cretaceous phosphate deposits. In: *Phosphate Deposits of the World, Volume 2: Phosphate Rock Resources.* A.J.G. Notholt, R.P. Sheldon and D.F. Davidson (Eds). pp. 187-193. Cambridge University Press, Cambridge (1989).
72. Y. Nathan and Y. Shiloni. The phosphate fields of the Negev (southern Israel). In: *Phosphate Deposits of the World, Volume 2: Phosphate Rock Resources.* A.J.G. Notholt, R.P. Sheldon and D.F. Davidson (Eds). pp. 352-356. Cambridge University Press, Cambridge (1989).
73. I.S. Jallad, O.S. Abu Murrey and R.M. Sadaqah. Upper Cretaceous phosphorites of Jordan. In: *Phosphate Deposits of the World, Volume 2: Phosphate Rock Resources.* A.J.G. Notholt, R.P. Sheldon and D.F. Davidson (Eds). pp. 344-351. Cambridge University Press, Cambridge (1989).
74. Office Chérifien des Phosphates, Casablanca, Morocco. The phosphate basins of Morocco. In: *Phosphate Deposits of the World, Volume 2: Phosphate Rock Resources.* A.J.G. Notholt, R.P. Sheldon and D.F. Davidson (Eds). pp. 301-311. Cambridge University Press, Cambridge (1989).
75. J.W. Berge and J. Jack. The phosphorites of West Thaniyat, Saudi Arabia. In: *Phosphate Deposits of the World, Volume 2: Phosphate Rock Resources.* A.J.G. Notholt, R.P. Sheldon and D.F. Davidson (Eds). pp. 340-343. Cambridge University Press, Cambridge (1989).
76. S. Atfeh. The phosphorite deposits of Syria. In: *Phosphate Deposits of the World, Volume 2: Phosphate Rock Resources.* A.J.G. Notholt, R.P. Sheldon and D.F. Davidson (Eds). pp. 357-362. Cambridge University Press, Cambridge (1989).
77. R.W. Swanson, V.E. McKelvey and R.P. Sheldon. Progress report on investigations of western phosphate deposits. *U.S. Geol. Surv. Circ.* **297** (1953).
78. R.P. Sheldon. Phosphorite deposits of the Phosphoria Formation, western United States. In: *Phosphate Deposits of the World, Volume 2: Phosphate Rock Resources.* A.J.G. Notholt, R.P. Sheldon and D.F. Davidson (Eds). pp. 53-61. Cambridge University Press, Cambridge (1989).
79. R.E. Garrison and M. Kastner. Phosphatic sediments and rocks recovered from the Peru margin during ODP Leg 112. In: *Proceedings of the Ocean Drilling Program, Scientific Results, 112.* E. Suess, R. von Huene, et al. (Eds.). pp. 111-134. Ocean Drilling Program, College Station (1990).

Shallow marine phosphorite–chert–palygorskite association, Upper Cretaceous Amman Formation, Jordan.

A. M. ABED
Department of Geology, University of Jordan, Amman, Jordan

Abstract-- Deposition of the Late Cretaceous phosphorites (Maastrichtian) seems to have been controlled by the paleotopography of this shallow epeiric shelf dominated by basins and swells. Upwelling currents from the north and west preferentially deposited phosphorites on the swells in a shallow subtidal environment. Similarly, oyster buildups formed patch reefs that dominated relatively shallower eastern central Jordan within tens of small basins where huge minable phosphorite was deposited. In both cases, and during a lower sea level stand (shallow subtidal to supratidal), phosphorites were reworked and accumulated on the flanks of the highs or the buildups. Chert increases towards the highs. This model clearly show that the Jordanian and subsequently the Tethyan phosphorites are much shallower than the Miocene phosphorite exemplified by the the Monterey deposits.

Key words: phosphorite, chert, palygorskite, Jordan, epeiric seas, Tethyan, depositional environments.

INTRODUCTION

Phosphorites and phosphatic rocks are widespread in Jordan and occur also nearby countries in Syria, Iraq, Saudi Arabia and Israel. They are an important and inseparable part of the Upper Cretaceous—Eocene Tethyan phosphorite province extending from India to northern Latin America [1,2]. Published works show that paleogeographical setting for the Tethyan phosphorites [3,4,5,6,7] was much shallower than that of the Miocene—Recent (N-S) or Trade Wind phosphorite province in western and eastern North America, western Latin America and eastern Australia [8,9,10,11,12]. In the former province, deposition of the phosphorites was in nearshore environments, in water depths not exceeding a few tens of meters, while in the Miocene—Recent province they were open shelf to slope sediments, and generally deposited at depths of few hundreds to more than 1000 m.
The aims of this paper are to describe the depositional environments of the Jordanian phosphorites and to investigate the effects of the paleotopography of the epeiric Late Cretaceous shelf on the deposition of these deposits.

GEOLOGICAL SETTING

Phosphatic deposits in Jordan form the upper part of the Amman Formation [13] or the Phosphorite Unit of Bender [14]. More recently the Natural Resources Authority (NRA) has the phosphatic deposits a formation status and named it Hesa Phosphorite Formation [15]. In this paper, the term Amman Formation will be used (Table 1).
These deposits are of Maastrichtian age [16,17,18], although others believe they are Campanian [14]. Eocene phosphorites are present in a few places in Jordan, especially in the extreme east, but they are of minor economic value.
Late Cretaceous phosphatic deposits occur extensively in Jordan, but minable deposits are found in only four areas (Fig. 1): North Jordan, Ruseifa, Central Jordan (Abyad and Hasa) and Esh-Shidiya in the extreme southeast. Commercial production is now concentrated in the

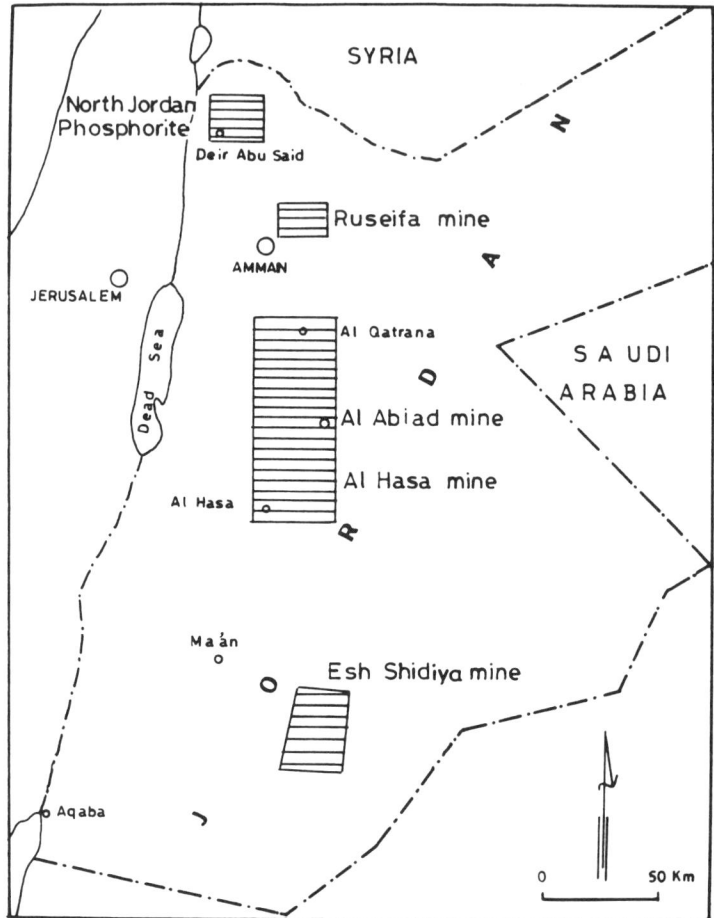

Figure 1. Location map of the four economic phosphorite deposits in Jordan.

latter two localities, Ruseifa mines stopped operation due to environmental problems. Each of the four areas will be treated separately.

Depositional Environments

There are ample evidence that the epeiric Tethyan platform in the Levant was not flat, but instead was dominated by highs and lows that affected sedimentation in the Late Cretaceous in general [19-25]. It appears that this paleotopography of the platform floor had played an important role in the sedimentation and accumulation of the phosphorite.

North Jordan Phosphorite (NJP)
The Ajlun—Irbid area in north Jordan (Fig. 2) was a high since at least the late Turonian. This area was named by Quennell [26] as the Ajlun dome, whose uplift was associated with the opening of the Dead Sea Rift in the post-middle Miocene. However, detailed sedimentological and microfacies analysis has shown that this area was a high since at least the late Turonian

Table 1.
Stratigraphic position of the phosphorites in Jordan.

System	Series	Units [14]	Formation [13]	
Early Paleogene	Paleocene	Ckalk-Marl	Muwaggar	B3
Late Cretaceous	Maastrichtian	Phosphorite	Amman	B2b
	Campanian	Silicified Limestone		B2a
	Santonian			
	Coniacian	Massive Limestone	El Ghudran	B1
	Turonian		Wadi Sir	A7
	Cenomanian	Echinoidal Limestone	Shueib	A5-6
			Hummar	A4
		Nodular Limestone	Fuheis	A3
			Naur	A1-2
Early Cretaceous		Kurnub (Hathira) Sandstone		

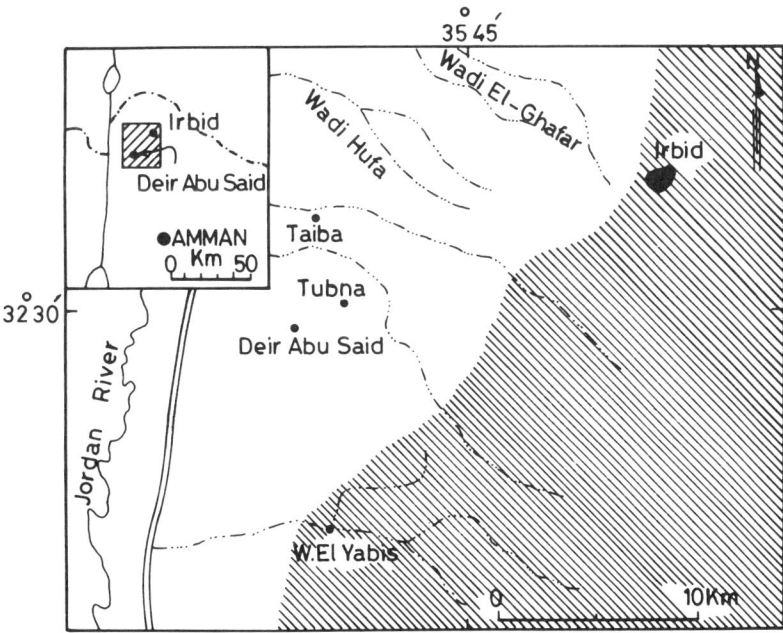

Figure 2. Location of the north Jordan phosphorites and the paleogeography of the this area during the Upper Turonian—Maastrichtian. Hatched area: Ajlun—Irbid High, open area: Al-Kora Low.

[27].

The topmost Wadi Sir Formation (Turonian) in this area consists of two rudist reef horizons separated by restricted lagoonal Milliolid packstones. Above the second rudist reef, a shallowing cycle (subtidal to supratidal) with mud cracks at the top indicates emergence of that area [27]. Farther to the west and northwest, in the Tubna area (Fig. 2), these facies are absent and are replaced by an open marine environment with diverse fauna. It is thus clear that since the late Turonian, northwestern Jordan was occupied by a high in the Ajlun-Irbid area and a low in the Tubna and the nearby areas. It is in these areas that phosphorites were deposited later. It is worth mentioning that the Turonian sediments are dominantly carbonates: limestone, marl and dolomite without phosphorite and chert.

During the Campanian—middle Maastrichtian, mixed mineralogy dominated by chert, phosphorite and porcelanite with limestone, marl and dolomite were deposited. This situation clearly indicates that upwelling from the deeper Tethyan ocean was situated to the north and west of this area and all of the Levant. The distribution of lithofacies and their thicknesses are shown in Figure 3. Without going into the details of each section, here are some generalized remarks.

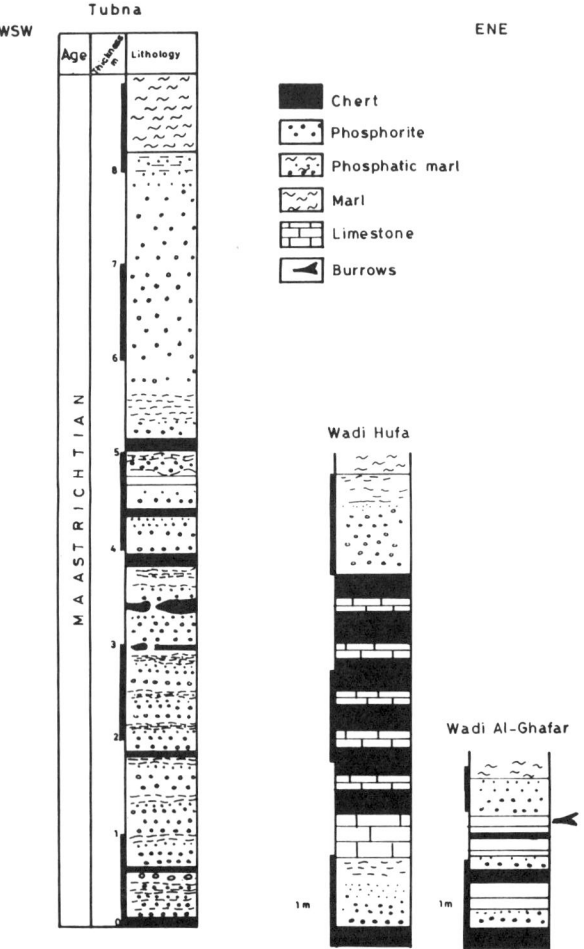

Figure 3. Columnar sections showing the variation in lithology and thickness in the basin (Tubna) and the high (Wadi Al-Gafar). For locations see Figure 2.

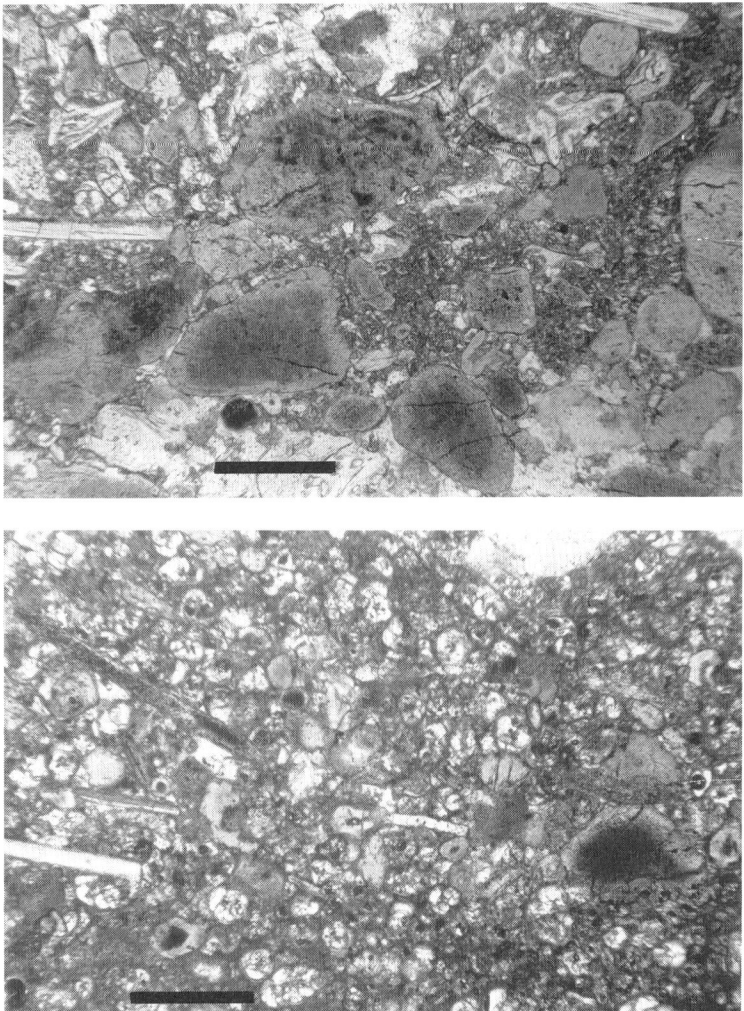

Figure 4. Upper: Packstone phosphorite from the basinal Tubna section. Lower: Same phosphorite with abundant foraminiferal tests. Bar = 0.5 mm.

1. Phosphorites are thinner on the paleohigh and increase in thickness toward the basin in the west and northwest (Fig. 3). In the Tubna area within the basin, they are 8-10 m thick compared with 1.75 m in Wadi El-Ghafar, the Irbid area eastwards on the high. They are grainstone phosphorites on or near the high, but grade to packstone phosphorites towards the basin. Also, some phosphatic intraclasts range up to 12 cm on the high, but are sand-size in the depression (Fig. 4).
2. Chert and the chert/phosphorite ratio increase towards the paleohigh, i.e. towards the shallower parts of the basin. This ratio is about 10 for four sections measured in the depression compared with more than 30 on the high in the east. This is contrary to the findings of Sheldon [2] in the Miocene phosphorites of the western and eastern USA where chert is a deeper facies than phosphorite. Usually, the phosphorites near and on the high are hard due to silica cementation, compared to friable deposits away from the paleohigh.
3. Marl and phosphatic marl form an important part of the section away from the paleohigh.

They are foraminiferal packstones with benthic foraminifera dominating over planktonic. They are frequently finely laminated, thus indicating a calm water environment below wave base for these marly facies which are not present on the high.

The foregoing clearly shows that economic phosphorites are preferentially found away from the high. But towards the deeper parts of the basin, marl and limestone become important as dilutants to the phosphorites.

Ruseifa Phosphorite

The general situation is rather similar to NJP. However, southern Ruseifa was uplifted during the Campanian—Maastrichtian as a part of the linear fault and fold belt known as the Amman—Hallabat structure [28,25] (Fig. 5). Abed [29] interpreted this structure as a part of the Syrian Arc System [19]. It is on the northern and northwestern flanks of this high that the Ruseifa economic phosphorite was deposited. On the high, the phosphatic sediments are made up of four three-membered cycles ranging from shallow subtidal to supratidal. (Fig. 6). Each cycle starts with chert conglomerates, followed by lensoidal, massive, fining upwards phosphorite grainstone with occasional oyster fragments and rare highly-broken foraminiferal tests. Chert fragments are rather abundant with the phosphorite (Fig. 7). These two members are high energy shallow subtidal to intertidal.

The third member is a mixed mineralogy horizon composed of algal-laminated chert, brecciated chert, and limestone with repeated small scale erosional surfaces, calcrete horizons and shrinkage cracks. Laterally, this member varies from completely eroded to fully preserved depending on the preceding phosphorite geometry (Fig. 8). Clearly, this mixed mineralogy member is a high energy intertidal to supratidal facies that records total emergence. Similar

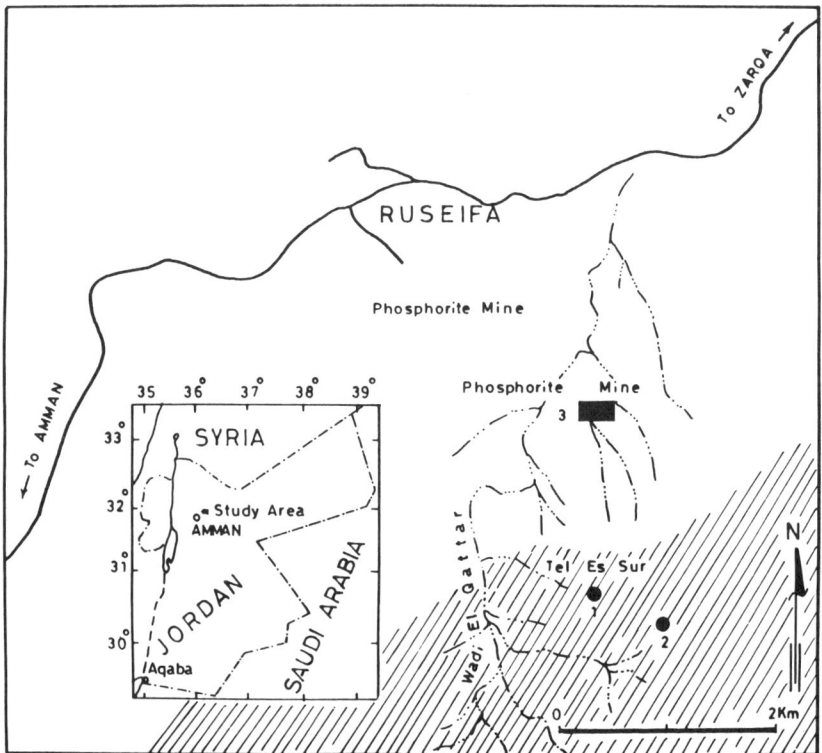

Figure 5. Location map of the Ruseifa phosphorites 1 and 2 are on the high, 3 on the flank of Ruseifa basin. Hatched area is part of the Amman—Hallabat structure.

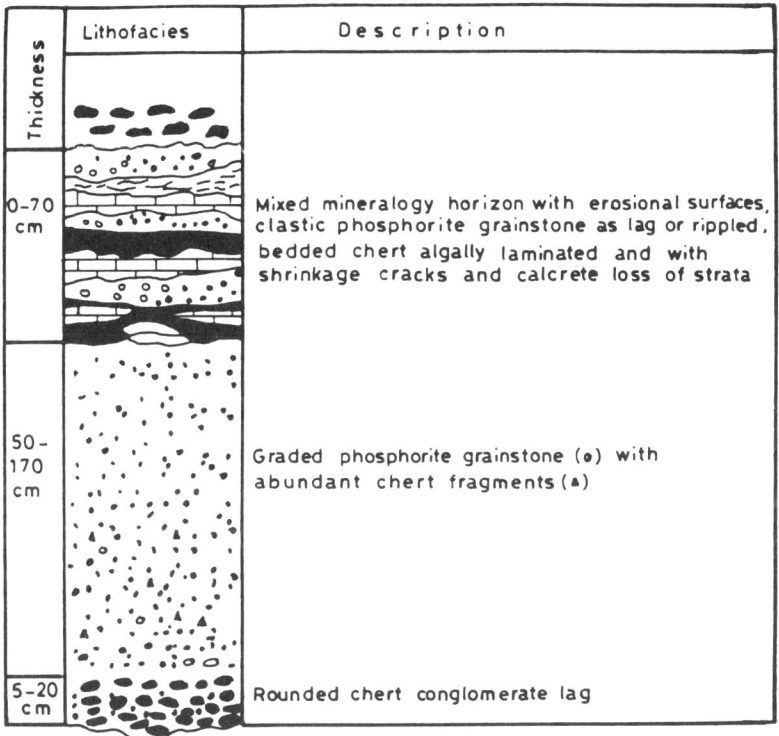

Figure 6. A typical shallowing upward cycle in Tel es Sur phosphorite (high), Ruseifa. See Figure 5 for location.

facies were reported in the Middle Cambrian phosphorite of Australia [30,31].
The topmost cycle is overlain by a coarsening upwards chert conglomerate of possible beach origin (Fig. 9). Then the whole sequence is capped by an oyster buildup, some 20 m thick (Fig. 10).
The basinal phosphorites (some 1 km to the north) are also made up of four cycles, but the nature of the material and cycles are different. The section is much thicker (Fig. 10). The phosphorites are packstone (Fig. 11) with abundant foraminifera especially in the carbonate beds. No chert conglomerate or chert fragments are present. Each cycle starts with a highly bioturbated surface occupied by large Ophiomorpha and Thallasinidea. This is followed by a friable fining upwards phosphorite packstone. Then the cycle is terminated by a relatively thick fossiliferous marl, marly limestone to limestone horizon (Fig. 10). One ammonite species was also encountered in these carbonates.
It is here believed that when sea level was at a high stand, phosphorite was deposited on the high while fossiliferous marly material was deposited in the low. At a low sea level stand, phosphorite of the high was reworked and redeposited on the flanks of the high. Chert/phosphorite ratios and total section thicknesses on the high and the depression are 35 versus 10 and 8 m versus 28 m, respectively (Fig. 10).
The foregoing clearly shows that phosphorites of the high are high energy tidal deposits, as evidenced by the presence of onshore oyster buildups, chert conglomerate, chert rock fragments within grainstone phosphorite, erosional surfaces, calcrete, brecciated chert and the rarity of fossils and fine sediments. All these characteristics are not present in the basinal section. Figure 12 shows a schematic cross-section of the Ruseifa phosphorite.

Central Jordan Phosphorites

These deposits extend for about 100 km in the N-S direction in central Jordan (Figs. 1 and 13). Several isolated ore bodies are located around Al-Abyad and Al-Hasa villages where most of the production in this area is from these two mines. There is a subtle difference between the paleogeographical setting of central Jordan during the phosphorite deposition compared to conditions in the Ruseifa and NJP areas.

There is abundant evidence that east central Jordan was shallower during the later half of the Late Cretaceous than west central Jordan: (1) The Wadi Sir Formation (Turonian) consists completely of carbonates with diverse fauna in the west in the Mujib area and around Karak City, whereas in the east its top third is dominated by sandstone with rare or complete absence of fossils near Al-Hasa mines [14,15]. (2) Two chalk horizons, 30 m thick each, within the

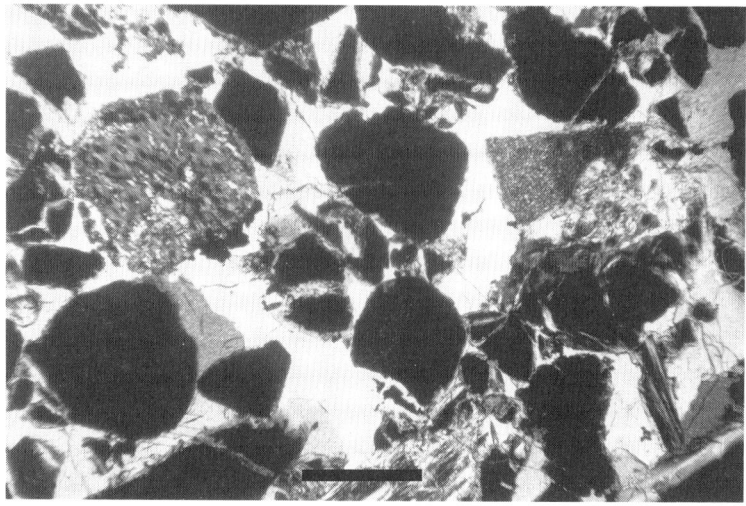

Figure 7. Grainstone phosphorite from Tel es Sur. Note the chert fragment on the left of center. Black particles are phosphatic intraclasts; white areas are sparry calcite cement. Cross nicols. Bar = 0.5 mm.

Figure 8. Laterally eroded top member in the shallowing cycles in Tel es Sur. Under hammer head.

Figure 9. Coarsening upward chert conglomerate of possible beach origin overlain by oyster buildup, Tel es Sur.

Ghudran Formation (Coniacian—lower Campanian) are present in the western area and completely absent in the east. They are made up of coccolith and planktonic foraminiferal packstone. It is in this rather very shallow epeiric platform that the oysters grew and formed the above described patchy buildups. Westwards, their debris is found in repeated beds in the Amman Formation as coquinoidal limestone.

East central Jordan is characterized by the presence of thick oyster buildups in the middle of the phosphorite unit. These buildups attain a thickness of 40 m and consist completely of oyster shells (Fig. 14 Upper). Often they show megacross-bedding (Fig. 14 Lower). The economic phosphatic deposits are almost all associated with these buildups. It should be emphasized here that the oysters and the buildups they construct are of shallow water origin. They can hardly exceed 40 m in depth [25]. Field investigation has shown that the buildups are very much like patch reefs with relatively deeper waters in between. These latter deeper areas acted as small basins some 2 to 4 km in diameter and were the sites for phosphorite deposition. This explains the isolated nature of the ore bodies in the area (Fig. 13).

Several facies relationships are evident (Fig. 15): (1) Economic phosphatic deposits are almost always overlie the buildups. They are relatively thin where the buildups are thickest. They then increase to a maximum thickness on the flanks of the buildups before becoming thin again in the center of the basin. (2) Buildups thin towards the basin and laterally change from boundstone through rudstone to grainstone. Finally they give way to marl and marlstone in the center of the basin. It should be noted that the area of maximum phosphorite thickness is roughly where the buildups interfinger with marl. Also, the base of the small basins is made up of the buildups (Fig. 15). (3) In several instances, the marl in the center of the basin is black due to high content of organic matter. They are a bituminous limestone or oil shale, suggesting a restricted euxinic environment in the center of the basins.

Esh-Shidiya Phosphorites

These deposits are situated in the extreme southeast (Fig. 16). More than 1000 million tons of minable phosphorites are present in this basin [32]. It is in this area that the phosphatic industry in Jordan is concentrated. The Esh-Shidiya basin is bordered on the southeast to southwest by the Paleozoic Nubian sandstone. Thus, quartz grains are an important mineralogical component especially in the lowermost phosphorite bed. Northwards, lies the much deeper Jafr basin. (Fig. 16). The relationship between the two basins is not clear. The area in between both is a flat desert with scarce outcrops, thus, the northward extension and nature of the phosphatic deposits there are not clearly understood.

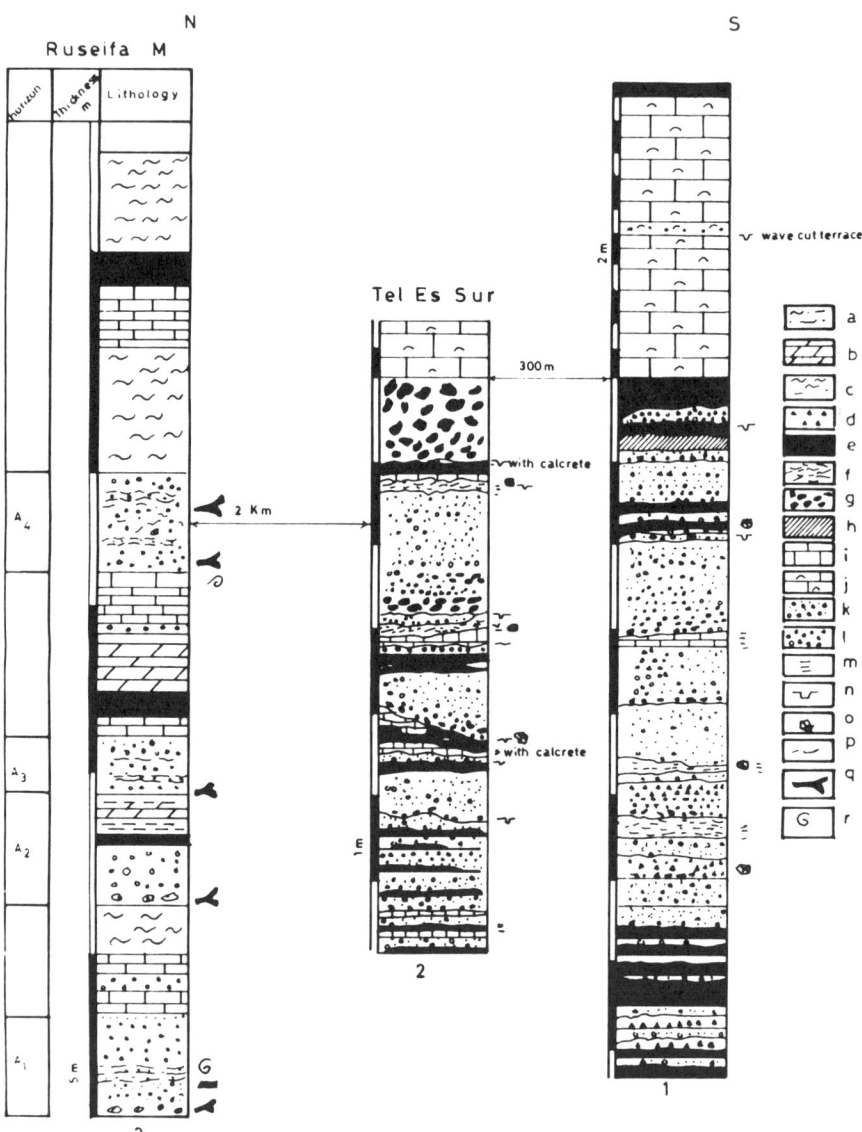

Figure 10. Columnar sections in the Ruseifa phosphorites. 1 and 2 are on the high, and 3 on the flank of the depression. a: marly phosphorite, b: dolomite, c: chalk—marl, d: fractured chert with angular fragments, e: chert bed, f: laminated chert, g: chert conglomerate, h: porcelanite, i: limestone, j: oyster buildup, k: phosphorite grainstone, l: phosphorite grainstone with chert fragments, m: lamination mostly algal, n: erosional surface, o: shrinkage cracks, p: ripple marks, q: burrows, and r: fossils. For location, see Figure 5.

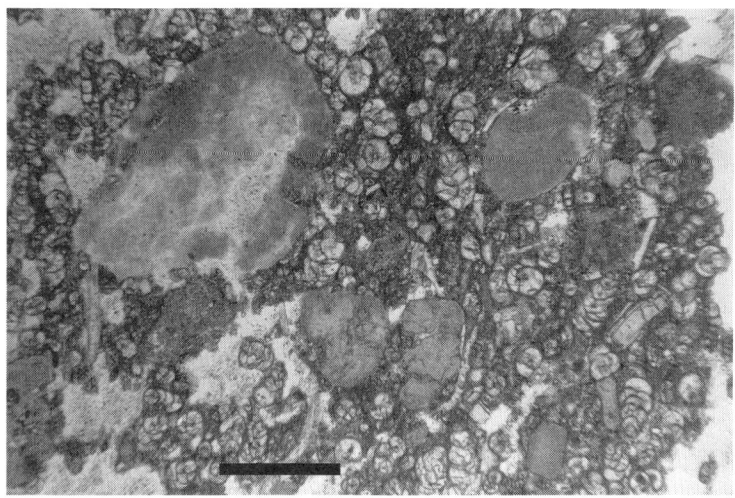

Figure 11. Packstone phosphorite with abundant foraminiferal tests in the basinal Ruseifa deposits. Bar = 0.5 mm.

The Esh-Shidiya section consists of four phosphatic horizons (A3 to A0); each overlain by marl, chert, limestone or dolomite (Fig. 17). Two beds of porcelanite are present in the A1 horizon. They are composed of opal-CT with abundant dolomite. The section is capped by a 2-3 m thick, thinly bedded, yellow marl horizon with up to 30% palygorskite (Fig. 18) with rare fossils such as foraminifera, oysters, gastropods and echinoderms. In central and north Jordan, this marl horizon consist entirely of foraminiferal tests which are hardly seen in Esh-Shidiya. Oyster fragments form a coquinoidal limestone bed, 2-3 m thick, in the mining area. This bed thins and disappears southwestwards.

Approaching the Nubian sandstone outcrop area in the southeast, the section becomes thinner and dominated by sandstone (Fig. 17). Further south and southeast, almost all the Upper

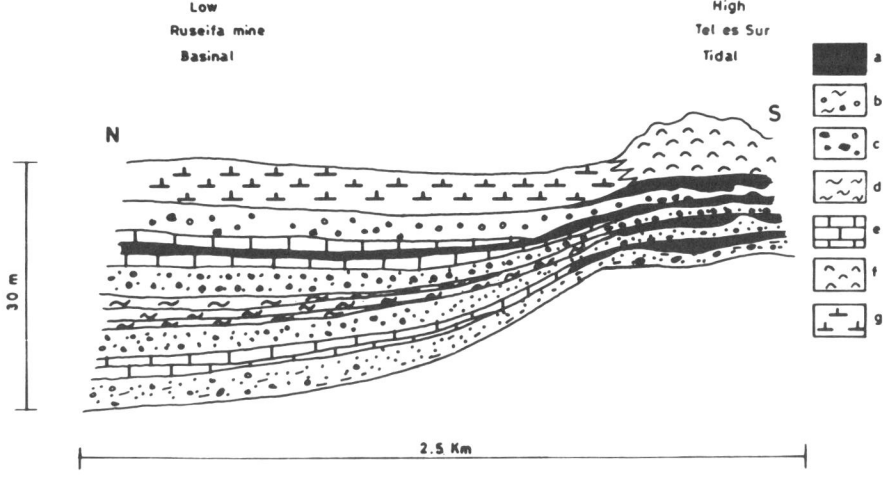

Figure 12. Schematic cross-section of the mineralogy distribution in the Ruseifa basin. a: chert, b: calcareous phosphorite, c: siliceous phosphorite, d: marl or marly limestone, e: limestone, f: oyster buildup, and g: chalk.

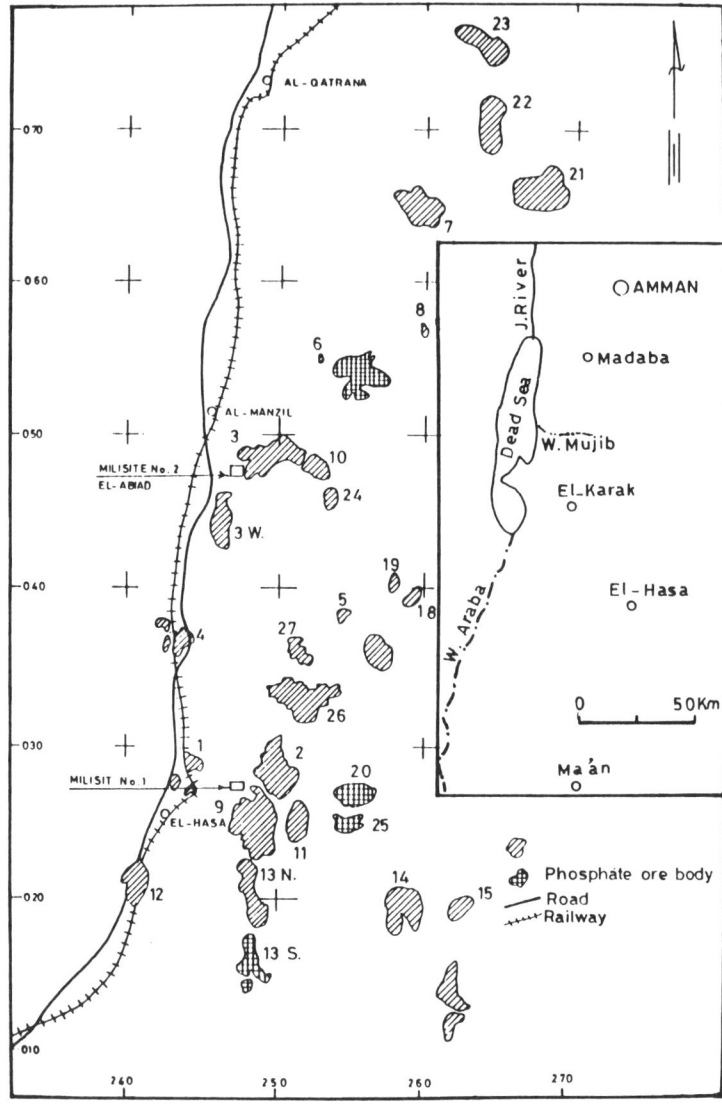

Figure 13. Distribution of phosphorite ore bodies in central Jordan. Note the isolated nature of these bodies. Areas in between are oyster buildups.

Cretaceous material is eroded due to uplift along a regional fault.

We believe that Esh-Shidiya basin is not fully marine as it is the case in Ruseifa and NJP basins. This is indicated by the proximity to the Nubian continent, the abundance of sandstone and rarity of fossils especially in the carbonate horizons which are highly fossiliferous in the northern basins.

The facies association of phosphorite, chert, porcelanite, dolomite and palygorskite is similar to those present in the much deeper and organic-rich Miocene Monterey Formation of California and the Pungo River Formation of North Carolina. This suggests that this facies association is not a depth or an environmental indicators.

Figure 14. Upper: Intact oyster shells making the buildups. For a scale, see lense cover on the right of center. Lower: Megacross-bedding in the buildups in central Jordan.

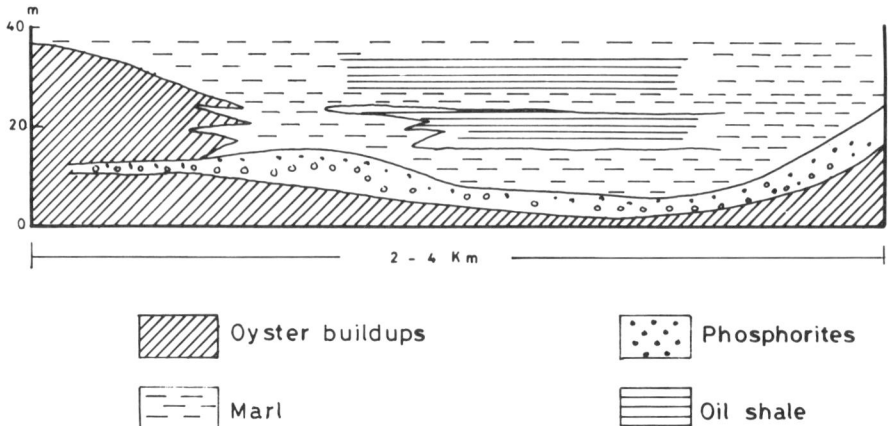

Figure 15. Schematic cross-section of the small basins produced by oyster buildups, Al-Hasa, Central Jordan.

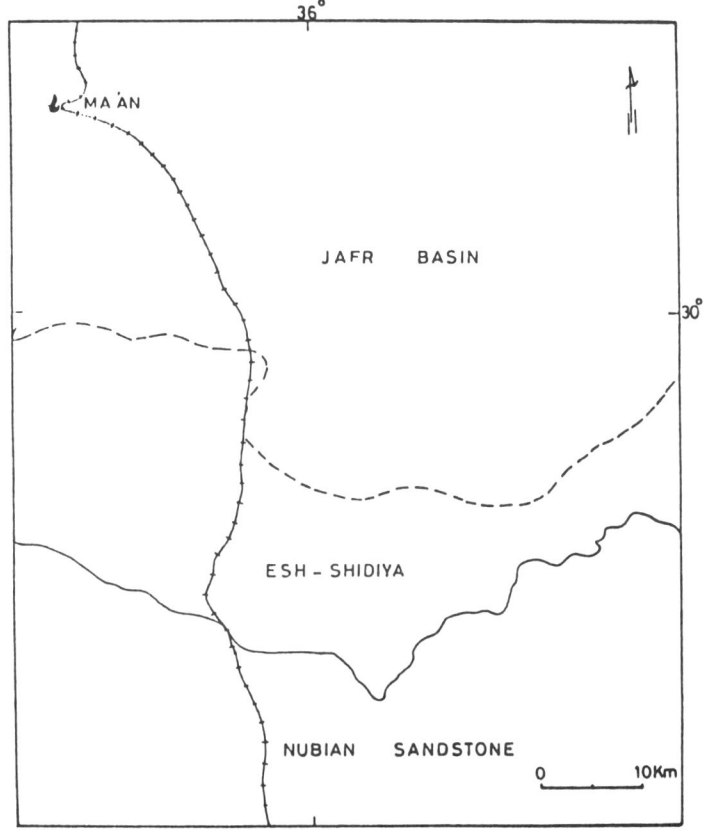

Figure 16. Location map of Esh-Shidiya and its relationship with the Jafr basin and Nubian sandstone.

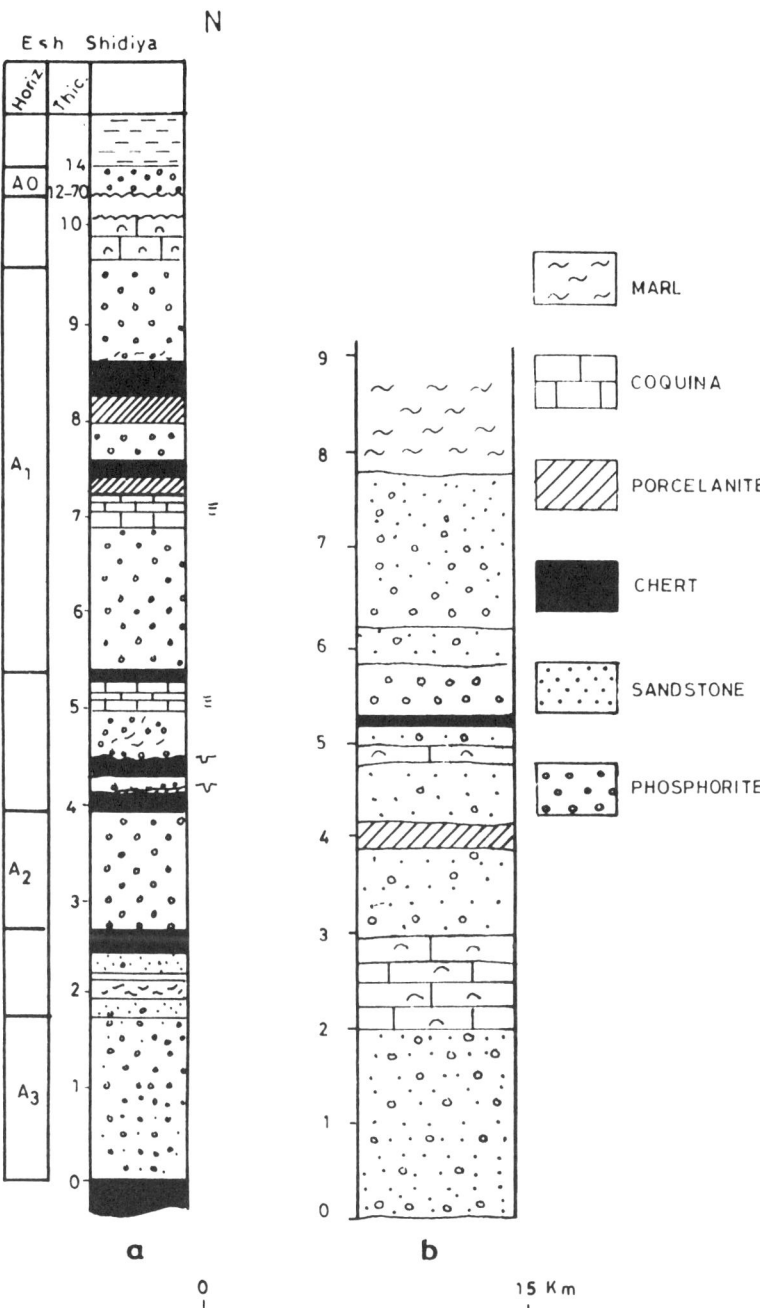

Figure 17. Esh-Shidiya phosphorite. a: in the mining area, b: nearer to the Nubian sandstone.

Figure 18. SEM micrograph showing palygorskite in the yellow marl horizon overlying the Esh-Shidiya phosphorite.

DISCUSSION

From the foregoing, it is rather clear that the Jordanian phosphorites were deposited in a shallow nearshore environment. In the depressions it was a subtidal environment below wave base, while on the highs it ranged from shallow subtidal (above wave base) to supratidal. This conclusion is in full agreement with the depositional environments of the Tethyan phosphorites in the Levant and North Africa. See, for example, Al-Maleh and Mouty [7] for the phosphorite in southeast Syria, Al-Bassam et al. [33] for western Iraq, and Shawly [6] and Al-Khattabi [34] for northwestern Saudi Arabia. In the Negev, Israel, Steinitz [35] demonstrated the deposition in coastal sabkha due to the presence of relics of evaporites in the associated chert. Evaporites associated with phosphorites are well developed in central Tunisia [4]. In Egypt, Glenn and Arthur [5] described a similar situation for the phosphorite belt in southern Egypt postulating that the source of P was from rivers rather than due to upwelling. Their main evidence for this interpretation is the abundance of Fe present as glauconite, a mineral completely absent in the Jordanian deposits.

The early Cambrian phosphorite of Georgina basin in northern Australia seems to have had a similar paleogeographical setting ranging from shallow subtidal to supratidal [30,31]. Cook and Elgueta [36] have shown that the depth of the basin did not exceed 50 m.

It is worth emphasizing that there is a subtle difference between the E-W Late Cretaceous—Eocene Tethyan phosphorite province and the N-S Miocene phosphorite province along the eastern and western margins of North America, known also as the trade wind province [2]. Although the lithofacies are rather similar in both provinces, i.e. phosphorite, chert, porcelanite, dolomite, Mg-clay minerals and sometimes glauconite, the paleogeographical settings are different. The Miocene deposits of the Monterey Formation of the western USA are much deeper water sediments, with paleodepths of as much as 2000 m in lower slope environments [9,37]. The Pungo River Formation in North Carolina and the equivalent formations in Florida are still much deeper than the Tethyan phosphorites, although they are generally shallower water deposits than the Monterey Formation [38,39].

Also Recent phosphorites off Peru are much deeper water sediments than the Tethyan deposits. They occur at water depths of a few hundred meters [40]. However, the quantities of Recent phosphorites off Peru are trivial compared with Tethyan deposits.

It is believed that the paleoceanographic setting was responsible for these differences. The

Miocene—Recent deposits of western North America are situated on a plate boundary, while the Tethyan phosphorites are the deposits of a broad epeiric shelf. The former are certainly much less important economically than the letter. A large proportion of the huge deposits in the eastern United States (not on a plate boundary) were shallow marine in origin. They were reworked to a deeper position and preserved, while most of the shallow facies were lost due to erosion [38,39].

What is the role of the paleotopography of the Late Cretaceous—Eocene Tethyan epeiric shelf on the accumulation of phosphatic deposits? Many studies indicate the presence of basins and swells during the Late Cretaceous in the Levant [20,21,22,24,41]. The phosphorite time, Campanian—Maastrichtian, is not an exception.

Most of the highs in Jordan face west and north, the presumed direction of the prevailing upwelling currents. These highs, being shallower, provide better sites for the deposition of phosphorite [12]. We believed that it is on those highs that the cold upwelled water rich in P and Si would authigenically produce the phosphorite and siliceous sediments during a high sea level stand. These sites are thus the phosphatic factories. Minor amounts of these sediments would be deposited in the depressions where they were diluted with other constituents, especially carbonates. In a regressive phase, the sediments on the highs were exposed to higher energy regimes and suffered winnowing, erosion and transportation towards the depression. Most of these reworked sediments did not reach the center of the depressions, but accumulated proximal to the highs on their flanks. Most probably, siliceous sediments resisted reworking more than phosphatic material. Subsequently the former deposits will be enriched on the highs at the expense of the latter. Such a model also explains the presence of uniformly interbedded oyster bioclastic limestone and phosphorites that extend westwards to the deeper water area in central Jordan, as well as their thinning in the same direction.

CONCLUSIONS

1. The Late Cretaceous sea in Jordan and the Levant was an epeiric shallow shelf dominated by basins and swells, extending to Campanian—Maastrichtian time, the time of maximum phosphorite deposition.
2. Phosphorites were deposited on the swells (highs) in shallow environments. During regressive phases they were reworked and redeposited preferentially on the flanks of the highs rather than in the center of the basins.
3. Chert/phosphorite ratios increase towards the shallower parts of the basin, i.e. towards the highs.
4. In central Jordan, oyster buildups created small basins in which phosphorites were deposited, with maximum accumulation on the flanks of the buildups.
5. Jordanian phosphorites as well as those of the Tethyan province in the Levant and North Africa, although having a similar mineralogy as those of the Miocene province, were much shallower water deposits.

REFERENCES

1. A.J. Notholt. Economic phosphatic deposits, mode of occurrence and stratigraphical distribution, *Jour. Geol. Soc. London* **137**, 793-805 (1980).
2. R.P. Sheldon. Association of phosphatic and siliceous marine sedimentary deposits. In:*Siliceous Sedimentary Rocks, Hosted Ores and Petroleum*. J.R. Hein (Ed.). pp.56-79. Nostrand Reinhold, New York (1987).
3. A.M. Abed and G. Kraishan. Evidence for shallow-marine origin of a 'Monterey Formation type' chert-phosphorite-dolomite sequence: Amman Formation (Late Cretaceous), central Jordan, *Facies* **24**, 25-38 (1991).
4. H. Belayouni and A. Beji-Sassi. Genesis of Tethyan phosphorites and associated petroleum source rocks. Phosphorites. In: *10th International field workshop and symposium, IGCP Project 156 Phosphorites*. Tunis (1987).
5. C.R. Glenn and H.A. Arthur. Anatomy and origin of a Cretaceous phosphorite-greensand giant, Egypt, *Sedimentology* **37**, 123-154 (1990).
6. S.J. Shawly. *Petrology and sedimentology of the Upper Cretaceous-Tertiary phosphate bearing sedimentary*

sequence, north of Saudi Arabia. M. Sc. Thesis, King Abdulaziz Univ., Jedda (1985).
7. K. Al-Maleh and K. Mouty. Sedimentological evolution and paleogeography of the Palmyrides, Syria, during the Cretaceous. In: *Proceeding of 3rd. Jordanian Geological Conference*. pp.213-244 (1990).
8. M.M. Bramlette. The Monterey Formation of California and the origin of its siliceous rocks, *U. S. Geol. Surv. Prof. Paper* **212**, 1-57 (1946).
9. K.A. Pisciotto and R.E. Garrison. Lithofacies and depositional environments of the Monterey Formation, California. In: *The Monterey Formation and Related Siliceous Rocks of California*. R.E. Garrison and R.G. Douglas (Eds). pp.97-123. Pacific Section, S.E.P.M., Los Angeles(1981).
10. S.R. Riggs and R.P. Sheldon. Plaeoceanographic and paleoclimatic controls of the temporal and geographic distribution of Upper Cenozoic continental margin phosphorites. In: *Phosphorite Deposits of the World Volume 3: Neogene to Modern phosphorites*. W.C. Burnett and S.R. Riggs (Eds). pp.207-222. Cambridge University Press, Cambridge (1990).
11. D.T. Heggie, G.W. Skyring, G.W. O'Brien, C. Reimers, A. Herczeg, D.J. Moriarty, W.C. Burnett and A.R. Milnes. Organic carbon cycling and modern phosphorite formation on the East Australian continental margin: An overview. In: *Phosphorite Research and Development*. A.J.G. Notholt and I. Jarvis (Eds). Spec. Publ. 52, pp.87-119. Geological Society, London (1990).
12. G.F. Birch. A model for penecontemporaneous phosphatization by diagenetic and authigenic mechanisms for the western margin of southern Africa. In: *Marine Phosphorites*. Y. K. Bentor (Ed.). Spec. Publ. 29, pp.70-101. SEPM, Tulsa (1980).
13. M. Masri. *Geology of the Amman-Zerqa area*. Central Water Authority, Amman (1963).
14. F. Bender. *Geology of Jordan*. Borntraeger, Berlin (1974).
15. J. Powell. *Stratigraphy and sedimentation of the Phanerozoic rocks in central and south Jordan. Part B, Kurnub, Ajlun and Belqa groups*. Geological Mapping Division, Bull. Geology Directorate, NRA, Amman (1989).
16. D.J. Burdon. *Handbook of the geology of Jordan*. Government of Jordan, Amman (1959).
17. K.A. Hamam. Foraminifera from the Maastrichtian phosphate-bearing strata of Al-Hasa area, Jordan, *Jour. Foram. Res.* **7**, 34-43 (1977).
18. A.M. Abed and M. Ashour. Petrography and age determination of the NW Jordan phosphorite, *Dirasat.* **14**, 247-263 (1987).
19. E. Krenkel. Der Syriasche Bogen, *Zentralb. Mineral. Geol. Palaont.* **9**, 274-281, 301-313 (1924).
20. Z. Garfunkel. The Negev: Regional synthesis of sedimentology in Israel, Cyprus and Turkey. In: Guidebook of 10th International Congress of Sedimentology. Part 1, pp.35-110. Jerusalem (1978).
21. Y. Bartov, Z. Lewy and I. Zak. Mesozoic and Tertiary stratigraphy, paleogeography and structural history of Gebel Areif en Naga area, eastern Sinai, *Israel J. Earth Sci.* **29**, 114-39 (1980).
22. Y. Bartov and G. Steinitz. Senonian oysteroid bioherm in the Negev, Israel. Implication of paleogeography and environments of deposition, *Israel J. Earth Sci.* **31**, 17-25 (1982).
23. A.M. Abed and W. Schneider. The Cenomanian Nodular Limestone Member, Jordan from subtidal to supratidal environment, *N. Jb. Geol. Palaont. Mh.* **9**, 513-22 (1982).
24. A.M. Abed. Emergence of Wadi Mujib (central Jordan) during the Lower Cenomanian time and its tectonic implications. In: *The Evolution of the Eastern Mediterranean*. J.E. Dixon and A.F. Robertson (Eds). pp. 213-217. Geological Society, London (1984).
25. K. Bandel and Sh. Mikbel. Origin and deposition of phosphate ores from the Upper Cretaceous, Ruseifa, Jordan, *Mitt. Geol. Palaont. Inst. Univ. Hamburg* **59**, 167-188 (1985).
26. A.M. Quennell . Tectonics of the Dead Sea rift. In: *Proceedings of 20th International Geological Conqress*. pp.285-305. Mexico (1956).
27. B.A. Mohammad. *Sedimentology of Wadi Sir Formation in North Jordan*. M. Sc. Thesis, Jordan University, Amman (1985).
28. Sh. Mikbel and W. Zacher. Fold structures in north Jordan, *N. Jb. Geol. Palaont. Mh.* **4**, 240-256 (1986).
29. A.M. Abed. On the genesis of the phosphorite-chert association of the Amman Formation in Tel es Sur area, Ruseifa, Jordan, *Sci. Geol. Bull. Strasbourg* **42**, 141-153 (1989).
30. P.M. Southgate. Cambrian phoscrete profiles, coated grains and microbial processes in phosphogenesis, Georgina Basin, Australia, *Jour. Sed. Petrol.* **56**, 429-41 (1986).
31. P.M. Southgate. Middle Cambrian phosphatic hardground, phoscrete profiles and stromatolites, and their implication for phosphogenesis. In: *Phosphate Deposits of the World Volume 1*. P.J. Cook and J.H. Shergold (Eds). pp.327-352. Cambridge University Press, Cambridge (1986).
32. Jordan Phosphate Mines Company (JMPC). Annual Report. Amman (1978).
33. K.S. Al-Bassam, A.A. Al-Dahan and A.K. Jamil. Campanian-Maastrichtian phosphorites of Iraq: petrology, geochemistry and genesis, *Mineral. Deposita* **18**, 215-233 (1983).
34. A.F. Al-Khattabi. *Geochemistry of phosphorite and associated rocks in Sirhan-Turayf basin*. M. Sc. Thesis, King Abdulaziz Univ., Jedda (1984).
35. G. Steinitz. Evaporite-chert association in Senonian bedded cherts, Israel, *Israel J. Earth sci.* **26**, 55-63 (1977).
36. P.J. Cook and S.A. Elgueta. Proterozoic and Cambrian phosphorites-deposits: Lady Annie, Queensland, Australia. In: *Phosphorite Deposits of the World Volume 3: Neogene to Modern Phosphorites*. P.J. Cook

and JH. Shergold (Eds.). pp.132-148. Cambridge University Press, Cambridge (1986).
37. R.E. Garrison, M. Kastner and C.E. Reimers. Miocene phosphogenesis in California. In: *Phosphorite Deposits of the World Volume 3: Neogene to Modern Phosphorites*. W.C. Burnett and S.R. Riggs (Eds). pp.285-299. Cambridge University Press. Cambridge (1990).
38. S.R. Riggs. Phosphorite deposition in Florida: A model of phosphogenic system, *Econ. Geol.* **74**, 285-314 (1979).
39. S.R. Riggs. Paleoceanographic model of Neogene phosphorite deposition, U. S. Atlantic continental margin, *Science* **223**, 123-131 (1984).
40. W.C. Burnett. Geochemistry and origin of phosphorite deposits from Peru and Chili, *Geol. Soc . Amer. Bull.* **88**, 813-823 (1977).
41. R. Bowen and U. Jux. *Afro-Arabian geology*. Chapman & Hall, London (1987).

Lithostratigraphy of Senonian phosphorite deposits in the Palmyridean region and their general sedimentological and paleogeographic framework

A. Kh. AL MALEH[1] and M. MOUTY[2]
[1]*General Est of Geology and Mineral Resources (GEGMR), P.O. Box 7645, Damascus, Syria and* [2]*Atomic Energy Commission (AECS), P.O. Box 6091, Damascus, Syria*

Abstract-- Phosphorite deposits in the Palmyrides Mountain Chain form a major portion of an Upper Campanian lithostratigraphic unit termed the Sawwaneh Formation which has a thickness that varies between 17 and 317 m. This region is subdivided into two lithologically differentiated zones, the northern ranges and the southern ranges. The Hamad Uplift, which bounds the Palmyridean Basin on the south, controlled phosphorite deposition. In the central part of the northern rim of the uplift only reduced phosphorite sedimentation occurred. The sedimentary milieus were enriched in phosphorus and plankton due to a regional paleogeographic evolution marked by the Senonian transgression and upwelling onto the Arabian platform.

Key words: phosphorite, Palmyrides Mountain Chain, Sawwaneh Formation, Palmyridean Basin, Hamad Uplift, Senonian transgression, Syria.

INTRODUCTION

Phosphorite deposits are encountered in some areas in Syria. The highest concentrations of phosphorites are located in the Palmyrides Mountain Chain, particularly in its southern central part where they have been mined in two mines for years.
Most of the studies that were carried out by Cayeux [1], Russian Technoexport Mission [2,3], Atfeh [4], Al Issa [5] and Abbas [6] on the phosphorite deposits focused mainly on the mineralogical and geochemical characteristics of the deposits exploited in the operating mines; in contrast, the lithologic and sedimentological characteristics were only restrictively and briefly reviewed. Accordingly, a clear perspective on the general paleogeographic framework of the phosphorite deposits and on their genesis has been lacking.
This study aims to specify the lithologic, paleontologic, petrographic and sedimentological characteristics of the phosphorite-bearing Senonian rocks in the Palmyrides Mountain Chain. We have constructed geologic cross-sections across the entire chain in order to delineate the paleogeographic framework within which the phosphorite deposits were formed.

LITHOSTRATIGRAPHY

The Palmyrides Mountain Chain, located in central Syria, is composed of a NE—SW elongated group, 350 km long, of mountains covering an area of 30,000 km². The Chain is segmented by a vast Neogene—Quaternary depression into northern ranges and southern ranges. The sedimentary series exposed in the Palmyrides Mountain Chain, which has ages ranging from the Upper Triassic up to Neogene, consists of several lithologic units (Figure 1); among these, the Soukhneh Group is characterized by its significant phosphorite deposits. This unit is composed mainly of two rock types: calcareous rocks and siliceous rocks. The first type is dominated by limestone, marly limestone, limy marl and marl with characteristic limy concretionary structures of few centimeters up to 2 m size. The siliceous rocks are

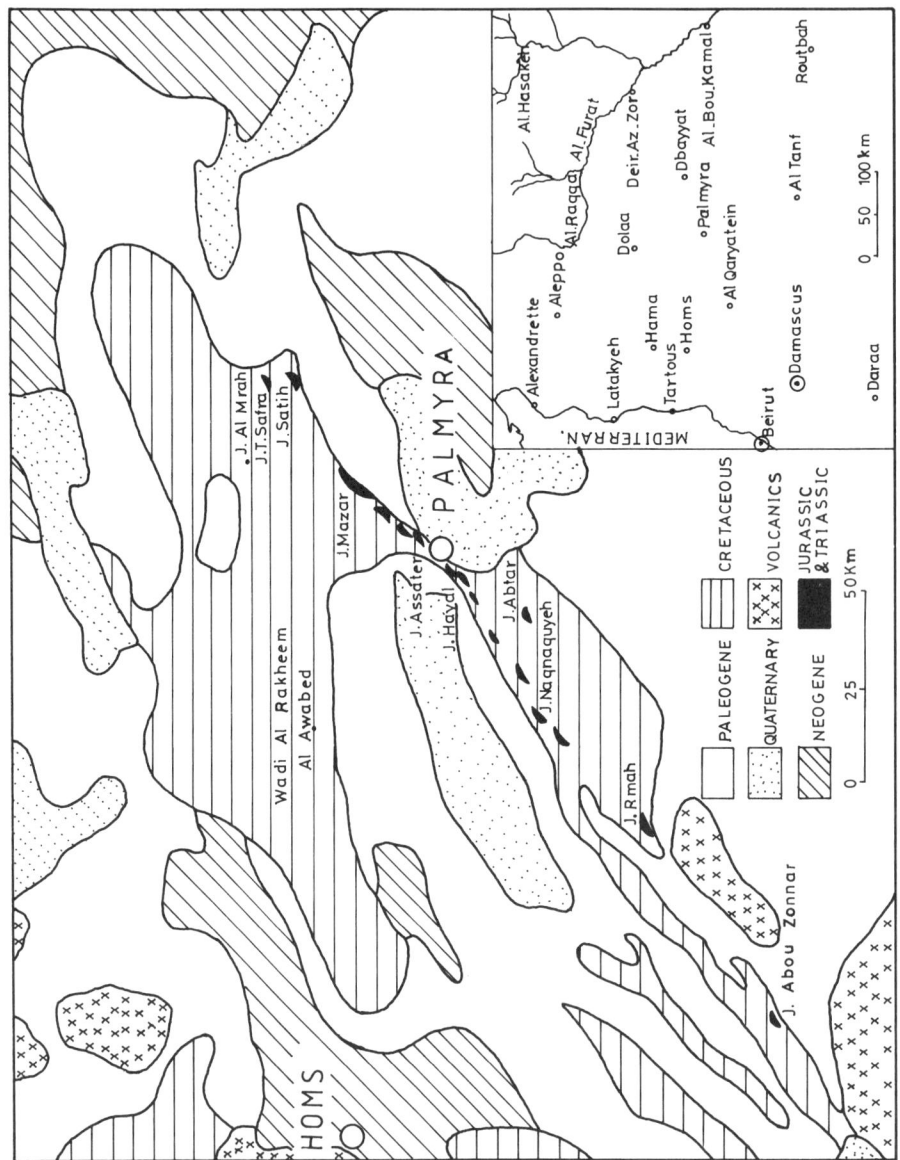

Figure 1. Simplified geological map of the Palmyrides Mountain Chain in central Syria.

composed generally of thin layered chert horizons or chert lenses and nodules. The Soukhneh Group is divided into two lithologic units: the Rmah Formation and the "Sawwaneh Formation" (Figure 3).

Rmah Formation
The lower part of the formation is composed of marly limestone and limy marl intercalated by thin limestone beds and concretions as well as by chert lenses and nodules. The upper part is composed mainly of marly limestone with concretionary structures distinguished by intercalating thin chert beds. The formation has an average thickness of 60—280 m,

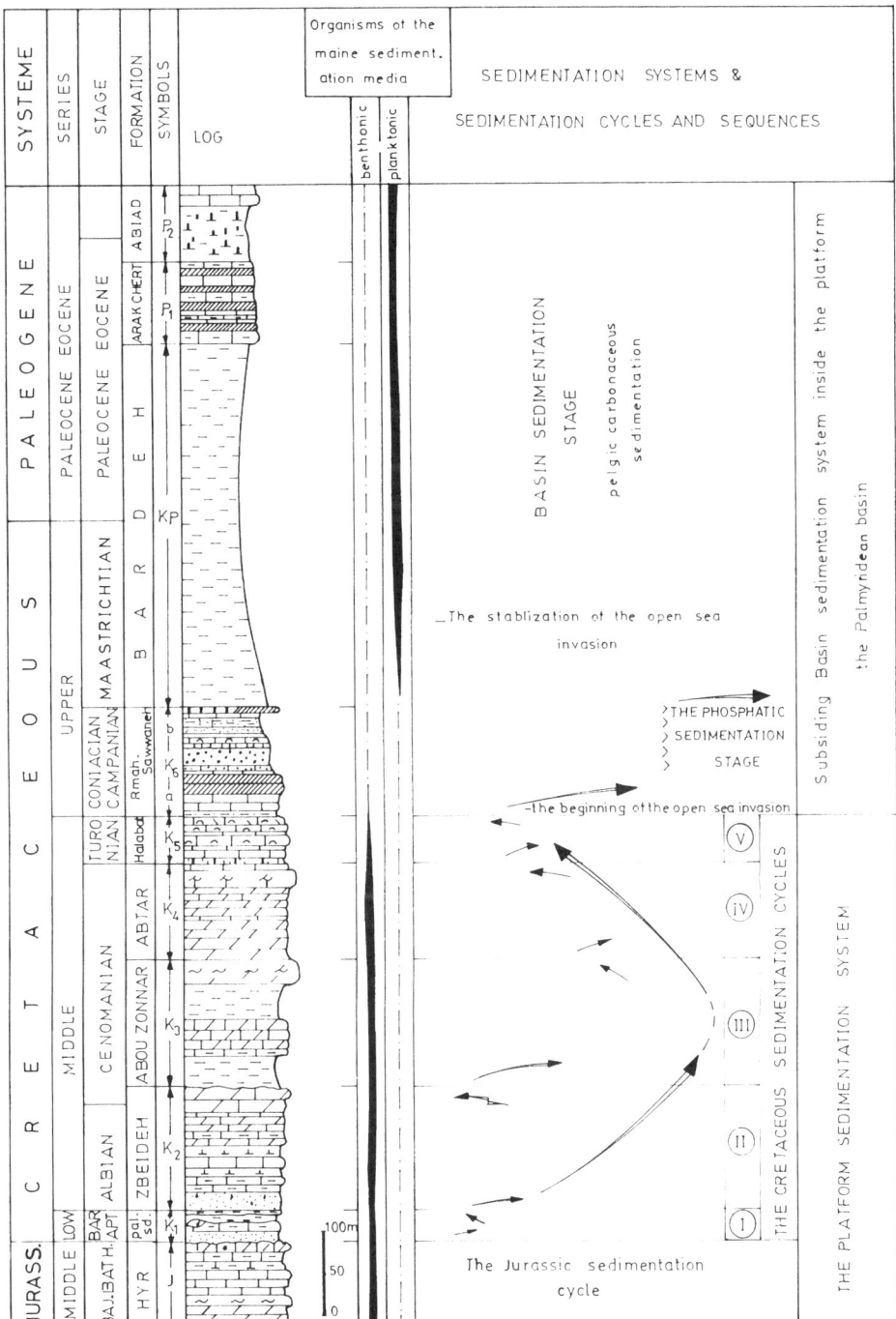

Figure 2. The general lithostratigraphic column in the Palmyrides Mountain Chain and the lithostratigraphic units.

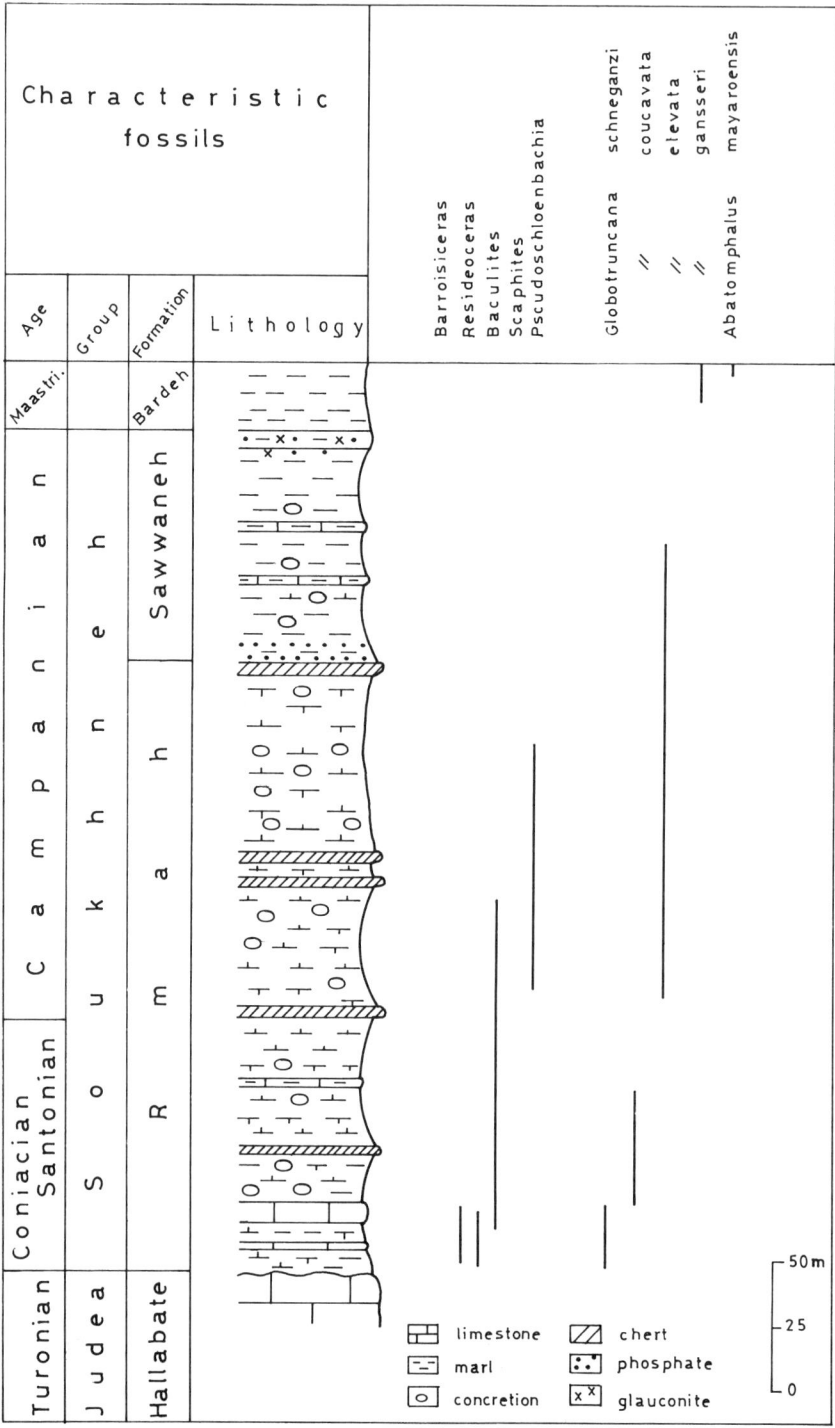

Figure 3. Stratigraphic section of the Senonian rocks in the Palmyrides Mountain Chain.

increasing northwards where it consists of a deeper water facies with a high content of ammonites, pelecypods and pelagic foraminifers. Toward the west and southwest the formation changes into a deep-water facies marl devoid of concretionary structures and is terminated by a cliff-forming chert horizon. According to the guide fossils present, the formation is given an age of Coniacian—Early Campanian (Figure 2).

Sawwaneh Formation

The basal part of the formation is marly limestone and limy marl intercalated with phosphorite beds varying in thickness and reaching a maximum of 10—12 m in the central southern range. The upper part of the formation is composed mainly of reddish yellow clayey marl intercalated with thin marly limestone beds with concretionary structures; this part of the unit is known as the"Erek Marl" and is overlain by a glauconite and phosphorite-bearing marl known as the "Tantour Marl". The latter marl is replaced locally by a siliceous horizon.

The thickness of the Sawwaneh Formation ranges from 17 m to 317 m. Generally, its thickness increases considerably northwards without significant facies changes, suggesting subsidence and deepening of the basin toward the north. The thick phosphorite horizons in the central part of the chain thin westwards and vanish totally below the overlying Erek Marl and Tantour Marl at the chain's westernmost extremities. Based on the occurrence of the microscopic pelagic fossils *Globotruncana calcarata* and *Globotruncana elevata*, the formation is assigned to the Upper Campanian.

The first listed fossil, defining the well-known *Globotruncana calcarata* zone, is characteristic for the Upper Campanian; its presence is restricted to the southwest extremity of the chain near Damascus. The second listed fossil, which is characteristic for Campanian in general, is present in the Sawwaneh Formation over the entire chain. Accordingly, the main phosphorites deposits in the Palmyrides Mountain Chain are assigned a Campanian age (Figure 3).

PETROGRAPHY

The macro- and microscopic facies of the Soukhneh Group show significant vertical and lateral variations. Among the main microscopic facies encountered are as follows:

In the limestone and marly limestone: Micrite and clayey biomicrite with pyrite, organic matter, clay minerals (sepiolite, attapulgite) and variable content of planktonic and benthonic foraminifers (the top horizons of mudstone and clay wackestone are rich in planktonic foraminifers).
Biomicrite—coquinoid microsparite with pelecypods of variable size.
Clayey glauconitic phosphatic biomicrite.
Laminated micrite with echinoids.
Organic biocalcarenite.
Biosparite with debris of mollusca, echinoids and benthonic foraminifers.
Biopelmicrosparite.

In the siliceous rocks: Siliceous lutite with micro- to cryptocrystalline quartz and chalcedony. It contains inclusions of micrite, microsparite, dolomicrite and fine planktonic foraminifers.

In the phosphorite rocks and phosphatic limestones; A wide spectrum of facies is encountered. The facies vary from biomicrite—microsparite with a small content of phosphatic grains to friable phosphatic arenite. The phosphatic components are generally granular with grain sizes of 0.05 to 0.5 mm; occasionally, the phosphate grains are millimetric and rarely centimetric to decimetric in size (remnants of vertebrae), and they form 10 to 80% of the rock. Some phosphatic rocks contain conglomerate-like chert pebbles and large shell horizons. The petrography of the phosphatic components reveals different kinds of phosphatic grains [7-9]; among these are phosphatic pellets, coated phosphatic grains composed of two or three aggregated grains, grains with non-phosphatic nuclei, irregular grains, coprolite grains resulting from the epigenesis of organic matter, grains with inclusions of non-organic, organic

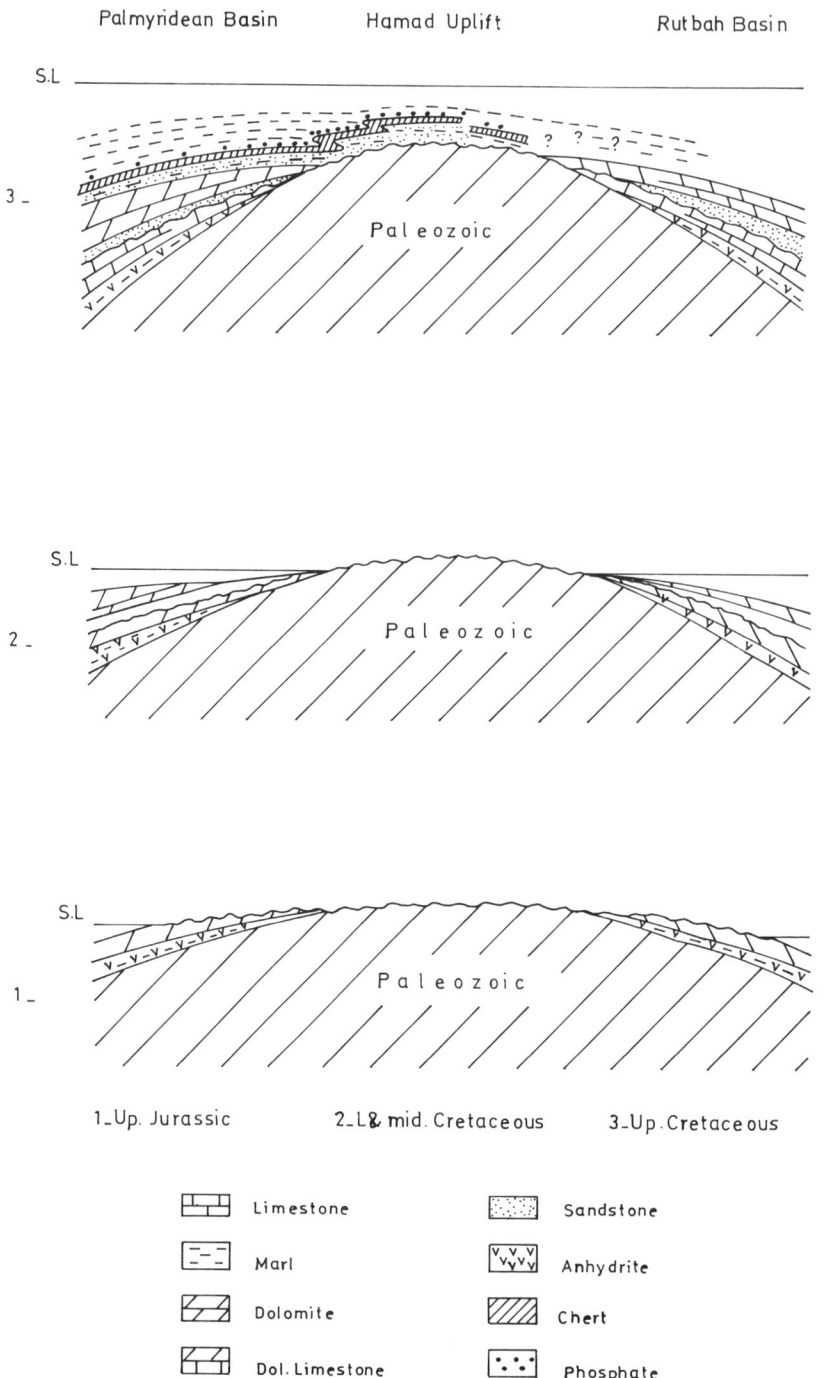

Figure 4. Lateral sedimentologic and paleogeographic effects of the Hamad Uplift from the Late Jurassic till the end of the Cretaceous.

and mineral substances, fish scales, teeth and vertebrae, irregularly-shaped and reworked grains. Phosphate beds may contain some limy and cherty clasts along with phosphatic grains which in turn contain organic matter in variable amounts.

SEDIMENTOLOGICAL AND PALEOGEOGRAPHIC FRAMEWORK OF THE FORMATION OF PHOSPHORITE DEPOSITS

During the deposition of the Soukhneh Group in the Palmyrides an important change in the sedimentary conditions throughout the entire region took place. For the first time since the onset of the Cretaceous, marine waters rich in planktonic foraminifers flooded the platform. Accordingly, the Hamad Uplift, which had remained emergent since the beginning of the Mesozoic, was submerged for the first time in its history by the Senonian sea. This submergence was modified by local features such as the Hamad Uplift which controlled phosphorite sedimentation in Senonian time [7-10] (Figure 4).

During the formation of the Rmah Chert Formation the Palmyridean region was initially shallow, in general, with lateral variations of sea floor morphology. The region was slightly but variably effected by the Senonian transgression. Consequently, it is marked by diversified sediments. Hence, the calcareous, siliceous and phosphatic limestones have vertical and lateral variations in structures and organic matter content. The calcareous sediments with clastic quartz at the base of this formation, encountered in the central part of the southern range of the Palmyrides, mark the onset of the Senonian transgression onto the Palmyridean region in general and the Hamad Uplift in particular. Unstable conditions during the phosphorite deposition were behind the presence of phosphatic limestone within this formation. The dominance of siliceous sedimentation at the end of the formation of the Rmah Chert Formation throughout the entire Palmyrides indicates general subsidence and paleoceanographic conditions suitable for silica abundance in the waters of the Palmyridean Sedimentary Basin. In part at least, this might be related to submarine acidic volcanism on the northern Arabian platform which enriched the sedimentary milieus by a regional silica supply. More likely, however, it reflects increased upwelling and higher biosiliceous productivity [11]. Such milieus in the northern range of the Palmyrides were deeper and more pelagic, for instance, at Jabal Al Mrah.

During the deposition of the Sawwaneh Formation, the entire southern range of the Palmyrides was characterized by the sedimentation of phosphorite around the rim of the Hamad Uplift. This structure was elongated to the south of the Palmyridean Basin between the localities of Bardeh and Soukhneh. The phosphorite sedimentation was concentrated most markedly in the central part of the northern rim of the mentioned uplift. The presence of associated calcareous and siliceous sediments confirms lateral variations in sedimentary conditions in the Palmyridean Basin. The marginal rim of the rising Hamad Uplift formed submarine highs which were especially suitable sites for phosphate sedimentation and concentration. Among the additional factors contributing to this sedimentation was the enrichment of phosphorus from the P-rich deep water on the northern edge of the Arabian Platform by currents which upwelled onto this extensive platform. The warming up of the platformal water on one hand, and the abundance of nutrients on the other, caused a proliferation of plankton which assimilated, stored and concentrated a large amount of phosphorus. After the deposition of the plankton, a huge amount of phosphorus dissolved and became concentrated in the sea-floor sediments, The contribution of the bacterial activity in such environments in the deposition of phosphorite is a possibility. Such environments existed in the central northern rim of the Hamad Uplift, in particular at variable depths ranging from few meters to fifty meters at maximum. The differences in the prevailing physico-chemical and energy conditions in these environments led to successive phosphatic, calcareous and siliceous deposits. The amount of planktonic and benthonic foraminifers in the sediments deposited varied considerably according to the prevailing conditions resulting from the Senonian transgression. Generally, the depth of the Palmyridean Basin increased from the central northern rim of the Hamad Uplift outwards, but the depths increased more rapidly northwards. Consequently, the northern range of the Palmyrides has experienced steady subsidence and deepening since the deposition of the Rmah Chert Formation and accordingly pelagic calcareous sedimentation, almost devoid of phosphorite, prevailed.

The reviewed paleogeographic and sedimentological characteristics confirm that the Palmyridean Basin started differentiating and continued developing throughout Coniacian—Santonian—Campanian time. This basin was bounded on the south by the Hamad Uplift where phosphorite sedimentation prevailed. The deep and subsiding central part of the basin, located in the northern range of the Palmyrides, was characterized by a steady deposition of pelagic calcareous sediments.

The presence of this paleogeographic structure (Hamad Uplift) is confirmed by the variations in thicknesses of the Soukhneh Group measured in sections across the entire Palmyrides and by the typical reduction in phosphorites located around the rising rim of the Hamad Uplift.

Phosphorite deposition diminished gradually towards the northern areas of the Palmyridean Basin where it was replaced by monotonous and thick calcareous sediments. Local phosphorite deposition at the northern margin of the Palmyridean Basin (Al Awabed, Wadi Al Rakheem) was most likely related to sea floor rises attached to the southern rim of the Aleppo Uplift; i.e. a distinctive paleogeographic structure bounded the Palmyridean Basin to the north. The remarkable thickness of the marly, pelagic Bardeh Group (Cretaceous—Paleogene), which overlies the Soukhneh Group, gives evidence for the belief that the Palmyridean Basin continued subsiding, deepening and developing a differentiated shape during Late Cretaceous—Paleogene time span.

Acknowledgments

We gratefully thank Prof. I. Haddad, the Director General of the AECS, for his great support during the performance of this research. We thank Mr. Y. Radwah, the staff geologist at the AECS, for his help in improving the manuscript. Our thanks are also due to Prof. R. E. Garrison and Prof. A. Iijima for their fruitful suggestions and for the reviewing and editing of the manuscript.

REFERENCES

1. L. Cayeux. Contribution sur les phosphates Senoniens de Syrie, *C. R. Acad. Sci.* **200**, 1553-1555 (1935).
2. Russian Technoexport Mission. *The geological map of Syria. Scale 1/1,000,000: Explanatory notes.* (Technoexport, U.S.S.R). Ministry of Industry - Syria (1966).
3. Russian Technoexport Mission. *The geological map of Syria. Scale 1/200,000: Explanatory notes, Sheet: XIV, XV, XVI, XX, XXII.* (Technoexport, U.S.S.R). Ministry of Industry - Syria (1966).
4. S. Atfeh. *The phosphate deposits in Syria.* Ph.D. Thesis, University of London, London (1967).
5. M. Al Issa. *The geological structure and the composition of phosphatic deposits in Palmyridean Basin.* Ph.D. Thesis, Institute of Geological Research, University of Moscow, Moscow (1972).
6. M. Abbas. *Geochimie de l'uranium des phosphorites des Palmyrides centrales, Syrie.* These d'Etat, Institut de Geologie, Universite de Strasbourg, Strasbourg (1987).
7. A.Kh. Al Maleh. Les depots phosphates du Senonien en Syrie et les conditions paleo-geographiques de leur genese. In: *27em Intern. Geol. Cong. Volume 11.* p.92. Moscow (1984).
8. A.Kh. Al Maleh. The general characteristics of Cretaceous sedimentation in the northern part of the Arabian Platform (the Syrian Platform). In: *8th IAS Regional Meeting of Sedimentology.* pp.34-35. Tunis (1987).
9. A.Kh. Al Maleh. The geology of the phosphatic deposits in Syria (Senonian and Palaeogene). In: *10th International Field Workshop and Symposium.* pp.xx-xx. Tunis (1987).
10. A.Kh. Al Maleh and M. Mouty. The sedimentary and paleogeographic evolution of the Palmyridean region during Cretaceous. In: *Proceedings 3rd Jordan Geological Conference.* pp.213-244 (1988).
11. Z. Reiss. Assemblages from a Senonian high-productivity sea, *Révue de Paleobiologie*, Vol. Spéc. no.2, 323-332 (1988).

Sedimentary petrographical, geochemical and sedimentological aspects of Triassic–Jurassic bedded cherts in Southwest Japan

Y. KAKUWA

Department of Astronomy and Earth Science, College of Arts and Sciences, University of Tokyo, Komaba 3-8-1, Meguro-ku, Tokyo 153, Japan

Abstbract -- Petrographical and geochemical characteristics of Triassic–Jurassic bedded cherts in Southwest Japan are summarized. The bedded chert formations record both world wide geological events and local differences of sedimentary environments. An example of the former is the earliest Triassic carbonaceous shale and succeeding early Triassic gray siliceous claystone with sparce radiolarians. The global Toarcian anoxic event is also recorded in the bedded chert formations. Examples of local environmental effects are the differences in colors, biogenic siliceous constituents and types of layering in chert beds in middle and upper Triassic cherts. Such local differences in chert beds become incospicuous in the Jurassic. The middle Jurassic section is characterized by radiolarian mudstone. Biogenic siliceous sediments are widespread during Triassic and Jurassic, but authigenic phosphorite is not associated with these bedded chert formations.

Keywords: bedded chert, Triassic–Jurassic, radiolarians, P/T boundary, phosphorite

INTRODUCTION

Unmetamorphosed Paleozoic and Mesozoic bedded cherts are widely distributed in the Ashio, Mino and Tamba Belts (AMT Belts) and in the Chichibu Belt in Southwest Japan (Figure 1). The age of bedded cherts range from Late Carboniferous to Late Jurassic. Bedded chert of these belts occurs as blocks in younger matrices or as repeated sheets of pile nappes, and are interpreted to be allochthonous blocks or sheets.

Slowly and continuously deposited bedded chert formations appear to record various geologic events and to represent deeper parts of the sea in contrast to carbonate rocks which typically represent shallower areas. I made detailed lithological and geochemical studies on eighteen well-exposed Triassic–Jurassic bedded chert sections of the AMT Belts which are well dated by conodonts and radiolarians [1, 2]. In the present paper, lithological and geochemical characters of Triassic–Jurassic bedded chert sections of Southwest Japan are summarized using published data [1-4] as well as new observations from the AMT Belts and the Chichibu Belt.

SUMMARY OF LITHOSTRATIGRAPHY AND GEOCHEMISTRY OF THE BEDDED CHERT SECTIONS

Early Triassic

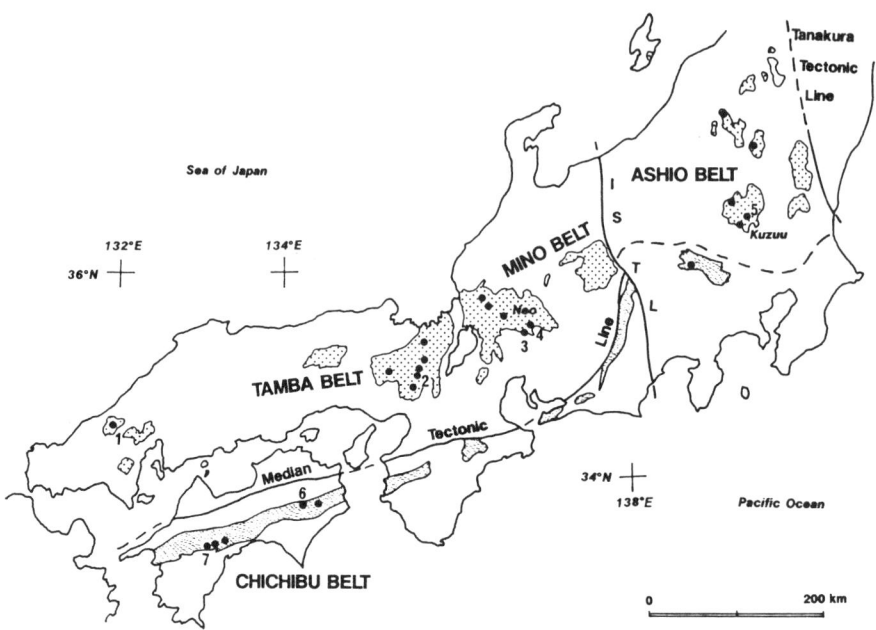

Figure 1. Index map. Dotted area: Ashio, Mino and Tamba Belts (AMT Belts). Shaded area: Chichibu Belt. Filled circle: studied section. 1: Koze, 2: Hozukyo, 3: Unuma, 4: Hisuikyo, 5: Karasawa, 6: Tenjinmaru, 7: Shiroishigawa, ISTL: Itoigawa-Shizuoka Tectonic Line.

The basal part of the bedded chert sections are characterized by the consistent occurrence of gray siliceous claystone (Figure 2). The gray siliceous claystone is composed of quartz, illite and chlorite with or without feldspar [3]. Authigenic pyrite is common. Siliceous skeletons are rare. Shale partings of bedded chert sequences are petrographically similar to the gray siliceous claystone, although pyrite is not common in shale partings. Chemical compositions of the siliceous claystone and shale partings are also similar to each other. Covariation diagrams of Al_2O_3 and lithogenous elements such as TiO_2, MgO, K_2O, Cr and V indicate that the detrital materials of the gray siliceous claystones and shale partings of the AMT Belts are almost identical to each other (Figure 3). The difference between the two argillaceous rocks is the content of Al_2O_3 and consequently SiO_2.

The average Mn contents in modern offshore muds and pelagic clays are plotted on the diagrams in Figure 3. The Mn contents are conspicuously higher in pelagic clays than in offshore muds and in the argillaceous rocks. On the other hand, the Mn content in offshore muds are lower than that of radiolarian mudstones and slightly higher than that of shale partings. Two possible origins are proposed for the high concentration of base metals in pelagic sediments; i.e., hydrogenous and hydrothermal [7-10]. Whether the origin of base metals in pelagic clays is hydrogenous or hydrothermal, the significantly higher concentration of Mn in pelagic clays compared to the argillaceous rocks suggests that the latter are equivalent to offshore muds rather than to pelagic clays [10, 12,13].

Chondrite-normalized rare earth element patterns of two chert beds and one siliceous claystone, as shown in Figure 4, generally resemble each other; i.e., light REEs (La, Ce, Sm and Eu) are much more enriched than heavy REEs (Tb, Yb and Lu). Such patterns are characteristic of the terrigenous fine sedimentary rocks such as the North American Shale

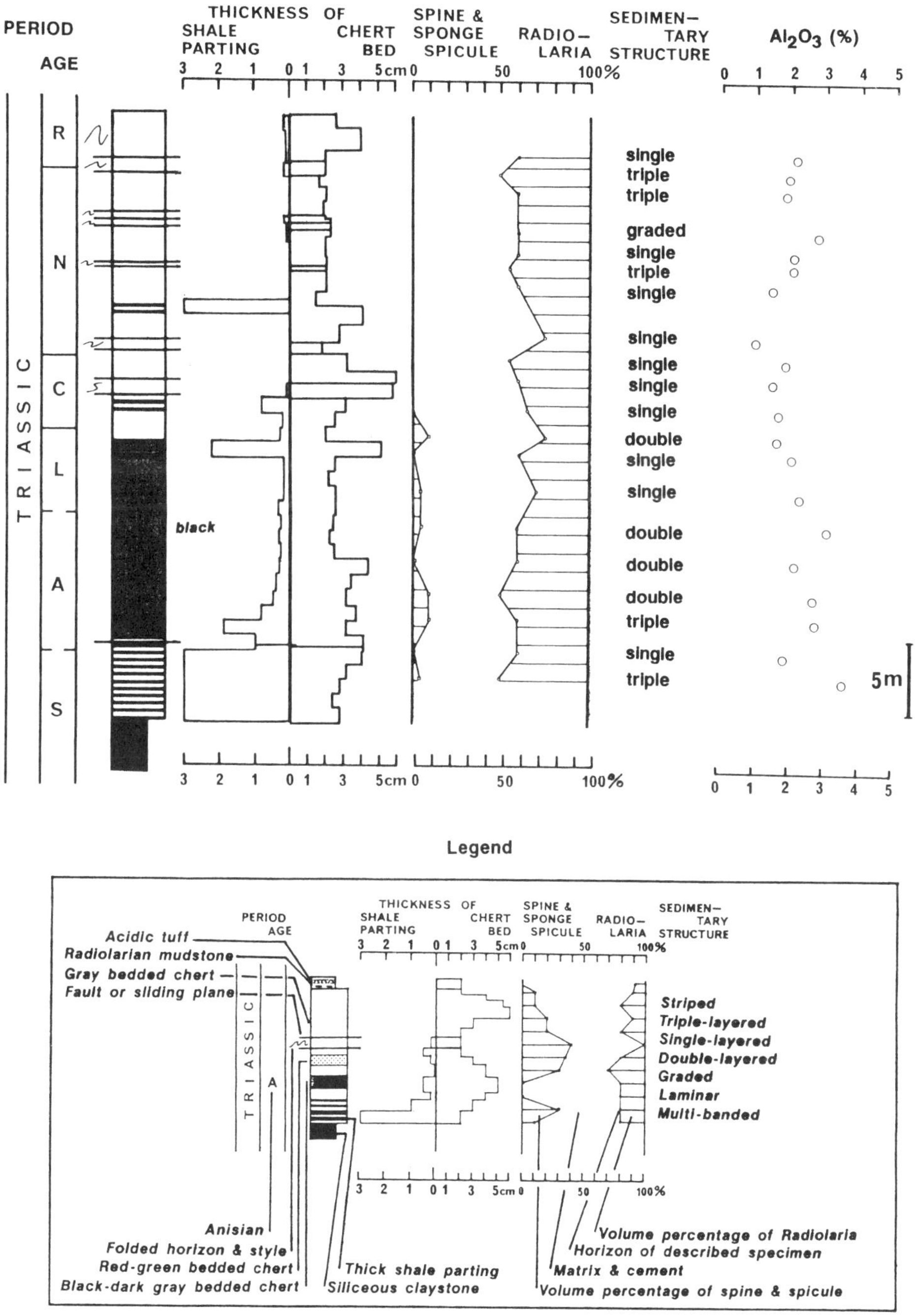

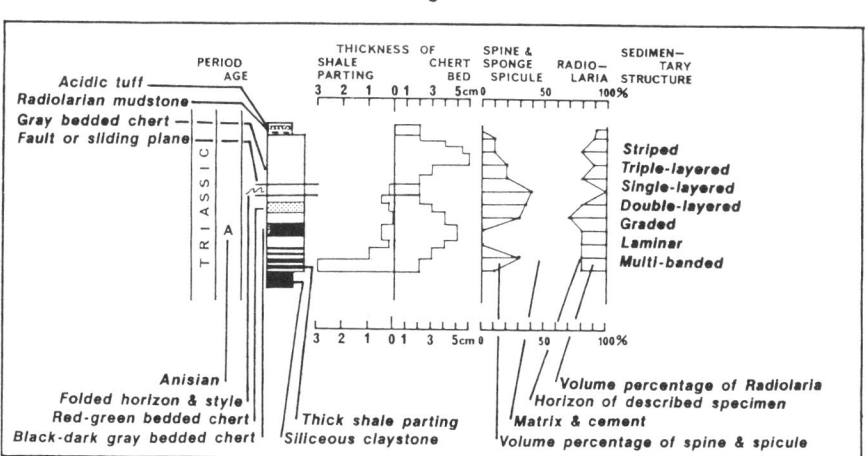

Figure 2. Description of the Hozukyo section in the Tamba Belt. The age is after [5,6]. S: Spathian, A: Anisian, L: Ladinian, C: Carnian, N: Norian, R: Rhaetian.

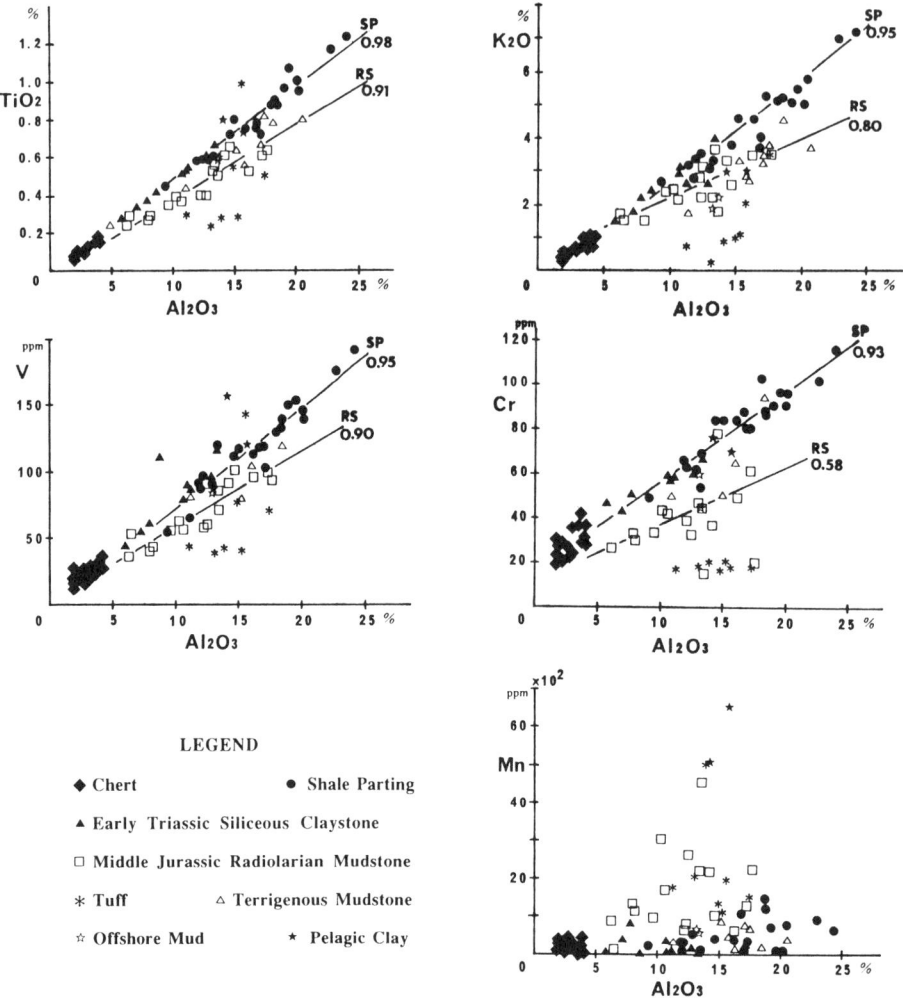

Figure 3. Relationship between the contents of Al2O3 and TiO2, K2O, V, Cr and Mn in the Triassic-Jurassic bedded cherts and argillaceous rocks of the AMT Belts: tuffs, middle Jurassic radiolarian mudstones [10,11] and averages of modern pelagic clays and offshore muds [10] are also plotted. SP and RS indicate the regression lines for shale partings of the bedded cherts and middle Jurassic radiolarian mudstones, respectively.

Composite (NASC) [14]. In addition, the REE patterns of NASC and siliceous claystone show no Ce and Eu anomaly. These geochemical analyses indicate that the siliceous claystone is not tuffaceous, but similar to the average shale [15].

A black carbonaceous bed, a few meters thick, is in some cases observed lying below the gray siliceous claystone (Figure 5). Age-determinable fossils have not been identified from the black carbonaceous bed. However, its age is suggested to be earliest Triassic (Griensbachian and Dienerian) and may be extend into latest Permian, because the underlying bedded chert contains late Permian radiolarians and conodonts, and the overlying gray siliceous claystone yields Spathian conodonts [16,17] or Smithian to Spathian radiolarians [18]. Partly-preserved, fine-parallel laminations of this carbonaceous bed indicate anoxic condition occurred occasionally during earliest Triassic and/or latest Permian time [4].

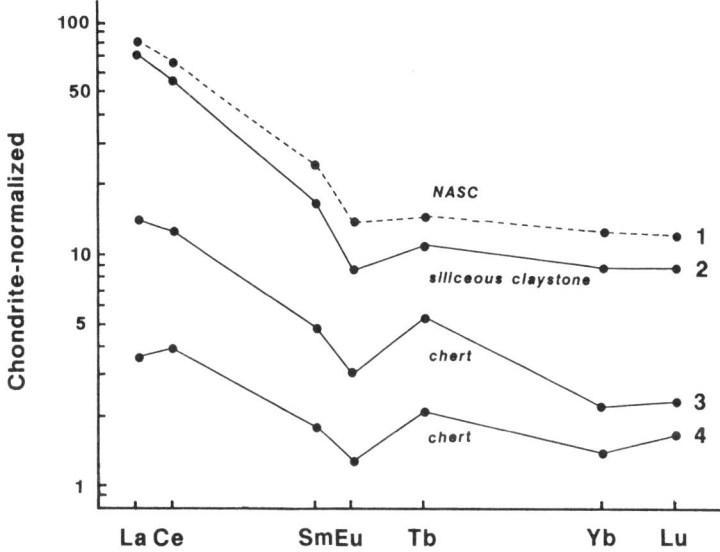

Figure 4. Chondrite-normalized rare earth elements pattern of the early Triassic siliceous claystone (2) and chert beds (3, 4). Broken line is chondrite-normalized pattern of North American Shale Composite (NASC) [14].

The black carbonaceous bed grades upwards into the gray siliceous claystone and then into gray or dark gray bedded chert with thick shale partings.

Middle Triassic
Middle Triassic bedded cherts are characterized by thick shale partings in the lower section. The shale partings become thinner upwards. Such thickness variations of shale partings are independent of the distributional areas of the bedded cherts. Other lithologic characters of the bedded cherts serve to define two different types of middle Triassic bedded chert formations in the AMT Belts; these two types are named the R and the G chert groups.

The R chert group is characterized by the occurrence of sponge spicule-bearing red chert (Figure 6). Layering in these chert beds is commonly multi-banded and graded. Multi-banded cherts are composed of sponge spicules with radiolarian lenses or laminations, with or without graded bedding [2]. Graded cherts show size grading of siliceous skeletons [2]. Those types of layering imply reworking of siliceous skeletons.

The R chert group of this age (Anisian to Ladinian) is asociated with thin sandstone beds or lamina in some localities in the Mino Belt. The sandstone-bearing chert beds are restricted to the multi-banded cherts. These sandstones are composed of lithic fragments of red, gray and vermillion cherts, gray siliceous claystone and minor amounts of calcitized, phosphatized or chloritized volcanic rock fragments. The average contents of P_2O_5 and CaO of five phosphatized volcanic rock fragments are 41.07% (standard deviation: 0.79) and 52.16 % (S.D.: 0.92), respectively, by EDS analysis.

Chemical composition of some vermillion radiolarian chert fragments in the sandstones was analyzed on an EDS. The result of the chemical analysis is plotted on the ternary Al-Fe-Mn diagram (Figure 7). For comparison, sandstone-bearing chert, red hematitic chert resting on

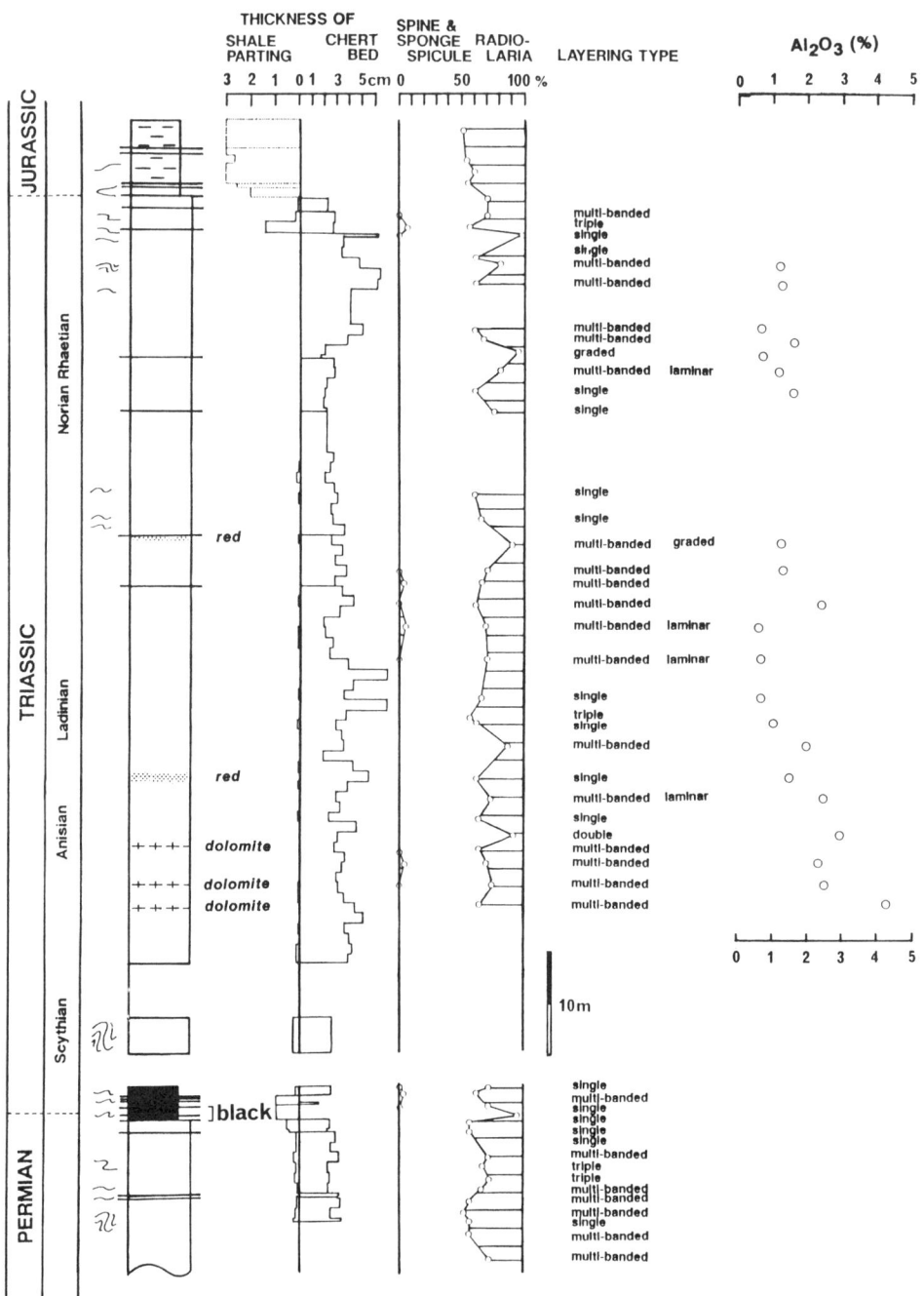

Figure 5. Description of the Tenjinmaru section and adjacent bedded chert section in the Chichibu Belt. The age is after [16, 17]. Legend is referred to figure 2.

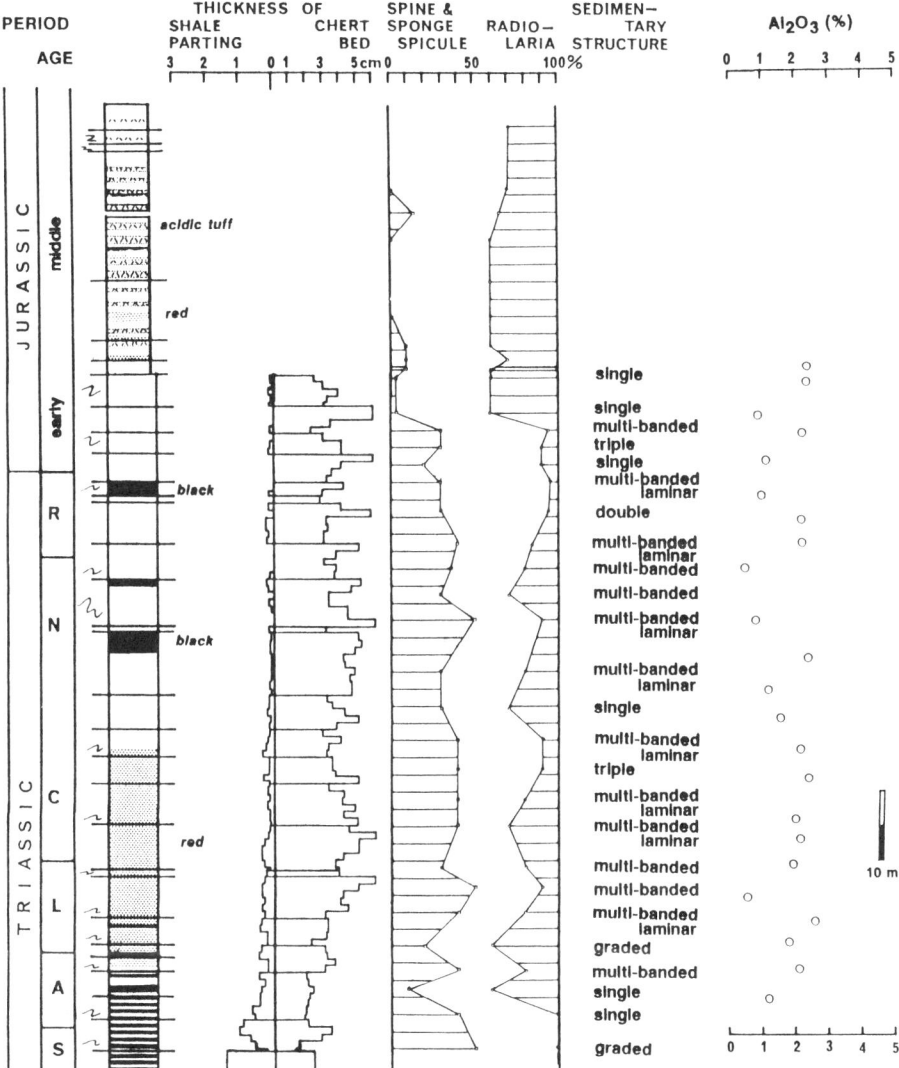

Figure 6. Description of the Hisuikyo section in the Mino Belt. The age is after [19-23]. Legend is referred to figure 2.

Permian pillow lava of the Mino Belt, eight cherts from the Anisian to Ladinian interval of the Hisuikyo section, and the ferruginous silicastones associated with Permian greenstones in the Tamba Belt [24-27] are also plotted on the diagram. The sandstone-bearing chert as well as the eight cherts of the Hisuikyo section lie within the "non-hydrothermal" domain, where as the vermillion chert fragment, red hematitic cherts and silicastones lie within the "hydrothermal" domain.

A possible interpretation of the depositional environment of the R chert group is shown in Figure 8. Bedded cherts of the R chert group were deposited near a topographic high formed by basic volcanic rocks. Both siliceous sponges, which lived on hard surface of such a

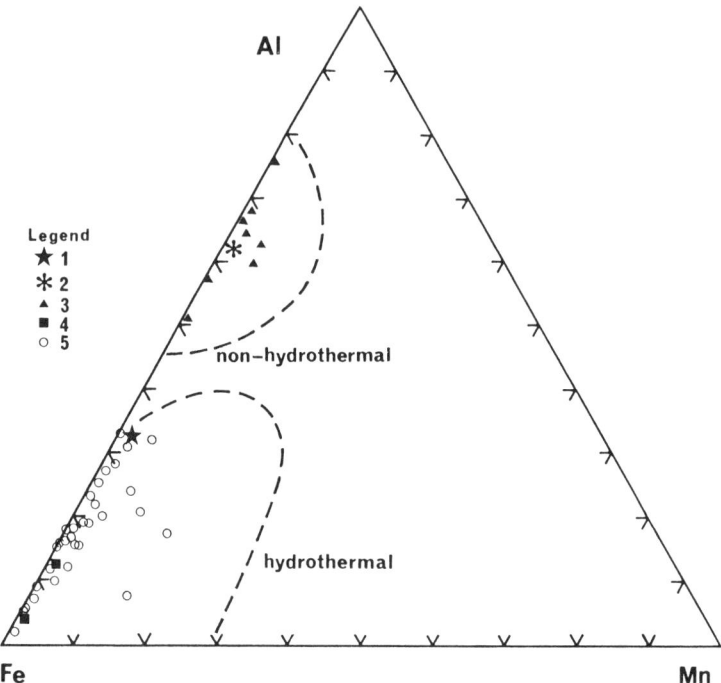

Figure 7. Al-Fe-Mn diagram showing compositional difference between the vermillion chert fragment of sandstone and both the sandstone-bearing chert and the Anisian to Ladinian chert of the Hisuikyo section. The areas of "non-hydrothermal" and "hydrothermal" encircled by broken lines are after [28]. 1: vermillion chert fragment, 2: sandstone-bearing chert, 3: Anisian to Ladinian cherts, 4: hematitic cherts rest on the Permian pillow lava, 5: silicastones which are closely associated with Permian greenstones in the Tamba Belt [24-27].

topographic high, and radiolarians, which accumulated on the high, were reworked and redeposited into a deeper basin. Volcanic rocks might be phosphatized on the shallow seamount, but in the deeper basin where the bedded chert was deposited, no *in situ* phosphorite was observed.

The G chert group is characterized by gray-colored radiolarian cherts (Figure 2). Layering in chert beds of the G chert group is mostly of the single and triple-layered types. The triple-layered cherts show a symmetrical, three-fold structure consisting of upper and lower argillaceous layers and a middle highly siliceous layer. The single-layered cherts, in contrast, are either homogeneous or structureless with or without faint clay lamination. Neither of these layering types show evidence of reworking of siliceous skeletons [2,29,30].

The R and G chert groups are not recognized in the bedded chert sections of the Chichibu Belt. In the Chichibu Belt, Triassic to Jurassic red cherts are rare, and sponge spicules are not common; both lithologic characters are representative of the G chert group. However, the multi-banded, graded and laminar types, which resemble the R chert group are common (Figure 5).

Late Triassic
The R and G chert groups persist within late Triassic sequences of the AMT Belts (Figures 2, 6). Sedimentation rates of the bedded cherts were estimated at four measured sections where detailed conodont and radiolarian biostratigraphic studies have been performed. The

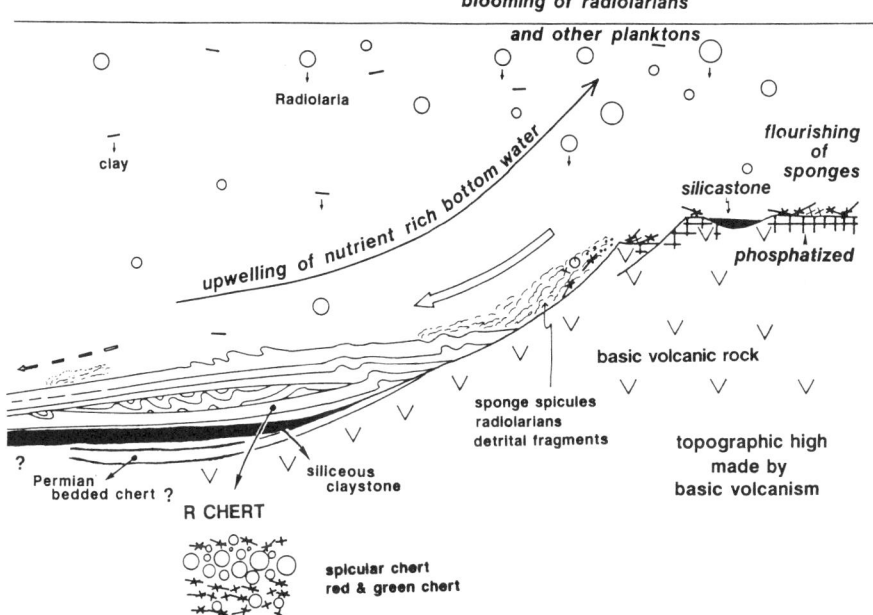

Figure 8. Schematic illustration of the depositional site of the Hisuikyo section. Bedded chert of the R cherts was deposited near a topographic high formed by basic volcanic rocks. Both biogenic siliceous constituents of R cherts such as radiolarians and sponge spicules and clastic materials such as silicastones and phosphatized volcanic rocks were redeposited into a deeper basin.

sedimentation rates are calculated within the limit of one or two geochronological stages which are defined by the occurrence of index fossils or assemblages and in which no faults are positively recognized. Correlation between stages and conodonts refers to [31] and radiolarians to [32]. The time intervals of ages in the Triassic are after [33].

The calculated sedimentation rates are shown in Table 1. Compacted sedimentation rates of the bedded cherts have a range from 0.51 to 0.36 mm/ky. Precompacted sedimentation rates are estimated to be a range of 2.55 to 18.4 mm/ky, if the compaction rate of bedded cherts is postulated to be 5, according to Iijima et al.[37]. Biostratigraphically estimated accumulation rate and probably the production rate of siliceous organisms were higher during the late Triassic interval than during the early and middle Triassic intervals.

The upper Triassic is characterized by the occurrence of white micritic limestones and/or gray dolostones in some sections of the Chichibu Belt and within the G chert group of the AMT Belts. The limestone interbeds usually have thicknesses of over a few meters, whereas the dolostone interbeds are less than a few tens of centimeters. White micritic limestone contain abundant calcitized radiolarians and conodonts. Conodont fossils suggest that the carbonate deposition is restricted almost entirely to early Carnian to late Norian.

Early Jurassic

Early Jurassic cherts, even the R chert group, have fewer sponge spicules as well as less multi-banded and graded types of red chert, and their petrographic characters are more similar to the G chert group (Figures 6, 9). No conspicuous petrographical difference was observed at or near the Triassic and Jurassic boundary, and the end-Triassic mass extinction event is not

Table 1.
Sedimentation rate of bedded cherts in the measured section of the AMT Belts. The age of the Hisuikyo, Unuma, Hozukyo and Koze sections is after [19-21], [34,35], [35,36] and [37], respectively. Time span is after [33]. S.R.: rate of sedimentation, Nor & Car: Norian and Carnian.

measured section	stage	time span (m.y.)	thickness (mm)	number of couple	S.R. (mm/k.y.)	frequency (year/couple)
Hisuikyo	Nor.& Car.	12	43238	1054	3.60	11385
	Ladinian	7	11850	236	1.69	29661
	Anisian	5	12161	381	2.43	13123
Unuma	Norian	6	22091	622	3.68	9646
Hozukyo	Norian	6	13645	529	2.27	11342
	Carnian	6	4552	149	0.76	40268
Koze	Ladinian	7	3572	299	0.51	23411
	Anisian	5	6433	274	1.29	18248

apparently recorded in the bedded chert sections. Highly carbonaceous shales and cherts occur in the late early Jurassic, which is correlated to the world-wide Toarcian anoxic event [38]. In the Karasawa section of the AMT Belts, petrified wood occurs in bedded cherts of the uppermost section (Figure 9) [37].

Middle-Late Jurassic
The middle-late Jurassic section is characterized by the common occurrence of bedded radiolarian mudstones. The radiolarian mudstones are green, gray, red and black in color, and contains well-preserved radiolarians, which form parallel laminations a few millimeters thick. These laminations are commonly disturbed by bioturbation. Bioturbated structures are also observed in the bedded cherts that persist until late middle Jurassic; such bioturbated structures have not been positively identified in the Triassic and early Jurassic bedded cherts. Sponge spicules are very rare. Acidic tuff interbeds commonly occur in the radiolarian mudstones. Manganese nodules composed of rhodochrosite are common and considered to have formed during early diagenesis, based on isotopic studies [40, 41].
The Al_2O_3 content in the cherts and radiolarian mudstones, which means the content of argillaceous material, gradually increases upward. The average thickness of acidic tuffs in each interval of one meter in the cherts and mudstones indicates that acidic tuff increases upward, excepting in the uppermost part (Figure 10). Some thick acidic tuff interbeds show graded bedding and dish structure. The uppermost part of the radiolarian mudstones is intercalated with epiclastic sandstones and overlain by thick sandstone with abundant plant debris.
Covariation diagrams of Al_2O_3 and lithogenous elements in Figure 3 indicate that the chemical composition and therefore the provenance of argillaceous materials in the middle to late Jurassic radiolarian mudstones were different from the early Triassic siliceous claystone.

DISCUSSION AND CONCLUSION

Site of deposition of the Triassic—Jurassic bedded cherts
Two theories have been proposed for the site of deposition of the Triassic—Jurassic bedded cherts in Southwest Japan: *i.e.*, the open ocean theory and the marginal sea theory.

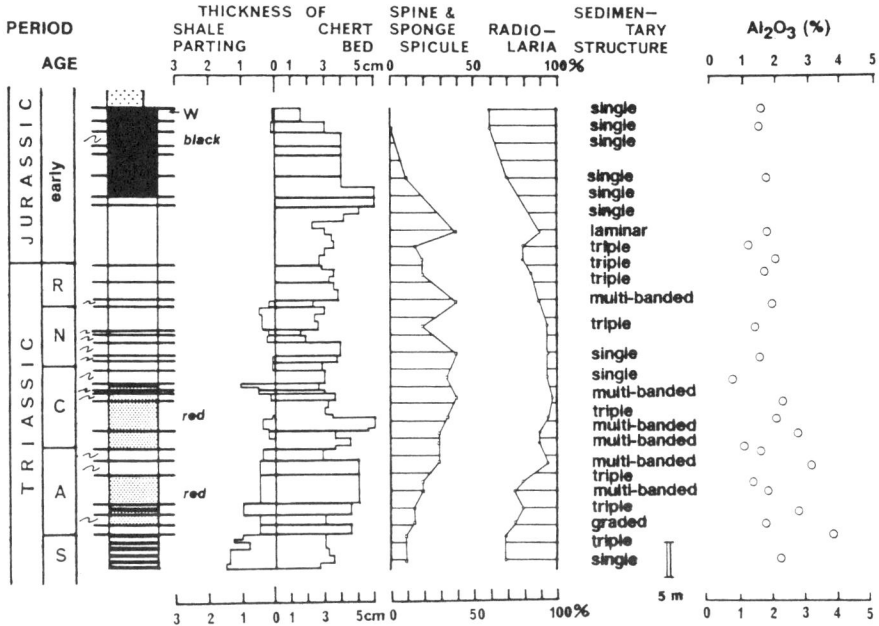

Figure 9. Description of the Karasawa section of the Ashio Belt. The age is after [37]. Legend is referred to figure 2. W: petrified wood.

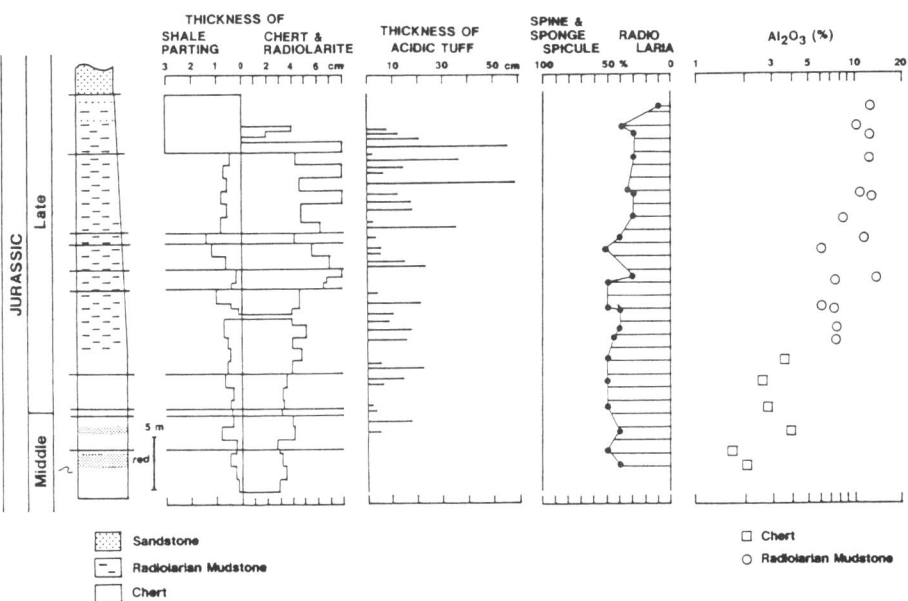

Figure 10. Description of the Shiroishigawa section of the Chichibu Belt. The age is after [39].

Matsuda and Isozaki [42] considered that the bedded cherts of the AMT Belts are comparable to modern pelagic sediments accumulated in an open ocean environment far away from continents such as the Pacific Ocean. Their speculation is based on the following general occurrence of bedded cherts: the low sedimentation rate, high purity of biogenic silica, long duration of continuous deposition (more than 50 m.y.), and wide along-strike extent (more than 1000 km) of bedded cherts.

Iijima, Matsumoto and Tada [29] and Iijima, Kakuwa and Matsuda [37] concluded that the Triassic—Jurassic bedded cherts of the AMT Belts were deposited in a back-arc marginal sea. Their conclusion is mainly based on the relation of terrigenous materials (shale partings) to the origin of rhythmic bedding, the finding of non-drifted silicified wood in the bedded cherts, and the regional sedimentological framework.

The geochemical data mentioned above, as well as all other geochemical studies [3,10, 12, 13], have revealed strong influences of detrital input from continental areas on the sedimentation of bedded cherts, and therefore support the marginal sea theory. In addition, middle Jurassic radiolarites recovered from the Pacific Ocean bottom have never displayed such strongly bedded and cherty ribbon radiolarites as distributed in the circum Pacific regions of the same age [43, 44]. Therefore, proponents of the open ocean theory must explain why bedded cherts are not discovered from the Pacific Ocean, and why chemical compositions of on-land bedded cherts are different from those of the sediments from the Pacific Ocean.

It has been asserted that marginal seas would be small and short-lived sedimentary basins, and would receive abundant terrigenous clastics from continents [42]. However, marginal seas are not always small and short lived. For example, Miocene bedded diatomaceous-siliceous sediments extend in the Sea of Japan for more than 1000 km along the Honshu Island Arc [44]. Pelagic calcareous sediments have been continuously deposited for as long as 70 m.y. without coarse clastic sediments on the Loard Howe Rise in the Tasman Sea [45], which occupies an area of 700 x 2000 km. Before the middle Jurassic, calcareous plankton were not typical pelagic organisms, therefore biogenic pelagic sediments would have been dominated by radiolarians.

Paleoceanography of the Triassic—Jurassic bedded cherts

Slowly-deposited bedded chert formations are commonly interpreted as records of variations of oceanic environments. Triassic—Jurassic bedded chert formations in Southwest Japan began with sedimentation of carbonaceous beds in an oxygen-deficient environment followed by rapid sedimentation of terrigenous argillaceous materials. Earliest Triassic carbonaceous deposits occur also in the Carnic Alps of the Mediterranean region [46] and South China [47] and in the western part of North America [48]. Such synchronous distribution of this lithofacies thus probably records not a local but a world-wide event during the transition from an environment hostile for living organisms resulting in the mass extinction event at the Permo—Triassic boundary to a more normal environment. Sedimentation of the early Triassic siliceous claystone represents proliferation of radiolarians which, however, had not sufficiently recovered to form chert beds during Spathian. Both the Al_2O_3 contents in chert beds and the thicknesses of shale partings decrease upward, suggesting that the supply of radiolarian skeletons gradually increased and terrigenous clay decreased. Both trends resulted in the formation of bedded chert by middle Triassic time.

Lithological differentiation of bedded cherts in the AMT Belts, which probably reflect the differences in environments of local sedimentary basins, became conspicuous in middle Triassic time and continued into late Triassic. The frequent occurrence of carbonate rocks in the bedded chert formations indicates that the CCD was locally deepened during late Triassic time.

The mass extinction event between the Triassic and Jurassic is not apparently recognized, but the Toarcian anoxic event is recorded in the bedded chert formations. The sedimentary environment characteristic of Triassic bedded chert deposition significantly changed from at least the middle Jurassic: Sponge spicules almost completely disappeared in the middle Jurassic and bioturbation starting by at least middle Jurassic time. The supply of terrigenous materials increased again from middle Jurassic to late Jurassic, in association with acidic tuff deposition.

Phosphogenesis and Triassic—Jurassic bedded cherts
Phosphatized volcanic rock fragments in the middle Triassic bedded cherts of the Mino Belt indicate phosphogenesis on the seamount previously mentioned (Figure 8). Where was the seamount situated? Tsukamoto [49] described the occurrence of lithic fragments of lutecite and quartz schist with volcanic rock fragments from Triassic bedded cherts in the Mino Belt, and suggested that continental crust covered with intermediate to basic volcanic rocks formed the background of the bedded chert basin. If such quartzitic detrital lithic fragments are not aeolian in origin, the seamount was situated not far from continental crust [37].
The age of the phosphogenesis on the above-stated seamount is uncertain. Basic volcanism was prevalent during late Carboniferous to middle Permian time in the AMT Belts. Middle to late Permian radiolarians, which are considered to be reworked, from the sandstone layers in the middle Triassic bedded cherts [49] suggest that the phosphorite might have formed sometime during the Permian Period. Other occurrences of phosphatic sediments from the AMT and Chichibu Belts are as follows: In the Kuzuu area of the Ashio Belt [50], phosphorite nodules are observed in middle Permian carbonate rocks which cover a Late Paleozoic seamount, detrital phosphorites occur in the Triassic limestone breccias on the karstified surface of the carbonate rocks. An allochthonous phosphorite block which contains late Permian radiolarians was discovered from the early Jurassic? muddy matrix within the Chichibu Belt [51]. All of these reports imply that the phosphogenetic episode was Permian and not Triassic to Jurassic.
Murchey and Jones [48] summarized the age-dependent chert formation during the Permian Period for the Circum-Pacific and Mediterranean regions, and discussed the genetic relation between Permian biosiliceous rocks and phosphorites. Triassic and/or Jurassic is also the age of biogenic siliceous sedimentation not only in Japan, as afore-mentioned, but also in other Circum-Pacific and Tethyan regions: Oman, Philippines, Malaysia, Alaska, the western part of Canada, the western part of North America (compiled by [52, 53]) and the eastern part of Russia [54]. But the Triassic—Jurassic bedded cherts of the AMT and Chichibu Belts are not associated with authigenic phosphatic deposits. In addition to their absence in Southwest Japan, phosphatic sediments are scarse in rocks of Triassic and Jurassic age on a global scale [55]. The reason why the "Triassic—Jurassic biosiliceous event" was not accompanied by phosphogenesis is unknown. The distribution of major continents during the Triassic and early Jurassic appears to have been not very different from that of middle Permian; the Pangea supercontinent was amalgamated and extended from pole to pole [56]. Consequently the major patterns of oceanic circulation during Triassic and Jurassic probably were very similar to that during the Permian major phosphogenic episode. The most significant difference between Permian and Triassic to Jurassic was existence of continental glaciation during Permian. The Carboniferous to Permian is known as one of the major glacial periods of earth history [57]. Oxygen and nutrient-rich cool water, which originated in polar regions, thus may have upwelled and formed phosphorites in the western part of the continental margins during Permian [55]. But such upwelling of nutrient-rich bottom water was probably weak during Triassic to Jurassic due to insignificant development of major continental glaciation in polar regions, and thus phosphatic sediments were not associated with Triassic—Jurassic biosiliceous sediments.

Acknowledgements

I express my sincere gratitude to Emeritus Professor Azuma Iijima of University of Tokyo for his guidance and encouragement during this study and critical reading of manuscript. I indebted to Professor Robert E. Garrison of University of California at Santa Cruz for reviewing of the manuscript. Thanks are due to Professor Ryo Matsumoto and Dr. Shigenori Ogihara of University of Tokyo who gave me helpful suggestions on the geochemical study.

REFERENCES

1. Y. Kakuwa. Geochemical study of Triassic to Jurassic bedded cherts in the Ashio, Mino and Tamba terranes in Japan, *Sci. Pap. Coll. Arts & Sci. Univ. Tokyo* **38**, 17-41 (1988).
2. Y. Kakuwa. Lithology and petrography of Triasso-Jurassic bedded cherts of the Ashio, Mino and Tamba Belts in Southwest Japan, *Sci. Pap. Coll. Arts & Sci. Univ. Tokyo* **41**, 7-57 (1991).
3. Y. Kakuwa. Petrography and geochemistry of argillaceous rocks associated with Triassic to Jurassic bedded chert of the Mino-Tamba terrane, *Sci. Pap. Coll. Arts & Sci. Univ. Tokyo.* **36**, 137-162 (1987).
4. Y. Kakuwa. Sedimentary petrographic study on bedded cherts of the Northern Chichibu Belt in eastern Shikoku — with special reference to the P/T boundary, *Bull. Geol. Surv. Japan* **44**, (1993). (in Japanese). (in press).
5. Y. Isozaki and T. Matsuda. Age of the Tamba Group along the Hozugawa "Anticline", western hills of Kyoto, southwest Japan, *J. Geosci. Osaka City Univ.* **23**, 115-134 (1980).
6. Y. Isozaki and T. Matsuda. Middle and late Triassic conodonts from bedded chert sequences in the Mino-Tamba Belt, southwest Japan. Part I: Epigondolella. *J. Geosci. Osaka City Univ.* **25**, 103-137 (1982).
7. M.L. Bender, K. Teh-Lung and W.S. Brocker. Accumulation rates of manganese in pelagic sediments and nodules, *Earth Planet. Sci. Letts.* **8**, 143-148 (1966).
8. E. Bonatti, M. Zerbi, R. Kay and H. Rydell. Metalliferous deposits from the Apennine ophiolites: Mesozoic equivalents of modern deposits from oceanic spreading centers, *Geol. Soc. Am. Bull.* **87**, 83-94 (1976).
9. S. Krishnaswami. Authigenic transition elements in Pacific pelagic clays, *Geochim. Cosmochim. Acta* **40**, 425-435 (1976).
10. R. Matsumoto and A. Iijima. Chemical sedimentology of some Permo—Jurassic and Tertiary bedded cherts in central Honshu, Japan. In: *Siliceous Deposits in the Pacific Region.* A. Iijima, J.R. Hein and R. Siever (Eds). pp.175-192. Elsevier Sci. Publ. Co., Amsterdam (1983).
11. A. Inazumi. Chemical composition of Mesozoic shales in Chugoku, southwest Japan, *Mem. Fac. Educ. Kagawa Univ., II* **2**. 43-55 (1975). (in Japanese).
12. R. Sugisaki, K. Yamamoto and M. Adachi. Triassic bedded cherts in central Japan are not pelagic, *Nature* **298**, 644-647 (1982).
13. K. Yamamoto. Geochemical study of Triassic bedded cherts from Kamiaso, Gifu Prefecture, *J. Geol. Soc. Japan* **89**, 143-162 (1983).
14. L.P. Gromet, R.F. Dymek, L.A. Haskin and R.L. Korotev. "The North American Shale Composite": Its compilation, major and trace element characteristics, *Geochim. Cosmochim Acta* **48**, 2469-2482 (1984).
15. R. Matsumoto, Y. Minai and A. Iijima. Manganese content, cerium anomaly, and rate of sedimentation as clues to characterize and classify deep-sea sediments. In: *Formation of Active Oceanic Margin.* N. Nasu, K. Kobayashi and H. Kagami (Eds). pp.913-939. Terra Science, Tokyo (1986).
16. S. Yamakita. Stratigraphic relationship between Permian and Triassic strata of chert facies in the Chichibu terranes in eastern Shikoku, *J. Geol. Soc. Japan* **93**, 145-148 (1987). (in Japanese).
17. S. Yamakita. Jurassic—earliest Cretaceous allochthonous complexes related to gravitational sliding in the Chichibu terrane in eastern and central Shikoku, southwest Japan, *J. Fac. Sci. Univ. Tokyo, Sec. II* **21**, 467-514 (1988).
18. K. Sugiyama. Lower and middle Triassic radiolarians from Mt. Kinkazan, Gifu Prefecture, central Japan, *Trans. Proc. Palaeont. Soc. Japan, N.S.* **167**, 1180-1223 (1992).
19. T. Matsuda and Y. Isozaki. Radiolarians around the Triassic—Jurassic boundary from the bedded chert in the Kamiaso area, southwest Japan. Appendix: "Anisian radiolarians", *News of Osaka Micropaleontologists, Special Volume* **5**, 93-102 (1982). (in Japanese).

20. S. Kido. Occurrence of Triassic chert and Jurassic siliceous shale at Kamiaso, Gifu Prefecture, central Japan, *News of Osaka Micropaleontologists, Special Volume* **5**, 135-145 (1982). (in Japanese).
21. S. Kido, I. Kawaguchi, M. Adachi and S. Mizutani. On the *Dictyomitrella (?) kamoensis-Pantanellium faveatum* assemblage in the Mino area, central Japan, *News of Osaka Micropaleontologists, Special Volume* **5**, 195-210 (1982). (in Japanese).
22. Y. Isozaki and T. Matsuda. Early Jurassic radiolarians from bedded chert in Kamiaso, Mino Belt, central Japan, *Chikyu Kagaku (Earth Science)* **39**, 429-442.
23. A. Matsuoka. Stratigraphic distribution of two species of *Tricolocapsa* in Hisuikyo section of the Kamiaso area, Mino Terrane, *News of Osaka Micropaleontologists, Special Volume* **7**, 411-420 (1985). (in Japanese).
24. S. Iwao. On some characters of the "Omine" brick silica stone, with special reference to chemical composition and to microscopic features, *Bull. Geol. Surv. Japan* **2**, 468-472 (1951). (in Japanese).
25. S. Iwao. Mg-enrichment around some ore-deposits in Japan — Particularly with reference to hydrothermal gypsum and silica deposits, *J. Geol. Soc. Japan* **61**, 543-555 (1955). (in Japanese).
26. S. Iwao. Geology of silica deposits in the Tamba district, Japan, *Mining Geol.* **12**, 334-345 (1962). (in Japanese).
27. S. Iwao, T. Anzai and T. Okano. Brick silica stone deposits in Tamba district, Kyoto and Hyogo Prefecture, *Bull. Geol. Surv. Japan* **2**, 138-157 (1951). (in Japanese).
28. A. Iijima, Y. Kakuwa and Y. Yanagimoto. Shallow-sea, organic origin of the Triassic bedded chert in central Japan, *J. Fac. Sci. Univ. Tokyo, Sec. II* **19**, 369-400 (1978).
29. A. Iijima, R. Matsumoto and R. Tada. Mechanism of sedimentation of rhythmically bedded chert, *Sedimentary Geology* **41**, 221-233 (1985).
30. K. Bostrom and M. N. A. Peterson. The origin of aluminum poor ferromanganoan sediments in areas of high heat flow on the East Pacific Rise, *Mar. Geol.* **7**, 427-447 (1969).
31. T. Koike. Biostratigraphy of Triassic conodonts in Japan, *Sci. Repts. Yokohama Natnl. Univ., Sec. II* **28**, 25-42 (1984).
32. A. Yao. Geologic age of Jurassic radiolarian zones in Japan and their international correlations, *News of Osaka Micropaleontologists, Special Volume* **7**, 63-74 (1986). (in Japanese).
33. W.B. Harland, A.V. Cox, P.G. Llewellyn, C.A.G. Pickston, A.G. Smith and R. Walters. *A Geologic Time Scale*. Cambridge University Press, Cambridge (1982).
34. A. Yao, T. Matsuda and Y. Isozaki. Triassic and Jurassic radiolarians from the Inuyama Area, central Japan, *J. Geosci. Osaka City Univ.* **23**, 135-154 (1980).
35. A. Yao. Middle Triassic to early Jurassic radiolarians from the Inuyama area, central Japan. *J. Geosci. Osaka City Univ.* **25**, 53-70 (1982).
36. K. Tanaka. Kanoashi Group, an olistostrome, in the Nichihara area, Shimane Prefecture, *J. Geol. Soc. Japan* **86**, 613-628 (1980). (in Japanese).
37. A. Iijima, Y. Kakuwa and H. Matsuda. Silicified wood from the Adoyama Chert, Kuzuh, central Honshu, and its bearing on compaction and depositional environment of radiolarian bedded chert. In: *Siliceous Deposits of the Tethys and Pacific Regions*. J.R. Hein and J. Obradovic (Eds). pp.151-168, Springer-Verlag, New York (1987).
38. R. Hori and H. Masuda. Late early Jurassic (Toarcian age?) oceanic event in bedded chert from SW Japan, *97th Ann. Meet. Geol. Soc. Japan, Abs.* p.152 (1990). (in Japanese).
39. A. Matsuoka. Middle and Late Jurassic radiolarian biostratigraphy in the Sakawa and adjacent areas, Shikoku, southwest Japan, *J. Geosci. Osaka City Univ.* **26**, 1-48 (1987).
40. R. Matsumoto. Origin of manganese nodules in the Jurassic siliceous rocks of the Inuyama district, central Japan. In: *Siliceous sedimentary rock-hosted ores and petroleum*. J.R. Hein (Ed.). pp.181-205, Van Nostrand Reinhold Co., New York (1987).
41. K. Minoura, S. Nakaya and A. Takemura. Origin of manganese carbonates in Jurassic red shale, central Japan, *Sedimentology* **38**, 137-152 (1991).
42. T. Matsuda and Y. Isozaki. Well documented travel history of Mesozoic pelagic chert in Japan: from remote ocean to subduction zone, *Tectonics* **10**, 475-499 (1991).
43. Shipboard Scientifc Party. Site 801. In: *Proc. Ocean Drilling Program, Init. Repts. Volume 129.* Y. Lancelot and R. Larson *et al.* (Eds). pp.91-170, A & M Univ. Texas (1990).

44. A. Iijima, J.R. Hein and R. Siever. An introduction to the siliceous deposits in the Pacific Region. In:*Siliceous Deposits in the Pacific Region.* A. Iijima, J.R. Hein and R. Siever (Eds). pp. 1-6, Elsevier Sci. Publ. Co., Amsterdam (1983).
45. Shipboard Scientific Party. Site 208. In: *Init. Repts, DSDP Volume 21.* R.E. Burns, T.E. Andrews *et al.* (Eds). pp. 271-331, U.S. Government Printing Office, Washington (1973).
46. P.B. Wignall and A. Hallam. Anoxia as a cause of the Permian/Triassic mass extinction: facies evidence from northern Italy and the western United States, *Paleogeogr. Paleoclimat. Paleoecol.* **93**, 21-46 (1991).
47. C. Chai, Y. Zhou, X. Mao, S. Ma, J. Ma, P. Kong and J. He. Geochemical constraints on the Permo—Triassic boundary event in South China. In: *Permo—Triassic events in the Earstern Tethys.* W.C. Sweet, Z. Young, J.M. Dickins and H. Yao (Eds). pp. 158-168, Cambridge Univ. Press, London (1992).
48. B.L. Murchey and D.L. Jones. A mid-Permian chert event: widespread deposition of biogenic siliceous sediments in coastal, island arc and oceanic basins, *Paleogeogr. Paleoclimat. Paleoecol.* **96**, 161-174 (1992).
49. H. Tsukamoto. Lutecite in Triassic bedded chert from the south Mino terrane, central Japan, *J. Fac. Sci. Nagoya Univ.* **36**, 1-14 (1989).
50. H. Matsuda and A. Iijima. Occurrence and genesis of Permian dolostone in the Kuzuu area, Tochigi Prefecture, central Japan, *J. Fac. Sci. Univ. Tokyo, Sec. II* **22**, 89-119 (1989).
51. S. Yamakita and A. Takemura. A Permian(?) radiolaria-bearing phosphate nodule in the Chichibu terrane in Shikoku. *99th Ann. Meet. Geol. Soc. Japan , Abs.* p.118 (1992). (in Japanese).
52. J.R. Hein and J.T. Parrish. Distribution of siliceous deposits in space and time. In: *Siliceous Sedimentary Rock-Hosted Ores and Petroleum.* J.R. Hein (Ed.). pp.10-57, Van Nostrand Reinhold Co., New York (1989).
53. P.De Weaver. Radiolarians, radiolarites, and Mesozoic Paleogeography of the Circum-Mediterranean Alpine Belts. In: *Siliceous Deposits of the Tethys and Pacific Regions.* J.R. Hein and J. Obradovic (Eds). pp. 31-50, Springer-Verlag, New York (1987).
54. S. Kojima. Mesozoic terrane accretion in northeast China, Shikhote Alin and Japan regions. *Paleogeogr. Paleoclimat. Paleoecol* **69**, 214-232 (1988).
55. R.P. Sheldon. Association of phosphatic and siliceous marine sedimentary deposits. In: S*iliceous Sedimentary Rock-Hosted Ores and Petroleum.* J.R. Hein (Ed.). pp. 58-80, Van Nostrand Reinhold Co., New York (1987).
56. C. R. Scotese, R.K. Bambach, R. Van der Voo and A.M. Ziegler. Paleozoic base maps, *Jour. Geol.* **87**, 217-277 (1979).
57. A.G. Fischer. The Phanerozoic super cycles. In: *Catastrophes and Earth History.* W.A. Berggren and J.A. Van Couvering (Eds). pp. 129-150. Princeton University Press, New Jersey (1984).